Health Monitoring Systems

Health Monitoring Systems
An Enabling Technology for Patient Care

Edited by
Rajarshi Gupta
and
Dwaipayan Biswas

CRC Press
Taylor & Francis Group
Boca Raton London New York

CRC Press is an imprint of the
Taylor & Francis Group, an **informa** business

CRC Press
Taylor & Francis Group
52 Vanderbilt Avenue,
New York, NY 10017

Printed on acid-free paper

International Standard Book Number-13: 978-1-4987-7582-3 (Hardback)

Visit the Taylor & Francis Web site at
http://www.taylorandfrancis.com

and the CRC Press Web site at
http://www.crcpress.com

Contents

Foreword

The field of remote healthcare technologies is not only booming, but it is also vast and extremely diverse. What could a new book on remote healthcare technology cover which has not been covered already? Should it concern new diagnostic or therapeutic technologies? Should it focus on acute or chronic disease applications? Should it be disease specific or disease agnostic? Should it focus on technologies for the patient or for the doctor? Should it cover early innovations or stick to validated technologies with regulatory approval? The good news is, it does not matter what the exact answers are to the above questions; there is value in sharing knowledge and sharing challenges with a broader community in all the above permutations.

This book looks at the remote healthcare technologies space from an engineering perspective and particularly focuses on aspects related to the gathering and processing of data: data compression, signal analysis, and security. These are essential ingredients to the application of digital healthcare services in the real world. By covering a few use cases in the cardiovascular and metabolic domain, this book will also satisfy the reader interested in applications of new technology.

Chris Van Hoof
Vice President R&D at IMEC,
General Manager OnePlanet Research Center
Kapeldreef 75, Heverlee, 3001, Belgium
Email: Chris.VanHoof@imec.be

Preface

Healthcare practices have undergone a sea change over the last few decades. Primarily this is driven by a general awareness among people for enjoying a good quality of life. In the United States, healthcare has been identified as one of the top areas of priority. According to World health Organization (WHO) statistics, the per capita total expenditure on healthcare in the USA has increased from US$6000 in 2005 to nearly US$9900 in 2016. The growing interest on healthcare and its peripheral services has given a high impetus to the research and contribution in multidisciplinary areas of science and technology. Undoubtedly, healthcare is one of the fastest growing areas of research and business in the current century. However, in spite of the tremendous growth of information and communication technology (ICT), embedded technology, computer engineering, sensing technology, biomedicine, and biotechnology in last century, the level of mortality rates due to illness has been a serious issue of concern for policy makers. For example, cardiovascular diseases (CVDs) contribute nearly 20% of the total annual mortality rates each year in the USA. For a developing nation like India, the figure is slightly higher at 22%. Healthcare scientists and policy makers attribute two prime factors for this. The first is inadequate doctor to patient ratio. The second is aggravation of minor health problems due to a lack of early detection and treatment. ICT has played a pivotal role in healthcare in the integration of healthcare components, their interoperability, and maintaining healthcare information records. The practice of health monitoring of remote patients was initiated in early the 1960s. Since then, development of embedded technology, communication and networking protocols, and computing has revolutionized the practice of health monitoring over the decades. From point-to-point communication based on a fixed station, we are now in the era of e-healthcare, a general term commonly used to collectively define a host of services. Innovations in healthcare research have been driven by some societal demands, viz., mobility of the caregiver (physician); wearable technologies for monitoring; use of handheld devices (mobile phone, PDAs); and technological developments in microelectronics, embedded systems, networking, and communications.

Health monitoring is contributed to by various specialized applications like *telehealth, telecare, telemedicine,* etc. Telehealth refers to the use of technology for connecting with a patient who is situated at a remote location. Telecare uses information and communication technology to monitor the vital signs of a patient and detects pathological conditions from the acquired signals for therapeutic actions. Telemedicine uses audiovisual consultation between two physicians for the treatment of a remote patient. Many modern health monitoring systems are enabled with Internet-of-Things (IoT) technology, which has paved the path for pervasive and ubiquitous monitoring and computing, resulting in strong cooperation between the various components of the complex healthcare setting. As a result, present-day health monitoring technology enables multi-patient monitoring using wearable sensors integrated with a wireless communication framework. On one hand, this facilitates ubiquitous monitoring of patients but, on the other, has demanded system-level requirements like managing huge medical data, preserving the security of data, and privacy of personal data.

The objective of this book is to provide a comprehensive discussion on different aspects of health monitoring systems. While discussing the basic technological areas related to the domain, it also focuses on the research carried out in pertinent areas and future research directions and challenges. Common medical signals used for cardiovascular and respiration measurement, such as electrocardiogram (ECG), photoplethysmogram (PPG or SpO_2), electroencephalogram (EEG), respiration, and blood pressure, have been considered for illustrations of circuits, algorithms, and systems in the various chapters of this book. Although the chapters are standalone in nature, detailed discussion on basic aspects of biomedical instrumentation, circuit analysis, signal processing concepts, soft computing tools, and communication technology are beyond the scope of this book. Adequate end-of-chapter references are provided for enthusiastic readers for further reading. We hope that this book will be useful for undergraduate and postgraduate students of engineering and medical physics, technicians, practicing engineers, and researchers.

Book Organization

This book discusses the core areas related to remote health monitoring systems, describing signal acquisition, compression, transmission, and analysis using different computing methods. Privacy and security of data collected/transmitted/processed using wireless sensor networks (WSN) have been discussed in detail. Examples of health monitoring applications have been presented to augment the theoretical details.

The book comprises 13 chapters, with focus primarily on cardiovascular monitoring. Chapter 1 introduces new paradigms of remote health monitoring systems. Chapter 2–9 focus on sensing, processing, pervasive computing, and security associated with remote health monitoring. Sensing and data acquisition have been discussed in Chapter 2, and details on compressing the acquired data is mentioned in Chapter 3. Research on IoT-based health monitoring has been presented in Chapter 4. Chapter 5 lists out state-of-the-art systems using telemedicine technology. Methods to analyze biomedical signals have been the focus of Chapter 6, and techniques for pervasive computation in a resource-constrained ambulant environment are mentioned in Chapter 7. Systems using big data and cloud computing have been discussed in Chapter 8, and issues related to cybersecurity in WSN have been detailed in Chapter 9. Chapters 10–13 present four examples of state-of-the-art health monitoring applications, namely, activity monitoring using wearable sensors; blood pressure monitoring using PPG and ECG sensors; wireless telecardiology; and diabetes monitoring.

MATLAB® is a registered trademark of The MathWorks, Inc. For product information, please contact:

The MathWorks, Inc.
3 Apple Hill Drive
Natick, MA 01760-2098 USA
Tel: 508-647-7000
Fax: 508-647-7001
E-mail: info@mathworks.com
Web: www.mathworks.com

Acknowledgments

We sincerely acknowledge the chapter authors for sharing their wide experience and expertise to enrich the book, and sparing their valuable time to contribute to this book. We are grateful to the CRC Press/Taylor & Francis Group team for their patience and tireless effort toward copyediting, typesetting, and publication of this book. We also thank the readers for selecting this book for advancement of their knowledge and technical skills.

Editors

Rajarshi Gupta received his MTech and PhD (Tech) in Instrumentation Engineering from University of Calcutta, India, in 2002 and 2012, respectively. He is currently an Associate Professor with the Electrical Engineering Section, Department of Applied Physics, University of Calcutta, India. His research interests include cardiovascular signal measurements and intelligent health monitoring. He has published 50 papers in peer-reviewed international journals and conferences and three book chapters with a Google Scholar citations of 523, h-index of 12, and i-10 index of 12. He is currently the principal investigator/co-principal investigator of five funded projects, with total funding of INR 62 lakhs in the area of biomedical devices and systems development. He has guided one PhD student and is currently guiding five PhD students. Rajarshi is an active volunteer in IEEE and IET professional activities and has been associated with many flagship conferences in key positions.

Dwaipayan Biswas received his MSc in System on Chip in 2011 and his PhD in Electrical Engineering in 2015 from the University of Southampton (UoS), UK. During 2011–2015, he was involved with three European Union funded projects, namely Chiron (https://artemis-ia.eu/project/17-chiron.html), StrokeBack (https://www.strokeback.eu/), and PLEASED (https://www.physense.eu/pleased-project/). His research focused on developing low-complexity algorithms and architecture design targeting low-power VLSI implementation. During 2015–2016, he went on to work as a post-doctoral research fellow at UoS in the projects funded by the Engineering and Physical Sciences Research Council (EPSRC), namely Refresh (http://www.refresh-project.org.uk/) and Prime (http://www.prime-project.org/). He worked on smart office design based on cognitive monitoring and energy optimization for multi-core embedded platforms. In November 2016, he joined IMEC, Belgium, as a researcher on digital IC design for biomedical applications. He has been working on digital front ends for low-power biomedical sensor readouts aimed at pervasive health monitoring. Further, he has been actively involved in embedded algorithm–architecture development for motion artifact reduction during ambulatory sensing. His research interests include low-power VLSI design, biomedical signal processing, machine learning, brain computer interface, and computer architecture. He has authored over 40 peer-reviewed articles including journals, conference publications, and book chapters.

Contributors

Amit Acharyya
Department of Electrical Engineering
Indian Institute of Technology (IIT) Hyderabad
Telangana, Hyderabad, India

Domenico Balsamo
Microsystems Design, School of Engineering
Newcastle University
NE1 7RU, United Kingdom

Simone Benatti
EEES Lab, Department of Electronic and Information Engineering (DEI)
University of Bologna
Bologna, Italy

Saptarshi Das
Department of Mathematics
University of Exeter, Penryn Campus
Penryn, United Kingdom

Elisabetta Farella
E3DA Unit, ICT Center
Fondazione Bruno Kessler
Trento, Italy

Victor Kartsch
EEES Lab, Department of Electronic and Information Engineering (DEI)
University of Bologna
Bologna, Italy

Velu P. Kumaravel
E3DA Unit, ICT Center
Fondazione Bruno Kessler
Trento, Italy

I. E. Lamprinos
Intracom S.A. Telecom Solutions
Paiania, Greece

Fabio Montagna
EEES Lab, Department of Electronic and Information Engineering (DEI)
University of Bologna
Bologna, Italy

Luca Monzo
Department of Cardiovascular, Respiratory, Nephrologic and Geriatric Sciences
Sapienza University, Policlinico "Umberto I"
Rome, Italy

Jayanta Mukhopadhyay
Indian Institute of Technology (IIT) Kharagpur
Kharagpur, West Bengal, India

Sourav Kumar Mukhopadhyay
Ryerson Universiy
Toronto, Ontario, Canada

Chandrajit Pal
Indian Institute of Technology (IIT) Hyderabad
Telangana, Hyderabad, India

Paolo Emilio Puddu
Department of Cardiovascular, Respiratory, Nephrologic and Geriatric Sciences
Sapienza University, Policlinico "Umberto I"
Rome, Italy

Michele Schiariti
Department of Cardiovascular, Respiratory, Nephrologic and Geriatric Sciences
Sapienza University, Policlinico "Umberto I"
Rome, Italy

Naresh Vemishetty
Indian Institute of Technology (IIT) Hyderabad
Telangana, Hyderabad, India

1

Remote Healthcare Technology: A New Paradigm

Dwaipayan Biswas
IMEC

CONTENTS

1.1 Introduction

Increased life expectancy due to better medical facilities in developed nations has also paradoxically augmented the prevalence of health impairments among ageing population. Cardiovascular disease (CVD), neurodegenerative disease, and motor impairments are some of the most common syndromes affecting majority of the world's population [1–5]. The effect of these diseases is not limited only to the elderly, but have also affected large sections of the society in their middle and young ages, mainly due to the stress prevalent in current living conditions (unrestricted working hours, eating disorders, etc.). CVD is one of the major reasons for mortality around the world, representing 31% of all global deaths in 2015, as reported by the World Health Organization (WHO) [2]. It is caused by disorders of the heart and blood vessels which result in coronary artery disease (CAD), heart failure, cardiac arrest, and sudden cardiac death [1]. Dementia, Alzheimer's disease, cerebrovascular accident, popularly termed as stroke, are some of the common neurodegenerative diseases plaguing the elderly population at large [6–9]. Stroke survivors often suffer from impairments of their various body parts, leading to a reduced quality of life [10]. Motor impairments are also common among patients suffering from Parkinson's disease having a Freezing of Gait (FoG) syndrome and survivors of epileptic seizures [11]. Treatment, rehabilitation, and care expenses associated with these diseases have an adverse impact on the financial aspect, in addition to the loss of quality of life causing societal disruption.

Technological advancements have aided the treatment of above-mentioned diseases and helped to deliver qualitative patient care within home environment, enabling proactive rather than reactive clinical intervention. The remote delivery of healthcare services to patients, using telecommunication infrastructure is termed as 'telemedicine' [12,13]. This negates the need for unnecessary patient visits to the clinic/rehabilitation centers and acts as a screening mechanism, enabling an interactive support system for patients, forming a closed loop with caregivers and clinicians. The terms, 'e-Health', 'remote healthcare', 'telemonitoring' are to a great extent synonymous with the concept of 'telemedicine' and are interchangeably used in relevant literature. Subject-specific behavioral and clinical traits, coupled with physiological differences across individuals, necessitate a personalized healthcare delivery model for round-the-clock monitoring within the home environment. There are several components of a remote health monitoring system, some of the fundamental ones being – (a) sensing mechanism proximal to the patient acquiring information on the vital/monitored parameters, (b) a remote server or a doctor's interface (e.g. personal data assistant, i.e. PDA, personal computer, etc.), and (c) secure communication gateway linking the above two infrastructures [14]. This book primarily focusses on the first stage, i.e. sensing, which encompasses data acquisition through relevant sensors, data processing and formulating clinical inference. Details on secure communication have also been presented.

Developments in microelectromechanical systems (MEMS) and wireless sensor networks (WSNs) have proliferated the use of miniaturized sensors capable of recording physiological parameters for long durations. These sensors having a small form factor can be positioned directly or indirectly on the human body, i.e. used as wearable sensors or placed in the environment proximal to the patient. They can capture vital signals that are reflective of the physiological state of the subject and are also used for acquiring information pertaining to patient movement, position, location and interaction with objects of interest within a zone. This leads to the collection of holistic information, which can be analyzed with respect to the application of interest. CVD monitoring is an illustrative application area where research has been instrumental in enabling a personalized closed-loop monitoring system, using wearable sensors. As an example, consider electrocardiogram (ECG) patches attached proximal to the chest, photoplethysmography (PPG) sensors attached to the wrist or other peripheral body parts, helping to gather cardiovascular information; accelerometers–gyroscopes–GPS present in a smartphone/smartwatch can act as an activity/location tracker; pressure sensors placed on the bed can help in tracking sleep; sensitized cutlery/radio frequency identification (RFID) tags on items of daily use can indicate food intake patterns [15]. Apart from sensing, front-end amplification, microcontroller functions, and radio transmissions have all been integrated into a single circuit thus resulting in a system-on-chip (SoC) implementation [16]. The availability of high-speed internet connection, large amount of data generated from such sensors active for long durations, data storage facilities in the cloud act as key elements in delivering round-the-clock high-quality health service to a wide population. This smart infrastructure gives rise to Internet of Things (IoT), enabling data acquisition through sensors, performing autonomous actions based on actuators, coordinate functions, and share information among multiple nodes [14,17]. International Data Corporation (IDC) predicts that IoT will reach about a US$ 1.7 trillion market by 2020. Gartner expects 25 billion connected devices by 2020 while Cisco mentions about 50 billion, and Harvard Business Review expects 28 billion 'things' connected to the Internet. IoT applied to health monitoring is also termed as Internet of Health things (IoHT), deemed promising for the near future, benefiting healthcare delivery and societal well-being. IoHT is predicted to

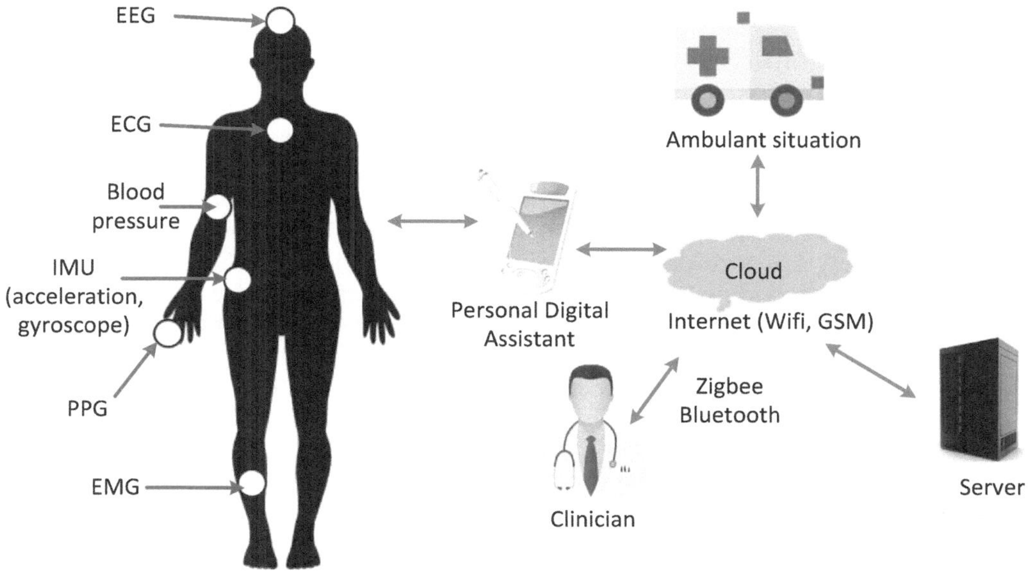

FIGURE 1.1
Overview of a typical remote health monitoring system. (Adapted from [30].)

be worth USD 163.2 billion by the year 2020, having a high economic impact across the world [18]. State-of-the-art IoHT systems exist for ECG [19,20], electroencephalogram (EEG) [21], diabetes [22], and vital signs including PPG [23], blood oxygenation (SPO_2), respiration rate, body temperature, glucometer, galvanic skin response, blood pressure, position (accelerometer, gyroscope, etc.), and electromyography (EMG) [24–28].

A holistic overview of a typical wireless body area network (BAN) system for remote healthcare monitoring framework is shown in Figure 1.1. In this figure, the central hub acts as a gateway and can be used for displaying the vital information on user interface or transmitting the clinically relevant information to a remote medical center or a clinician [29]. The sensor nodes comprise ECG, EMG, SpO_2, galvanic skin response (GSR), etc. among many other sensors.

The factors significantly contributing toward home-based/remote rehabilitation can be summarized as follows:

- logistical convenience, since it helps to reduce unnecessary patient transfer to the clinic or hospital,
- cost-effectiveness, since it helps to reduce expenses incurred toward hospitalization and human intervention,
- wireless sensing and computer-aided technology to provide proactive clinical intervention,
- maximizing number of patients being treated at home across a wide spectrum of the population [31].

In the following sections, we highlight a few significant state-of-the-art systems that have transformed the concept of remote health monitoring into reality.

1.2 IoT Applications

Technological developments such as 5G wireless networks, mobile edge computing (MEC), cloud computing, have enabled the development of 'smart healthcare', gaining attention from academia, governments, and industry. 5G wireless networks facilitate efficient communication between resource-constrained device, faster data generation and processing as well as high-quality data transmission to stakeholders [32]. With the growing emergence of IoT devices, privacy and security of data are also important issues as have been discussed in [33]. Example systems benefitting from such developments are EcoHealth, a middleware platform developed for IoT that connects patients, healthcare providers, and devices [34]. It allows data management and aims to simplify and standardize IoT application development, addressing issues like interoperability between different devices. Similarly, an IoT-based network reported in [35] was designed to monitor patients in rural areas and areas with low population density. Another solution, named U-Healthcare, is based on a mobile gateway for ubiquitous healthcare systems that collects data, processes it, and stores it in the cloud for remote access [36].

IoHT applications are targeted toward providing a holistic care for patients and can mainly be divided into two segments – (a) ambient assisted living (AAL), supporting elderly or incapacitated patients, in their home environment, providing a safer environment and increased autonomy toward an active life [37] and (b) systems for monitoring specific diseases. CVD monitoring, movement/location tracking, fall detection, food intake are some of the key monitoring components which ensure that subjects receive the proper assistance in ambient environment (i.e. away from clinic/hospital). State-of-the-art AAL systems, discussed in Section 1.2.1, incorporate variants of these monitoring components, in association with sensing, processing, storage, and transmission modalities. In addition, there exist IoT systems for tracking specific disease which have also been discussed in Section 1.2.2. Processing of sensor data for IoHT applications deserves special attention given the importance of energy efficient operation of battery-operated sensor nodes, ensuring long-term continuous monitoring, as discussed in Section 1.2.3. A few state-of-the-art remote health monitoring systems have been listed in Table 1.1 for further reference.

1.2.1 Ambient Assisted Living

According to the Global Health Observatory [49], healthy life expectancy at birth was 63.1 years in 2015, giving rise to a population needing assistance as they grow older. The products presented for remote health monitoring are used by patients who need to be continuously monitored but not necessarily inside a hospital. They can be at home using a device for monitoring their condition and send the data to be analyzed by a health professional remotely. Living outside the hospital environment aids patient comfort and reduces costs and the risk of hospital infection. As proposed in [50], remote healthcare monitoring frameworks may be an effective way to reduce hospital readmission rates. EarlySense is a system offering the most complete solution [42], monitoring heart and respiratory rate, has a fall prevention system, detects early patient deterioration, and prevents pressure ulcer. The results of the study are published on the website for further reference.

Wearables are smart devices that can be attached, for example, to the body, such as watches, shoes, or other body parts. A common application of wearable devices is for monitoring user's physical activity, such as the system proposed in [29], used to monitor

TABLE 1.1
State-of-the-Art Remote Health Monitoring Systems

Refs.	IoHT Service	Product	Description
[38]	AAL system	BeClose	Remote monitoring system from Assisted Living Technologies Inc., for both caregiver and subject.
[39]	AAL mobile application	Fade	Android application for detecting 'fall' and generating an alarm.
[40]	Mobile application	Medisafe	Medication reminder
[41]	Mobile application	OnTrack	Managing diabetes
[42]	Monitoring system	EarlySense	Continuously monitors heart rate, respiratory rate, fall prevention, and pressure ulcer prevention.
[43]	Monitoring system	AccuSom	Monitors sleep
[44]	Monitoring system	Proteus Discover	System comprises ingestible sensors, portable sensor patch, mobile application, manufactured by Proteus Digital Health Inc. The pill dissolved in the stomach produces a signal, detected by a body-worn sensor, which transmits the data to a mobile application.
[45]	Wearable sensor	Apple watch	Monitors cardiac parameters and kinematic information
[46]	Wearable sensor	Enterprise	A product from Bittium, providing customized solution for industrial, healthcare, and sports device manufacturers
[47]	Wearable sensor	QardioCore	Monitors cardiac parameters and transmits information to clinician
[48]	Wearable sensor	Monica AN24	Clinical and home monitoring system

the elderly population. In this context, a system developed for vital signs monitoring on elderly people in care homes is described in [51]. Sensors located on the patient's clothes collect data used to monitor their physiological parameters. Smart sensory furniture used for elderly people living alone was proposed in [52]. A new method toward achieving assistive smart homes based on an intention recognition mechanism incorporated into an intelligent agent architecture was proposed in [53]. The method provided a web interface and provided assistance during physical activities. Robotics and home automation was combined in [54] for AAL. RFID intelligent system was used to identify user–object interactions employing machine learning techniques to enable an AAL system in a retail store. A system developed in [55] allows AAL systems to identify and predict situations that may endanger users in their living environment, proposing a complementary and alternative way to acquire comprehension of a person's behavior providing this knowledge when cognitive impairments occur. A survey reported in [56] covers signal analysis and processing techniques employed with different types of sensors, such as pyro-electric infrared, and vibration sensors, accelerometers, cameras, depth sensors, and microphones, analyzing the increase of diseases and healthcare costs, shortage of caregivers, and a rise in the number of individuals unable to live independently.

The prevalence of smartphones and the use of fitness applications have gained immense popularity, as investigated in [57], presenting information from more than 14,000 cellular towers and 4,000 users. This work correlates key factors, such as temporal, location, mobility, and personal incomes, to the usage of such mobile applications. These findings are important for developing public policy and city planning. Healthcare solutions using smartphones has become increasingly common these days. These solutions range from a simple application to remind the patient to take a medicine or an oximeter. According to Saúde Business [58], it is estimated that there are more than a 100 million m-health Apps around the world. A review of using medical apps on smartphones, discussing their limitations and future trends, is presented in [59]. A review of all randomized clinical trials reporting the effects of psychological interventions delivered via smartphone on symptoms of anxiety, observing a reduction in total anxiety scores, is reported in [60]. M-hub, a system addressing issues of mobility and security in medical environments, uses a middleware in mobile devices that automatically connects smart objects [61]. Similarly, H3IoT is a five-layer architectural framework for monitoring health of elderly people [62]. There are some devices specifically built for the AAL market. BeClose monitoring system is an example of a solution that combines several sensors with artificial intelligence approaches providing more freedom to elderly people [63]. MC10 offers a state-of-the-art body-worn sensor capable of gathering complex physiological data [64]. Bittium's smartwatch takes the leading position capable of measuring vital signs, monitoring patient location, detecting body positioning and medicine dosage, and timing management [46]. As security is necessary in all IoHT solutions, Bittium also offers a solution that ensures secure data transfer between sensor devices and cloud services, being a company specialized in the development of reliable, secure communications and connectivity solutions. It is also important to mention the Apple Watch as a wearable development platform [45], allowing a large number of applications for healthcare. It is possible to obtain the body mass index (BMI), body surface area, estimated glomerular filtration rate (eGFR), and several scores used in CVDs. The popularity can be judged from the fact that nearly 6 million devices were sold in 2016.

Fall is one of the main causes of fatal injury in the elderly population, turning their lives more dependent on care [65]. The Fade App aims to solve the problem with fall and is easy to use [39]. A fall detection system for elderly patients is shown [66–68], monitoring and detecting activities such as walking, running, sitting or standing up, lying down, and falling. A system based on cloud computing for monitoring Parkinson patients was developed in [69]. An m-health wearable wireless sensing system for monitoring human motion disorders was proposed in [70], capable of generating timely rhythmic auditory stimulations to release the gait block during freezing of gait (FoG) episodes in Parkinson's disease. Tracking movements of visually impaired people, with an indoor anti-collision alarm system based on wearable sensors, is proposed in [71]. It identifies the distance between the RFID tag and antenna, keeping track of the degree at which the user is straying off the normal path. Information from tri-axial accelerometers and a heart-rate monitor were used to distinguish different physical activity using supervised machine learning algorithms [72]. A smartphone application was used to acquire, process, and store inertial sensor data and rotation matrices about device position for gait parameters estimation during daily living [73]. HealtheBrain is an application, proposed in [74], focussing on allowing the square stepping exercise done by elderly people without cognitive loss. A correlation-based sleep scheduling mechanism (LCSSM) was proposed in [75] to implement energy efficient wireless sensor networks in ambient-assisted homes (AAH).

A prediction system, Wanda [76], was developed to detect CVD risk factors, where subjects receive technical support and reinforcement. A patient health monitoring system integrated with cloud computing and IoT [77] was applied for real-time monitoring of congestive heart failure using ECG. Similarly, a wearable ECG monitoring system analyzing heart and respiratory rates was proposed in [78]. A smartphone-based real-time CVD monitoring app was developed in [79] for patients with heart disease, diabetes, and hypertension and other chronic diseases. Also, a ubiquitous system for monitoring elderly patients with Alzheimer's disease is described in [80], enabling the patient to notify the clinician on vital parameters such as oxygen saturation, blood pressure, and heart rate. More specific wearable devices are also found in literature, such as the system proposed in [81], enabling real-time data access in the cloud and monitoring blood pressure, temperature, ECG, and oxygen saturation. Ubiquitous cardiac care (UCC) is a ubiquitous health monitoring system for cardiac arrhythmias detection [82], where data is captured using wearable ECG sensors and sent to cardiologists for analysis. An ECG remote monitoring system dedicated to long-term residential health monitoring integrated with IoT was proposed in [83]. Majority of the systems presented here are capable of measuring vital parameters and ensuring a general well-being (also termed as 'wellness monitoring') of patients as well as proactively notifying respective clinicians in times of emergency.

1.2.2 Disease Monitoring

In this section, health monitoring systems, not necessarily restricted to the home and living environment, but targeting specific diseases, which are key toward societal well-being, have been discussed. A 5G-Smart Diabetes monitoring system has been proposed in [84], providing communication infrastructure for continuous physiological monitoring of diabetic patients. It targets achieving a cost-effective solution toward personalized monitoring. The number of people with diabetes has risen from 108 million (in 1980) to 422 million in 2014. According to the World Health Organization, these numbers are likely to increase considerably [4]. One good solution is OnTrack Diabetes [85], an application for people with type-2 diabetes. It helps users manage their conditions by performing blood glucose, blood pressure, exercise, food, medication, pulse, and weight checks. An approach described in [86] enables real-time location monitoring system for elderly people, ensuring patient safety, with respect to drug compliance in diabetes therapy management.

Assessing voice disorders using deep learned features of voice signals and their corresponding treatments is proposed in [87]. The processing, assessment, and treatment are managed by an application management system in the cloud, an edge computing (EC) application platform management system. An embedded system measuring blood flow with low-cost technology is described in [88]. A remote body pressure monitoring system is proposed in [89]. Self-monitoring of blood pressure for pregnant women was demonstrated in [90].

An integrated system design for monitoring of dementia patients has been proposed in [91]. It allows data collection from different sensors placed in a house and transmission over the internet. A smartphone-based accelerometer is used to quantify spatiotemporal gait parameters when attached to the body or in a bag, belt, hand, and pocket, allowing detection of behavior changes among patients diagnosed with bipolar disorder [92]. A smartphone-based AAL solution has been demonstrated in [93], supporting diagnosis, clinical communication, providing drug references or medical education.

An automatic detection of chronic wounds (using color and size as features) was proposed in [94], using image analysis and machine learning, developed as a mobile application. A StripTest reader was proposed in [95], a smartphone interpreter of biochemical tests based on paper-based strip color using the phone camera for acquisition, processing the images within the phone and producing comparable results with respect to gold standard. Skin monitoring and automatic detection of melanoma, using smartphones, showed that an early intervention at home could successfully treat this disease [96]. Girl Talk, a free smartphone app, provides comprehensive sexual health information, particularly targeting teenage girls [97]. A method to detect and monitor the outbreak of chikungunya virus based on IoT and fog computing was proposed in [98], diagnosing infected users and generating alerts during emergency.

The characteristics of wearable applications for IoT scenarios was analyzed in [99], describing the interaction between wearable or mobile devices and smart objects, and a Web of Things application for evaluating interaction patterns in a smart environment was presented. In USA, 15% of the healthcare consumers use wearable devices, such as smartwatches and fitness devices. It is predicted that 110 million of wearable devices will be sold by 2018–2019 [100]. The goal is always to make the user as comfortable as possible in order to provide quality healthcare. Therefore, the main aspects taken into consideration to analyze the previously presented wearable solutions were patient's comfort, data security, and usability. IoT is bringing innovations to healthcare, resulting in new companies taking interest, thereby helping to create an impact on product development and marketing.

1.2.3 Energy Efficient Operation

The physiological data collected from different sensors distributed over the human body or in the living environment are transmitted continuously to a remote server (or cloud), using standard communication protocols, i.e. wirelessly over mobile network or through the patient's Bluetooth-enabled mobile device. The analysis of the transmitted signal in the cloud/server takes place using 'computationally intensive signal processing/machine learning techniques' [101]. The fundamental problem with this process of continuous data transmission, followed in typical remote monitoring systems, is the energy expenditure at the radio-front end of the sensors. Hence, continuous transmission of data affects the operational lifetime of the battery-operated sensors, thereby rendering them ineffective for long-term remote monitoring applications. Therefore, in such wearable systems, the data analysis primarily needs to be carried out on the sensor node, yielding an energy efficient solution compared to the remote monitoring systems based on conventional continuous data transmission, provided the complexity of data analysis algorithms is reduced [101]. Since the energy consumption is proportional to the computational complexity of the data processing algorithm, the use of high-complexity algorithms (although they may be more 'accurate') will drain the battery faster, defying the objective of long-term monitoring. Hence, for long-term system operation using body-worn wireless sensor node comprising heterogeneous sensors, it is essential to select data analysis algorithms that have low computational complexity. Energy efficient operation has grown as a key factor in IoHT applications, with approaches [102] and [103] paying special focus to improve battery consumption of smartphones and using an event-driven design. On node sensor processing aimed at AAL was proposed in [104]. This usually takes the form of implementing the low-complexity processing methodology in hardware or software [105]. Hardware implementation involves dedicated application-specific integrated circuits (ASICs) or microcontrollers, ARM cores [106], field programmable gate arrays (FPGAs) [107]. Examples of software implementation include android/java applications on mobile platforms. Implementation choices are dictated by the application requirements for real-time operations.

1.3 Book Organization

In this book, we have primarily focussed on wearable sensing platforms for cardiovascular monitoring. We have presented an in-depth discussion on telemedicine technologies, sensing methodology, IoT infrastructure for health monitoring, signal processing techniques, pervasive computing, cybersecurity solutions, and lastly various state-of-the-art health monitoring example applications. Apart from the Introduction chapter, the book comprises 12 chapters, of which Chapters 2–9 focus on the theoretical aspects of sensing, processing, pervasive computing, and security associated with remote health monitoring. Chapters 10–13 present four key health applications which further compliment the methodological details presented in the previous eight chapters.

Chapter 2 presents an overview on medical sensing and associated data acquisition, focussing on sensors for cardiovascular (ECG, BP, phonocardiogram (PCG), PPG), respiratory (respiratory rate), nerve (EEG), and motor activity (EMG, bioimpedance) monitoring. Chapter 3 presents details on medical data compression, a key component in terms of the huge amount of data generated by smart sensing platforms and impending storage constraints. The compression modalities have been explained with respect to ECG and PPG signals.

The technological standards, challenges, and open research areas in IoT-based health monitoring have been discussed in Chapter 4. In particular, low-energy implementation, signal processing and machine learning challenges, communication constraints, and security concerns have been reviewed. State-of-the-art systems using telemedicine technology have been listed down in Chapter 5. Signal processing is a key aspect in biomedical applications, and hence the most popular time- or frequency-domain processing methodologies and machine learning techniques have been discussed in Chapter 6. The discussion is further substantiated by two illustrative use-case applications on prosthetic controller using EMG signals and a brain–computer interface speller application controlled by ECG. Chapter 7 presents a detailed discussion on pervasive computation for CVD monitoring, in particular focussing on an important research area of lead reconstruction for ambulant ECG monitoring. The chapter demonstrates solutions for reconstructing a 12-lead ECG system which is suitable for clinical diagnosis, from a two- to three-lead pervasive ECG recording setup.

Chapter 8 revisits major technological advances behind novel e-health and m-health services (both consumer-grade and medical-grade), with emphasis on big data and cloud computing. The role of IoT in healthcare is analyzed, particularly exploring the role of blockchain in health information sharing. The best practices for building modern, user-friendly health-related mobile applications are also summarized herein. In Chapter 9, the major driving force behind healthcare digitization is discussed, namely, the role of trust and cybersecurity. Key enabling technologies and approaches for secure and trusted applications such as homomorphic encryption, blockchain-based cybersecurity solutions, and sandboxing are presented in detail. Four key applications – human activity monitoring using wearable sensors, blood pressure monitoring using PPG and ECG measurements, wireless telecardiology, and lastly, diabetes monitoring – have been presented in Chapters 10, 11, 12, and 13, respectively.

We hope this book will act as a launching platform for readers interested in biomedical instrumentation and signal processing, presenting a holistic overview on sensing, processing, security, and state-of-the-art knowledge on key health monitoring applications.

1.4 Future Research Directions

Presently, wearable sensors claim to be doing more than simple monitoring; for example, there are headbands which alert the users when they become distracted during an important activity, wristbands administering shocks to smokers as a prohibitive alarm, navigation control for visually impaired and sensors that help with epileptic seizures and anxiety attacks. With rapid increase in prominence of wearable and smart sensors, handling the large amount of data generated is a critical research question. Data transmission to and from the wearable platforms and data privacy/information safety assume utmost significance [108].

1.4.1 Challenges

With majority gadgets connected to the internet, demands on a finite bandwidth are rapidly straining the system. By the end of 2014, global mobile-data traffic reached 2.5 exabytes (2.5 billion gigabytes) per month according to the networking-technology company Cisco Systems. Approximately, the 100 million odd wearable devices were generating 15 million gigabytes of monthly traffic on what is a physically finite portion of the electromagnetic spectrum, with their number expected to increase fivefold by 2019 [108]. Moreover, there will be occasions for gridlock, with an increased use of headsets which deliver data-hungry virtual and augmented reality experiences. All these devices clog up the airwaves, impairing performance and threatening essential internet traffic. Probable solutions include making efficient use of the spectrum; optimizing antenna designs to reduce interference and power consumption; and transforming wireless communications into the visible-light realm using light-emitting diodes (LEDs), acting as photoreceptors, to communicate either between wearables or directly to the internet. Cognitive radios is one such solution, which enables smart use of communication channels, identifying unused bandwidth regions, and opportunistically using them to speed up communication [109]. Optimum potential can be reached, when devices use a licensed frequency to communicate, and then it drops off the spectrum when someone with higher priority enters. Although techniques based on this principle have been used for decades, cognitive radio will make sharing available spectrum among devices more efficient.

Wearable sensors generate a wealth of personal data, causing data privacy concerns among potential users. Encryption is an effective way to deal with it; however, it is not always used in low-cost wearable devices. SensCrypt, an encryption protocol designed specifically for low-energy fitness trackers that reduces communications costs, is in practice [110]. There have been efforts to improve security of mobile and wearable devices by equipping them with biometrics such as fingerprint readers and iris scanners [111,112]. However, high-resolution cameras have been shown to be capable of capturing iris information from a distance and capture fingerprint using a phone's camera. Robust encryption mechanisms need to be used, which include heartbeat patterns, or other physiological signals such as DNA, paired with sensor platforms, to enhance security measures.

1.4.2 Wearable Biosensors

The face of wearable devices has changed rapidly in recent years, with researchers branching out from tracking physical activities for healthcare applications to the development of biosensors. These devices incorporate a biological recognition element into sensing

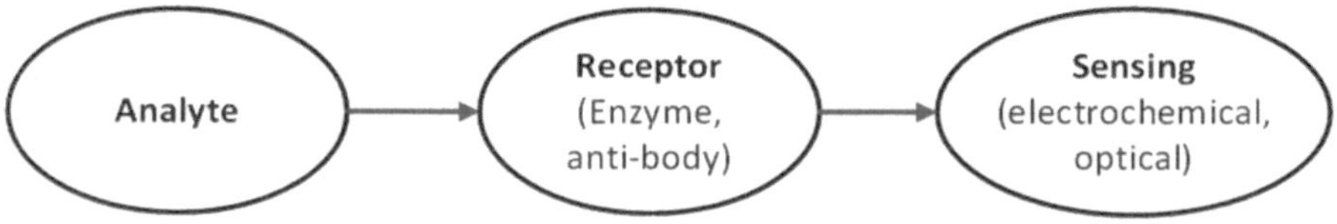

FIGURE 1.2
Generic overview of biosensing components. (Adapted from [113].)

mechanism (for example, enzyme, antibody, cell receptor). The potential utility of wearable biosensors is evident from the large number of reported studies. A typical biosensor comprises (a) bioreceptor (for example, enzyme, antibody) responsible for selective recognition of the target analyte and (b) physico-chemical transducer (for example, electrochemical, optical, or mechanical) which translates the biorecognition event into a signal of interest, as shown in Figure 1.2 [113]. These devices have been used in point-of-care clinical/home settings (for example, test strips for blood glucose measurement). These wearable sensors allow noninvasive chemical analysis of biofluids, such as sweat, saliva, or interstitial fluid (ISF), which fall under the category of epidermal wearable biosensors, or tears which represent ocular wearable biosensors. They hold promise due to their high specificity, speed, portability, and low-cost and low-power requirements. Epidermal devices rely on sweat or ISF sampling at the skin surface, along with transport of these biofluids over the biosensor surface. Biomarker molecules in tears diffuse directly from the blood and exhibit close correlation between tear and blood concentrations and present an opportunity for diagnosis of ocular disease.

The wide acceptance of such wearable biosensing technology requires a deep understanding of the biochemical composition of bodily fluids, such as sweat or tears, and its relation to blood chemistry. Wearable monitoring platforms can lend insights into dynamic biochemical processes in these biofluids by enabling continuous, real-time monitoring of biomarkers that can be related to a wearer's health and performance. Such real-time monitoring can provide information on wellness and enhance the management of chronic diseases by acting as a proactive mechanism for disease detection. They obviate painful and risky blood sampling procedures and can be readily blended with a wearer's daily routine. Multidisciplinary development of new biosensing technologies has led to numerous proof-of concept demonstrations and has driven growing efforts toward the commercialization of wearable sensors. However, these products still require further large-scale validation studies. Minimally invasive glucose monitoring devices is one such success story which has attracted a large-scale commercial interest. Some of the popular commercial prototypes are (a) smart contact lens from Google and Novartis, measuring glucose in tears [114,115], and (b) GlucoWatch, Freestyle Libre (in the form of a patch) [116], Eversense (subcutaneous implant), all measuring glucose in ISF [117].

References

1. N. D. Wong, Epidemiological studies of CHD and the evolution of preventive cardiology, *Nature Reviews Cardiology*, vol. 11, no. 5, 276, 2014.
2. WHO Fact Sheet, CVD. [Online]. Available: www.who.int/news-room/fact-sheets/detail/cardiovascular-diseases-(cvds).

3. WHO Fact Sheet, ASD. [Online]. Available: www.who.int/news-room/fact-sheets/detail/autism-spectrum-disorders.
4. WHO Fact Sheet, Diabetes. [Online]. Available: www.who.int/news-room/fact-sheets/detail/diabetes.
5. WHO Fact Sheet, Epilepsy. [Online]. Available: www.who.int/news-room/fact-sheets/detail/epilepsy.
6. Alzheimer's Association and others, 2017 Alzheimer's disease facts and figures, *Alzheimer's & Dementia*, vol. 13, no. 4, 325–373, 2017.
7. M. G. Austrom, M. Boustani, and M. A. LaMantia, Ongoing medical management to maximize health and well-being for persons living with dementia, *The Gerontologist*, vol. 58, no. suppl_1, S48–S57, 2018.
8. M. P. Calkins, From research to application: Supportive and therapeutic environments for people living with dementia, *The Gerontologist*, vol. 58, no. suppl_1, S114–S128, 2018.
9. E. J. Benjamin, S. S. Virani, C. W. Callaway, A. M. Chamberlain, A. R. Chang, S. Cheng, S. E. Chiuve, M. Cushman, F. N. Delling, R. Deo, and S. D. de Ferranti, Heart disease and stroke statistics—2018 update: A report from the American Heart Association, *Circulation*, vol. 137, no. 12, e67–e492, 2018.
10. D. Biswas, Recognition of elementary upper limb movements in nomadic environment, Doctoral Dissertation, University of Southampton, UK, 2015.
11. Z. Iqbal and M. Toft, TMEM230 variants in Parkinson's disease, *Nature Genetics*, 1, 2019.
12. R. A. Benschoter, M. T. Eaton, and P. Smith, Use of videotape to provide individual instruction in techniques of psychotherapy, *Academic Medicine*, vol. 40, no. 12, 1159–1161, 1965.
13. C. Combi, G. Pozzani, and G. Pozzi, Telemedicine for developing countries, *Applied Clinical Informatics*, vol. 7, no. 04, 1025–1050, 2016.
14. A. Al-Fuqaha, M. Guizani, M. Mohammadi, M. Aledhari, and M. Ayyash, Internet of Things: A survey on enabling technologies, protocols, and applications, *IEEE Communications Surveys & Tutorials*, vol. 17, no. 4, 2347–2376, 2015.
15. S. R. Islam, D. Kwak, M. H. Kabir, M. Hossain, and K.-S. Kwak, The Internet of Things for health care: A comprehensive survey, *IEEE Access*, vol. 3, 678–708, 2015.
16. O. Brand, Microsensor integration into systems-on-chip, *Proceedings of the IEEE*, vol. 94, no. 6, 1160–1176, 2006.
17. L. Atzori, A. Iera, and G. Morabito, The Internet of Things: A survey, *Computer Networks*, vol. 54, no. 15, 2787–2805, 2010.
18. A. M. Elmisery, S. Rho, and D. Botvich, A fog based middleware for automated compliance with OECD privacy principles in internet of healthcare things, *IEEE Access*, vol. 4, 8418–8441, 2016.
19. A. Khairuddin, K. F. K. Azir, and P. E. Kan, Limitations and future of electrocardiography devices: A review and the perspective from the Internet of Things, in *International Conference on Research and Innovation in Information Systems* (*ICRIIS*), 2017, pp. 1–7.
20. C. Li, X. Hu, and L. Zhang, The IoT-based heart disease monitoring system for pervasive healthcare service, *Procedia Computer Science*, vol. 112, 2328–2334, 2017.
21. P. M. Vergara, E. de la Cal, J. R. Villar, V. M. González, and J. Sedano, An IoT platform for epilepsy monitoring and supervising, *Journal of Sensors*, vol. 2017, 18, 2017.
22. S. Deshkar, R. Thanseeh, and V. G. Menon, A review on IoT based m-Health systems for diabetes, *International Journal of Computer Science and Telecommunications*, vol. 8, no. 1, 13–18, 2017.
23. D. Biswas, L. Everson, M. Liu, M. Panwar, B. Verhoef, S. Patrika, C. H. Kim, A. Acharyya, C. Van Hoof, M. Konijnenburg, and N. Van Helleputte, CorNET: Deep learning framework for PPG based heart rate estimation and biometric identification in ambulant environment, *IEEE Transactions on Biomedical Circuits and Systems*, vol. 13, no. 2, 282–291, 2019.
24. L. Catarinucci, D. De Donno, L. Mainetti, L. Palano, L. Patrono, M. L. Stefanizzi, and L. Tarricone, An IoT-aware architecture for smart healthcare systems, *IEEE Internet of Things Journal*, vol. 2, no. 6, 515–526, 2015.

25. Y. Yuehong, Y. Zeng, X. Chen, and Y. Fan, The Internet of Things in healthcare: An overview, *Journal of Industrial Information Integration*, vol. 1, 3–13, 2016.
26. M. W. Woo, J. Lee, and K. Park, A reliable IoT system for personal healthcare devices, *Future Generation Computer Systems*, vol. 78, 626–640, 2018.
27. B. Farahani, F. Firouzi, V. Chang, M. Badaroglu, N. Constant, and K. Mankodiya, Towards fog-driven IoT eHealth: Promises and challenges of IoT in medicine and healthcare, *Future Generation Computer Systems*, vol. 78, 659–676, 2018.
28. F. Firouzi, A. M. Rahmani, K. Mankodiya, M. Badaroglu, G. V. Merrett, P. Wong, and B. Farahani, Internet-of-Things and big data for smarter healthcare: From device to architecture, applications and analytics, *Future Generation Computer Systems*, vol. 78, 583–586, 2018.
29. D. Biswas, A. Cranny, and K. Maharatna, Body area sensing networks for remote health monitoring, in E. Vogiatzaki and A. Krukowski (eds.) *Modern Stroke Rehabilitation through e-Health-Based Entertainment*, Springer, Cham, 2016, pp. 85–136.
30. J. J. Rodrigues, D. B. D. R. Segundo, H. A. Junqueira, M. H. Sabino, R. M. Prince, J. Al-Muhtadi, and V. H. C. De Albuquerque, Enabling technologies for the internet of health things, *IEEE Access*, vol. 6, 13129–13141, 2018.
31. S. R. Levine and M. Gorman, 'Telestroke' the application of telemedicine for stroke, *Stroke*, vol. 30, no. 2, 464–469, 1999.
32. M. S. Hossain, C. Xu, Y. Li, J. Bilbao, and A. El Saddik, Advances in next-generation networking technologies for smart healthcare, *IEEE Communications Magazine*, vol. 56, no. 4, 14–15, 2018.
33. D. He, R. Ye, S. Chan, M. Guizani, and Y. Xu, Privacy in the Internet of Things for smart healthcare, *IEEE Communications Magazine*, vol. 56, no. 4, 38–44, 2018.
34. P. Maia, T. Batista, E. Cavalcante, A. Baffa, F. C. Delicato, P. F. Pires, and A. Zomaya, A Web platform for interconnecting body sensors and improving health care, *Procedia Computer Science*, vol. 40, 135–142, 2014.
35. E. Serafim and S. Motoyama, Estrutura de rede baseada em tecnologia IoT para atendimento médico em áreas urbanas e rurais, Doctoral dissertation, Master's thesis, Programa de Mestrado em Ciência da Computação, Faculdade Campo Limpo Paulista, 2014.
36. T. H. Laine, C. Lee, and H. Suk, Mobile gateway for ubiquitous health care system using zigbee and bluetooth, in *Eighth International Conference on Innovative Mobile and Internet Services in Ubiquitous Computing*, 2014, pp. 139–145.
37. N. M. Garcia and J. J. P. Rodrigues, *Ambient Assisted Living*, CRC Press, Boca Raton, FL, 2015.
38. Assisted Living Technologies. [Online]. Available: www.assistedlivingtechnologies.com/remote-monitoring-elderly/11-beclose.html. [Accessed: Apr-2019].
39. Fade: Fall Detector. [Online]. Available: http://fade.iter.es/. [Accessed: Apr-2019].
40. Medisafe. [Online]. Available: www.medisafeapp.com/. [Accessed: Apr-2019].
41. OnTrack Diabetes App. [Online]. Available: www.ontrack.org.au/diabetes/. [Accessed: Apr-2019].
42. Early Sense One. [Online]. Available: www.earlysense.com/earlysense-one/. [Accessed: Apr-2019].
43. NovaSom. [Online]. Available: www.novasom.com/. [Accessed: Apr-2019].
44. Proteus Discover. [Online]. Available: www.proteus.com/discover/. [Accessed: Apr-2019].
45. Apple Watch Series 4. [Online]. Available: www.apple.com/br/watch/. [Accessed: Apr-2019].
46. IoT and Wearable Solutions. [Online]. Available: www.bittium.com/products__services/iot_and_wearable_solutions/. [Accessed: Apr-2019].
47. QardioCore. [Online]. Available: www.getqardio.com/qardiocore-wearable-ecg-ekg-monitor-iphone/. [Accessed: Apr-2019].
48. Monica AN24. [Online]. Available: http://www.monicahealthcare.com/. [Accessed: Apr-2019].
49. Global Health Observatory, World Health Organization, Geneva, Switzerland, 2017. [Online]. Available: www.who.int/gho/publications/world_health_statistics/2017/en/. [Accessed: Apr-2019].

50. M. Pourhomayoun, N. Alshurafa, F. Dabiri, E. Ardestani, A. Samiee, H. Ghasemzadeh, and M. Sarrafzadeh, Why do we need a remote health monitoring system? A study on predictive analytics for heart failure patients, in *Proceedings of the 11th EAI International Conference on Body Area Networks*, 2016, pp. 171–172.
51. T. de Souza Rios and R. M. da Silva Bezerra, WHMS4: An integrated model for health remote monitoring: A case study in nursing homes for the elderly, in *10th Iberian Conference on Information Systems and Technologies (CISTI)*, 2015, pp. 1–6.
52. A. L. Bleda, F. J. Fernández-Luque, A. Rosa, J. Zapata, and R. Maestre, Smart sensory furniture based on WSN for ambient assisted living, *IEEE Sensors Journal*, vol. 17, no. 17, 5626–5636, 2017.
53. J. Rafferty, C. D. Nugent, J. Liu, and L. Chen, From activity recognition to intention recognition for assisted living within smart homes, *IEEE Transactions on Human-Machine Systems*, vol. 47, no. 3, 368–379, 2017.
54. F. Schwiegelshohn, M. Hubner, P. Wehner, and D. Gohringer, Tackling the new health-care paradigm through service robotics: Unobtrusive, efficient, reliable, and modular solutions for assisted-living environments, *IEEE Consumer Electronics Magazine*, vol. 6, no. 3, 34–41, 2017.
55. A. Machado, V. Maran, I. Augustin, L. K. Wives, and J. P. M. de Oliveira, Reactive, proactive, and extensible situation-awareness in ambient assisted living, *Expert Systems with Applications*, vol. 76, 21–35, 2017.
56. F. Erden, S. Velipasalar, A. Z. Alkar, and A. E. Cetin, Sensors in assisted living: A survey of signal and image processing methods, *IEEE Signal Processing Magazine*, vol. 33, no. 2, 36–44, 2016.
57. X. Chen, Z. Zhu, M. Chen, and Y. Li, Large-scale mobile fitness app usage analysis for smart health, *IEEE Communications Magazine*, vol. 56, no. 4, 46–52, 2018.
58. Que é Mobile Health 2017. [Online]. Available: http://saudebusiness.com/noticias/o-que-e-mobile-health-infogra_co/. [Accessed: Apr-2019].
59. J. P. Higgins, Smartphone applications for patients' health and fitness, *The American Journal of Medicine*, vol. 129, no. 1, 11–19, 2016.
60. J. Firth, J. Torous, J. Nicholas, R. Carney, S. Rosenbaum, and J. Sarris, Can smartphone mental health interventions reduce symptoms of anxiety? A meta-analysis of randomized controlled trials, *Journal of Affective Disorders*, vol. 218, 15–22, 2017.
61. A. J. J. Valera, M. A. Zamora, and A. F. Skarmeta, An architecture based on Internet of Things to support mobility and security in medical environments, in *7th IEEE Consumer Communications and Networking Conference*, 2010, pp. 1–5.
62. P. P. Ray, Home Health Hub Internet of Things (H3IoT): An architectural framework for monitoring health of elderly people, in *International Conference on Science Engineering and Management Research (ICSEMR)*, 2014, pp. 1–3.
63. Assisted Living Technologies. (2017). BeClose Remote Monitoring. [Online]. Available: www.assistedlivingtechnologies.com/. [Accessed: Apr-2019].
64. MC10. (2017). BioStampMD. [Online]. Available: www.mc10inc.com/our-products#BioStampMD. [Accessed: Apr-2019].
65. J. A. Stevens and R. A. Rudd, Circumstances and contributing causes of fall deaths among persons aged 65 and older: United States, 2010, *Journal of the American Geriatrics Society*, vol. 62, no. 3, 470–475, 2014.
66. I. C. Lopes, B. Vaidya, and J. J. Rodrigues, Towards an autonomous fall detection and alerting system on a mobile and pervasive environment, *Telecommunication Systems*, vol. 52, no. 4, 2299–2310, 2013.
67. E. T. Horta, I. C. Lopes, and J. J. Rodrigues, Ubiquitous mHealth approach for biofeedback monitoring with falls detection techniques and falls prevention methodologies, in S. Adibi (ed.) *Mobile Health*, Springer, Cham, 2015, pp. 43–75.
68. L. Y. Mano, M. M. Funes, T. Volpato, and J. R. T. Neto, Explorando tecnologias de IoT no contexto de Health Smart Home: uma abordagem para detecção de quedas em pessoas idosas, *Journal on Advances in Theoretical and Applied Informatics*, vol. 2, no. 1, 46–57, 2016.

69. K. A. Al Mamun, M. Alhussein, K. Sailunaz, and M. S. Islam, Cloud based framework for Parkinson's disease diagnosis and monitoring system for remote healthcare applications, *Future Generation Computer Systems*, vol. 66, 36–47, 2017.
70. P. Lorenzi, R. Rao, G. Romano, A. Kita, and F. Irrera, Mobile devices for the real-time detection of specific human motion disorders, *IEEE Sensors Journal*, vol. 16, no. 23, 8220–8227, 2016.
71. F. Xiao, Q. Miao, X. Xie, L. Sun, and R. Wang, Indoor anti-collision alarm system based on wearable Internet of Things for smart healthcare, *IEEE Communications Magazine*, vol. 56, no. 4, 53–59, 2018.
72. C. Dobbins, R. Rawassizadeh, and E. Momeni, Detecting physical activity within lifelogs towards preventing obesity and aiding ambient assisted living, *Neurocomputing*, vol. 230, 110–132, 2017.
73. L. Pepa, F. Verdini, and L. Spalazzi, Gait parameter and event estimation using smartphones, *Gait & Posture*, vol. 57, 217–223, 2017.
74. E. M. Shellington, T. Felfeli, R. Shigematsu, D. P. Gill, and R. J. Petrella, HealtheBrain: An innovative smartphone application to improve cognitive function in older adults, *Mhealth*, vol. 3, 7, 2017.
75. W. Liu, Y. Shoji, and R. Shinkuma, Logical correlation-based sleep scheduling for WSNs in ambient-assisted homes, *IEEE Sensors Journal*, vol. 17, no. 10, 3207–3218, 2017.
76. N. Alshurafa, C. Sideris, M. Pourhomayoun, H. Kalantarian, M. Sarrafzadeh, and J.-A. Eastwood, Remote health monitoring outcome success prediction using baseline and first month intervention data, *IEEE Journal of Biomedical and Health Informatics*, vol. 21, no. 2, 507–514, 2017.
77. J. H. Abawajy and M. M. Hassan, Federated Internet of Things and cloud computing pervasive patient health monitoring system, *IEEE Communications Magazine*, vol. 55, no. 1, 48–53, 2017.
78. C. Crema, A. Depari, A. Flammini, E. Sisinni, A. Vezzoli, and P. Bellagente, Virtual respiratory rate sensors: An example of a smartphone-based integrated and multiparametric mHealth gateway, *IEEE Transactions on Instrumentation and Measurement*, vol. 66, no. 9, 2456–2463, 2017.
79. D. Aranki, G. Kurillo, P. Yan, D. M. Liebovitz, and R. Bajcsy, Real-time tele-monitoring of patients with chronic heart-failure using a smartphone: Lessons learned, *IEEE Transactions on Affective Computing*, vol. 7, no. 3, 206–219, 2016.
80. M. W. Raad, T. Sheltami, and E. Shakshuki, Ubiquitous tele-health system for elderly patients with Alzheimer's, *Procedia Computer Science*, vol. 52, 685–689, 2015.
81. M. Chen, Y. Ma, Y. Li, D. Wu, Y. Zhang, and C.-H. Youn, Wearable 2.0: Enabling human-cloud integration in next generation healthcare systems, *IEEE Communications Magazine*, vol. 55, no. 1, 54–61, 2017.
82. J. Li, H. Zhou, D. Zuo, K.-M. Hou, and C. De Vaulx, Ubiquitous health monitoring and real-time cardiac arrhythmias detection: A case study, *Bio-Medical Materials and Engineering*, vol. 24, no. 1, 1027–1033, 2014.
83. E. Spanò, S. Di Pascoli, and G. Iannaccone, Low-power wearable ECG monitoring system for multiple-patient remote monitoring, *IEEE Sensors Journal*, vol. 16, no. 13, 5452–5462, 2016.
84. M. Chen, J. Yang, J. Zhou, Y. Hao, J. Zhang, and C.-H. Youn, 5G-smart diabetes: Toward personalized diabetes diagnosis with healthcare big data clouds, *IEEE Communications Magazine*, vol. 56, no. 4, 16–23, 2018.
85. M. Cassimatis, D. J. Kavanagh, A. P. Hills, A. C. Smith, P. A. Scuffham, S. Edge, J. Gibson, and C. Gericke, Development of the OnTrack diabetes program, *JMIR Research Protocols*, vol. 4, no. 2, 24, 2015.
86. L. Mainetti, L. Patrono, A. Secco, and I. Sergi, An IoT-aware AAL system for elderly people, in *International Multidisciplinary Conference on Computer and Energy Science (SpliTech)*, 2016, pp. 1–6.
87. G. Muhammad, M. F. Alhamid, M. Alsulaiman, and B. Gupta, Edge computing with cloud for voice disorder assessment and treatment, *IEEE Communications Magazine*, vol. 56, no. 4, 60–65, 2018.

88. E. Dantas, Low cost embedded hardware for blood pressure measurement based on Internet of Things, Master's thesis, Programa de Pós-Graduação em Engenharia Elétrica, Inst. Federal de Educação, Ciência e Tecnologia da Paraiba, 2016.
89. G. Matar, J.-M. Lina, J. Carrier, A. Riley, and G. Kaddoum, Internet of Things in sleep monitoring: An application for posture recognition using supervised learning, in *IEEE 18th International Conference on e-Health Networking, Applications and Services (Healthcom)*, 2016, pp. 1–6.
90. R. Ganapathy, A. Grewal, and J. Castleman, Remote monitoring of blood pressure to reduce the risk of preeclampsia related complications with an innovative use of mobile technology, *Pregnancy Hypertension: An International Journal of Women's Cardiovascular Health*, vol. 6, no. 4, 263–265, 2016.
91. E. Demir, E. Köseouglu, R. Sokullu, and B. Seker, Smart home assistant for ambient assisted living of elderly people with dementia, *Procedia Computer Science*, vol. 113, 609–614, 2017.
92. P. Silsupadol, K. Teja, and V. Lugade, Reliability and validity of a smartphone-based assessment of gait parameters across walking speed and smartphone locations: Body, bag, belt, hand, and pocket, *Gait & Posture*, vol. 58, 516–522, 2017.
93. B. M. Silva, J. J. Rodrigues, I. de la Torre Diez, M. López-Coronado, and K. Saleem, Mobile-health: A review of current state in 2015, *Journal of Biomedical Informatics*, vol. 56, 265–272, 2015.
94. L. Wang, P. C. Pedersen, D. M. Strong, B. Tulu, E. Agu, and R. Ignotz, Smartphone-based wound assessment system for patients with diabetes, *IEEE Transactions on Biomedical Engineering*, vol. 62, no. 2, 477–488, 2015.
95. M. Velikova, R. L. Smeets, J. T. van Scheltinga, P. J. Lucas, and M. Spaanderman, Smartphone-based analysis of biochemical tests for health monitoring support at home, *Healthcare Technology Letters*, vol. 1, no. 3, 92–97, 2014.
96. E. Chao, C. K. Meenan, and L. K. Ferris, Smartphone-based applications for skin monitoring and melanoma detection, *Dermatologic Clinics*, vol. 35, no. 4, 551–557, 2017.
97. L. M. Brayboy, A. Sepolen, T. Mezoian, L. Schultz, B. S. Landgren-Mills, N. Spencer, C. Wheeler, and M. A. Clark, Girl talk: A smartphone application to teach sexual health education to adolescent girls, *Journal of Pediatric and Adolescent Gynecology*, vol. 30, no. 1, 23–28, 2017.
98. S. K. Sood and I. Mahajan, Wearable IoT sensor based healthcare system for identifying and controlling chikungunya virus, *Computers in Industry*, vol. 91, 33–44, 2017.
99. S. Cirani and M. Picone, Wearable computing for the Internet of Things, *It Professional*, vol. 17, no. 5, 35–41, 2015.
100. L. Piwek, D. A. Ellis, S. Andrews, and A. Joinson, The rise of consumer health wearables: Promises and barriers, *PLoS Medicine*, vol. 13, no. 2, e1001953, 2016.
101. K. Maharatna, E. B. Mazomenos, J. Morgan, and S. Bonfiglio, Towards the development of next-generation remote healthcare system: Some practical considerations, in *IEEE International Symposium on Circuits and Systems (ISCAS)*, 2012, pp. 1–4.
102. N. Alshurafa, J.-A. Eastwood, S. Nyamathi, J. J. Liu, W. Xu, H. Ghasemzadeh, M. Pourhomayoun, and M. Sarrafzadeh, Improving compliance in remote healthcare systems through smartphone battery optimization, *IEEE Journal of Biomedical and Health Informatics*, vol. 19, no. 1, 57–63, 2015.
103. C. Seeger, K. Van Laerhoven, and A. Buchmann, My Health Assistant: An event-driven middleware for multiple medical applications on a smartphone-mediated body sensor network, *IEEE Journal of Biomedical and Health Informatics*, vol. 19, no. 2, 752–760, 2015.
104. E. Zdravevski, P. Lameski, V. Trajkovik, A. Kulakov, I. Chorbev, R. Goleva, N. Pombo, and N. Garcia, Improving activity recognition accuracy in ambient-assisted living systems by automated feature engineering, *IEEE Access*, vol. 5, 5262–5280, 2017.
105. D. Biswas, K. Maharatna, G. Panic, E. B. Mazomenos, J. Achner, J. Klemke, M. Jöbges, and S. Ortmann, Low-complexity framework for movement classification using body-worn sensors, *IEEE Transactions on Very Large Scale Integration (VLSI) Systems*, vol. 25, no. 4, 1537–1548, 2017.
106. M. Konijnenburg, R. van Wegberg, S. Song, H. Ha, W. Sijbers, J. Xu, S. Stanzione, C. van Liempd, D. Biswas, A. Breeschoten, and P. Vis, 22.1 A 769μW battery-powered single-chip SoC with BLE for multi-modal vital sign health patches, in *IEEE International Solid-State Circuits Conference-(ISSCC)*, 2019, pp. 360–362.

107. D. Biswas, G. J. Ajiwibawa, K. Maharatna, A. Cranny, J. Achner, J. Klemke, and M. Jöbges, Real-time arm movement recognition using FPGA, in *IEEE International Symposium on Circuits and Systems (ISCAS)*, 2015, pp. 766–769.
108. K. Austen, What could derail the wearables revolution? *Nature News*, vol. 525, no. 7567, 22, 2015.
109. Q. Dong, Y. Chen, X. Li, and K. Zeng, Position paper: A theoretical framework for general cognition evaluation of cognitive radios, in *IEEE International Smart Cities Conference (ISC2)*, 2018, pp. 1–8.
110. M. Rahman, B. Carbunar, and U. Topkara, SensCrypt: A secure protocol for managing low power fitness trackers, in *IEEE 22nd International Conference on Network Protocols*, 2014, pp. 191–196.
111. S. Pirbhulal, P. Shang, W. Wu, A. K. Sangaiah, O. W. Samuel, and G. Li, Fuzzy vault-based biometric security method for tele-health monitoring systems, *Computers & Electrical Engineering*, vol. 71, 546–557, 2018.
112. S. Ravindran, V. Loseu, M. Polley, M. Goel, K. H. Lee, and S. Lee, Method and apparatus for a wearable based authentication for improved user experience, U.S. Patent Application 10/142,332. 2018.
113. J. Kim, A. S. Campbell, B. E.-F. de Ávila, and J. Wang, Wearable biosensors for healthcare monitoring, *Nature Biotechnology*, vol. 37, no. 4, 1, 2019. doi:10.1038/s41587-019-0045-y.
114. D. L. Chandler, The eye as a window to health: Albeit slow, research is progressing on contact lenses for medical diagnostics, *IEEE Pulse*, vol. 9, no. 6, 20–23, 2018.
115. M. Senior, Novartis signs up for Google smart lens, 2014.
116. Freestyle Libre, Abbott. [Online]. Available: www.freestylelibre.us/. [Accessed: Apr-2019].
117. Eversense, Senseonics. [Online]. Available: www.eversensediabetes.com/. [Accessed: Apr-2019].

2

Biomedical Sensors and Data Acquisition

Rajarshi Gupta
University of Calcutta

CONTENTS

2.1 General Principles of Biomedical Measurements

Human body consists of many interconnected complex physio-psychological systems, many of which generate of tiny electrical potentials within certain body tissues. Measurement of these small potentials can provide valuable information of the functioning of the organs. For important body organs like heart, brain, muscles, and eyes, these electrical potentials can be related to their pathological conditions. These biomedical signals are broadly categorized into two broad classes, viz., endogenous signals and exogenous signals. Endogenous signals (often called physiological signals) are generated on specific body surface area or group of tissues due to the underlying physiological phenomena. Examples are electrocardiogram (ECG), electroencephalogram (EEG). On the other hand, exogenous signals are the result of modulations on some externally applied energy (ultrasound) or radiation (X-ray) non-invasively while passing through some body tissues and collected by some sensors. Examples are computed tomography (CT), magnetic resonance imaging (MRI), photoelectric plethysmography (PPG) [1]. Over the years, various engineering techniques have been

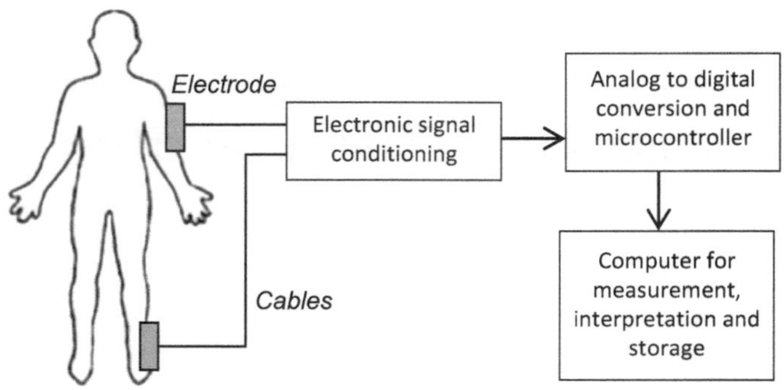

FIGURE 2.1
Basic schematic of a biomedical measurement system.

developed to collect the endogenous and exogenous signals for diagnostic actions. The general structure and measurement principle of an advanced biomedical measurement system can be shown in Figure 2.1.

The biomedical signals are picked up through the metal electrodes, and cables connect them to the biomedical equipment (BME). The first stage of the BME is an electronic circuit which is primarily responsible for refinement of the weak biopotential. This function includes noise reduction, amplification, scaling, and patient isolation, collectively known as signal conditioning. This emphasizes (boosts up) the clinical information from the weak biopotential, discarding the non-relevant information. The next stage is discretization of the continuous domain signal into discrete samples and their representation using finite bit-lengths using analog to digital converter (ADC). A microcontroller transfers the quantized samples to a computer which is the final data storage, interpretation, and measurement device. Often, for standalone biomedical instruments, a low- to medium-end processor is provided for elementary signal analysis and direct digital readout of the physiological parameter. As such, these principles are very similar to any typical process measurement system with the additional constraints of maintaining the safety of the human subject (patient) from electrical, radiation power levels, and other toxic hazards.

Over the last three decades or more, the development of various signal processing algorithms has enabled computerized measurement and interpretation of human physiological processes, equally correct and consistent with that of a medical doctor. As a result, in recent years, the tendency to use computers in hospital setting for automated measurement and analysis of medical signals has been increased for rapid processing of high volume biomedical data collected from a number of patients, with the additional facility of data archival and sharing with authorized users. The scope of this chapter, however, is to provide a brief overview of origin, sensing, and interfacing principles of some prime biomedical signals which are related to cardiovascular, respiratory, and nervous system of human physiologic system.

2.2 Bioelectric Phenomena and Biopotential Electrodes

The genesis of all physiological signals may be attributed to the transients in intercellular membrane potentials associated with excitable living cells, known as action potentials (APs) [2] in response to an external stimulation. The electrophysiological basis of generation of small electric potentials is due to the fact these cells are polarized in normal condition due to distribution of Na^+, K^+, Ca^{++}, Cl^- ions in the intra- and extracellular fluids and semi-permeable nature of the cell membrane. This potential, named resting membrane potential (RMP) of most human living cells, is around –20 to –70 mV while measured inside to outside of the cell membrane. The steady-state RMP is expressed by Goldman–Hodgkin–Katz equation [3]:

$$V_{mo} = -\frac{RT}{F}\left\{\frac{[Na^+]_i P_{Na^+} + [K^+]_i P_{K^+} + [Cl^-]_i P_{Cl^-}}{[Na^+]_e P_{Na^+} + [K^+]_e P_{K^+} + [Cl^-]_e P_{Cl^-}}\right\}, \tag{2.1}$$

where T is the Kelvin temperature, R is the gas constant [8.314 J/(mol K)], F is the Faraday number, 96,500 Cb/mol, and P_X is the permeability for ion species X.

This 'polarized' behavior cell of the cell undergoes a reversible process due to the application of short duration external stimulus. The semi-permeable nature of the cell membrane is governed by a Na–K pump theory, which describes the transfer of Na^+ and K^+ ions between intra- and extracellular fluids through the membrane. This makes the cell potential positive for sometime, called 'depolarization' of the cell. After the stimulus is withdrawn, it slowly recovers to the initial polarized state (negative polarity) through a slow 'repolarization' phase. The recording of the membrane potential during these reversible changes is called AP of the cell. Each type of excitable living cells have their specific signatures (amplitude, morphology, and frequency character). Figure 2.2 shows the typical waveform morphology of typical AP of a human cell. As soon as the external stimulus is applied, the cell membrane potential rapidly changes to slightly positive state, indicating depolarization (time duration D). When the stimulus is withdrawn, the membrane potential slowly recovers to the repolarized state (time duration R), after which some afterpotentials (state A) may be observed.

The recording of any physiological signal is the collective representation of these tiny APs resulted from a group of tissues over space and time at predefined body location(s).

Sensing of small bioelectric potentials involve interaction with the ions and convert them suitably into an electric current using electrodes and instrumentation electronics. Metal electrodes of different configurations are used for measuring medical signals. Very often, an electrode gel is used for better contact and chemical stability of the

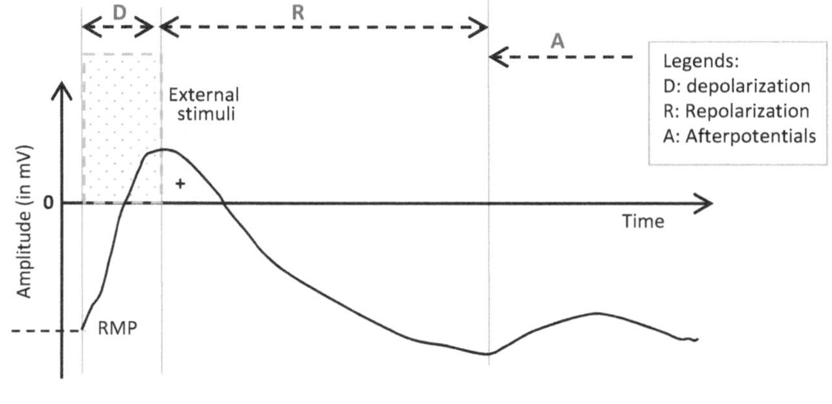

FIGURE 2.2
Typical AP waveform of a human excitable living cell.

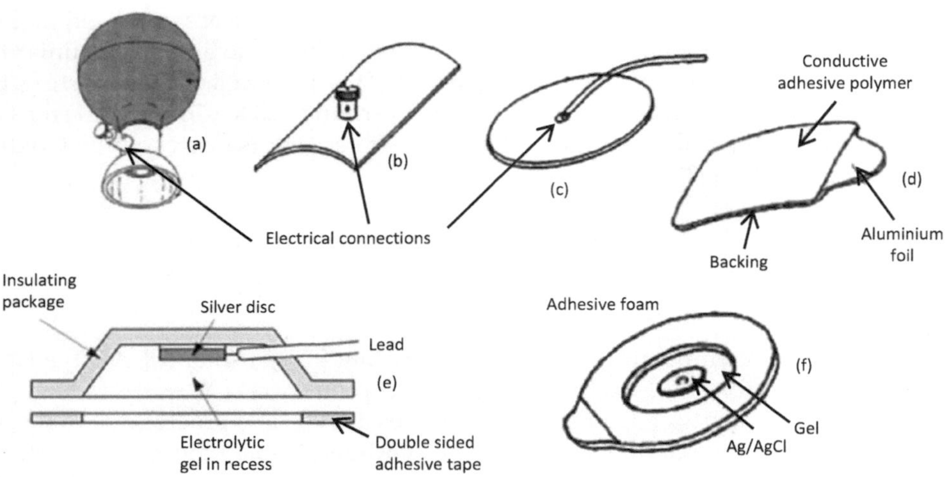

FIGURE 2.3
Common biopotential electrodes used in clinical setting: (a–c) metal electrodes; (d and e) disposable electrode; (f) reusable Ag/AgCl electrode.

electrode–electrolyte combination. It additionally provides protection against slippage of the electrodes at the body surface. Choice of materials for designing biopotential electrodes requires an understanding of the electrochemistry of the electrode–electrolyte combination and their behavioral model [4]. Electrodes are categorized into polarized and non-polarized according to their model. Silver–silver chloride electrodes are found to be most stable. They are made of silver disks coated electrolytically with silver chloride. Commercially, two types of electrodes are used in biopotential recordings, viz., reusable and disposable. Figure 2.3 shows some of the commonly used electrode configurations used in clinical settings [3]. Here, configurations (a–c) are metal reusable electrodes used in common clinical settings. Materials used for these types of electrodes include German silver (a nickel–silver alloy), silver, gold, and platinum. Among these, Figure 2.3a is a suction cup electrode for precordial or chest leads used for ECG recording. The Figure 2.3b type is used as limb electrode. The type in Figure 2.3c is a general metal disc electrode which is normally used in long-term ECG recording. A lead wire is soldered into the back surface for electrical connection to recorder. A thin insulating material epoxy or polyvinylchloride protects the conducting part from lead wire. Figure 2.3d shows special conductive polymer electrodes, exploiting the properties of conductive and adhesive properties. The polymer is attached to a metallic backing made of silver or aluminium foil, allowing electric contact to external amplifier. Figure 2.3e shows disposable electrodes used in long-term monitoring in the intensive care unit (ICU). Here, the electrodes must have a stable interface with the skin. Mechanical stability of the interface between the electrode and the skin can help to reduce motion artifacts. Here, a large foam pad is attached to the electrode body with adhesive coating on one side. The cup is then securely fastened to the skin surface using a double-sided adhesive ring. The gel is trapped in the packet, between the electrode contact surface and the skin. Movement of the skin with respect to the electrode may affect the electrolyte near the skin–electrolyte interface, but the electrode–electrolyte interface can be several millimeters away from this location, since it is recessed in the cup. The fluid movement is unlikely to affect the recessed electrode–electrolyte interface as compared to what would happen if the electrode was

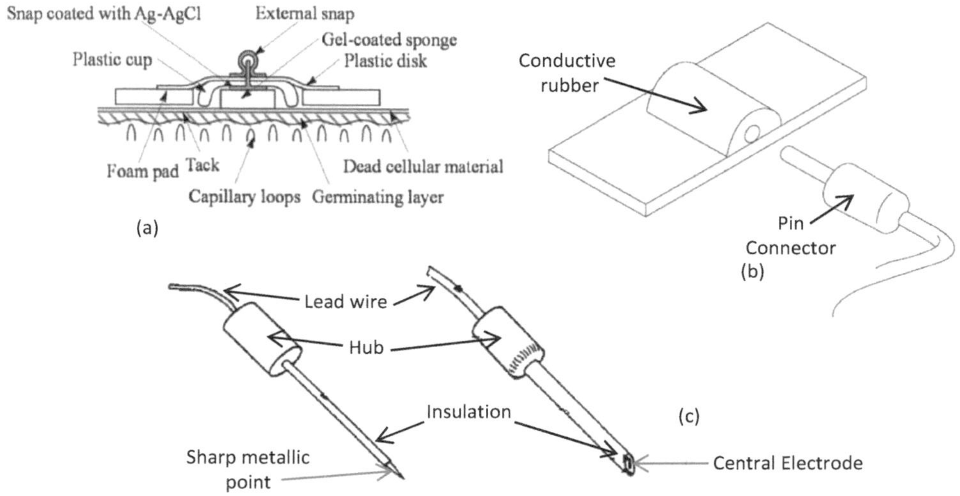

FIGURE 2.4
Some specialized biomedical electrodes: (a) basic recessed electrode; (b) flexible electrode; and (c) needle electrode.

separated from the skin by just a thin layer of electrolyte. Figure 2.3f shows another reusable silver–silver chloride electrode attached to the skin by non-allergic adhesive tape and provides high-quality measurement.

Some special electrode configurations used in patient monitoring are shown in Figure 2.4. A recessed electrode basic configuration is shown in Figure 2.4a. The electrolyte layer now consists of a sponge saturated with a thickened electrolytic solution. The sponge serves the same function as the recess in the cup electrodes and is coupled directly to a silver–silver chloride electrode. For neonatal monitoring, flexible electrodes, as shown in Figure 2.4b, are used. They are basically similar to the metal plate electrodes. A carbon-filled silicon rubber compound is used as the active material. The other one is a Mylar electrode, consisting of a 13 μm thick Mylar film, on which Ag and AgCl is deposited. The lead wire is attached to the Mylar substrate through a conductive adhesive. A thin layer of Ag is deposited on the adhesive layer and the Mylar. The additional advantages of flexible electrodes are being penetrable by X-rays and relief from repeated use and detachment for clinical tests. A second group of special electrodes are used to collect the biopotential from within the body tissues. These do not require electrolytic gel and can be broadly discussed under the two heads, viz., needle electrodes and microelectrodes. The basic needle electrode, as shown in Figure 2.4c, consists of a sharp stainless steel pointed needle, with the shank insulated with a coating. When this structure is placed in tissue such as skeletal muscle, electrical signals can be picked up by the exposed tip. One can also make needle electrodes by running one or more insulated wires down the lumen of a standard hypodermic needle.

For EEG measurements, special types of electrodes are required since these are to be placed on the cortex. The standard electrodes consist of a metal plate covering an area of a few square millimeters. The electrolyte, normally a saline solution, is either contained in liquid form in a pad wrapping the electrode or is bound in a gel or paste filling the gap between electrode and skin. Different methods of fixing the electrodes on the cortex are available: a grid of rubber strips, special caps with the electrodes being integrated into the cap textile, gluing using

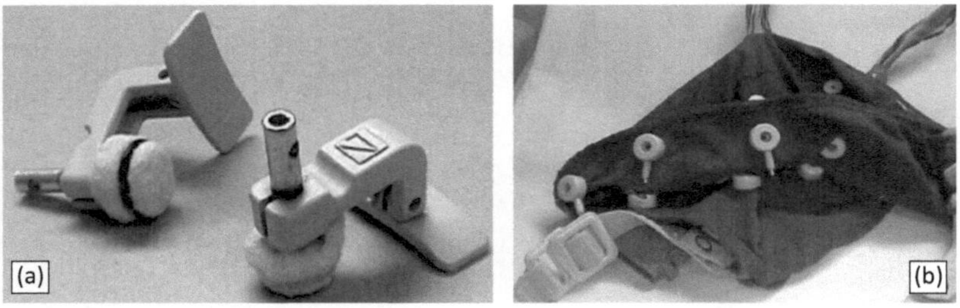

FIGURE 2.5
Electrodes used for EEG measurements: (a) metal electrodes; (b) electrodes fitted in a cap for automated positioning.

a collodium–acetone solution, paste (i.e., the paste providing the electrolyte also serves as a fixation aide). Ag/AgCl electrodes (silver electrodes coated with a thin silver chloride layer) are the current standard for routine applications. The chloride layer prevents polarization. A special variant is made by sintering a mixed powder of Ag/AgCl onto the electrode rather than applying it as a discrete metal coating. This electrode requires less maintenance because, with conventional Ag/AgCl electrodes, the AgCl layer is easily eroded with regular use.

Figure 2.5a shows metal electrodes used for EEG measurements. For specialized applications like ICU monitoring, needle electrodes made of refined steel are pierced into the skin. For regular use, modern electrode arrangements using caps (called 'electrocaps') are available (lower) where all the electrodes are automatically placed at their right position once the cap is applied.

2.3 Origin and Measurements of Cardiovascular and Respiration Signals

2.3.1 ECG, PPG, Respiration, Blood Pressure, and PCG

ECG is the time-averaged representation of tiny electrical potentials that are developed on cardiac muscles and propagate over specialized conducting fibers over the heart surface. These tiny electric potentials originate on the sinoatrial node (SA) of the right atrium and propagate to the atrioventricular (AV) node first, and then spread over the left and right ventricles over the Purkinje fibers. As a result, the atrial and ventricular cells sequentially depolarize and repolarize, generating the pumping action of the heart. One complete cardiac cycle comprises of following events: (a) collection of impure blood from inferior and superior vena cava in the right atrium, with simultaneous collection of oxygen-rich blood at left atrium to fill the atria; (b) atrial contraction to force fill blood into the ventricles from atria; (c) ventricular contraction to send the impure blood from right ventricles to lungs for purification, with simultaneous pumping of blood to body parts through aorta [5]. This generates the P-QRS-T waveform of the ECG, given by Einthoven, the inventor. A typical ECG waveform and associated electrophysiological events are shown in Figure 2.6.

ECG signal is collected from body extremities, viz., both hands, left leg, and chest. Each collection requires two electrodes and a reference. Each of these measurements is referred to as 'lead'. Universally, 12 lead ECG is the clinically accepted standard for cardiac function evaluation. Among these, six leads are collected from frontal plane, out of which three are

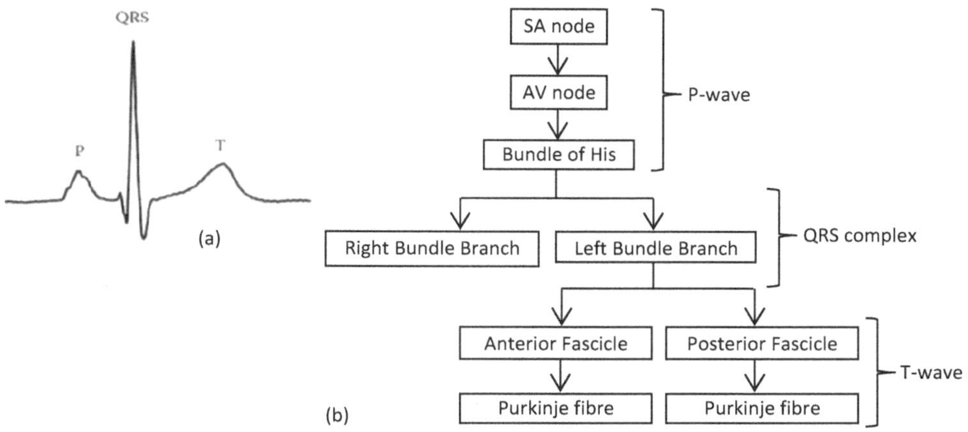

FIGURE 2.6
(a) A typical ECG waveform; (b) associated events with heart chambers

called bipolar leads of Einthoven, and the other three are unipolar leads. The remaining six are collected from the transverse plane and are called precordial or chest leads. The standard leads provide bipolar measurements, designated as I, II, and III, indicating differential measurements between a pair of limb positions. The remaining three unipolar leads from frontal plane measurements are unipolar measurements, designated as aVR, aVL, and aVF, also named augmented limb leads. These potentials are measured with reference to a common point, Wilson Central Terminal (WCT), which also provides reference for six chest leads, V1, V2, V3, V4, V5, and V6. The lead positions are shown in Figure 2.7 and detailed in Table 2.1.

Most of the clinical information in ECG is available in the frequency band of 0.05–100 Hz. However, there are various artifacts which corrupt the ECG, few of which have overlapping spectra with the ECG. The major artifacts in ECG are electromyography (EMG) noises, power line interference (PLI), electrode pop or contact noise, baseline wander, motion artifact, electrosurgical noise, and amplifier noise [5,6]. The noise and artifacts of ECG are briefly described below:

1. **EMG noise:** Picked up due to muscular activity of the subject during the ECG collection. Their amplitude is in the range 0.1–1 mV, and frequency 5 Hz–1 kHz is partly overlapping with that of ECG signal. EMG noise, if not properly taken care of, may completely destroy the signal in morphology-based analysis. For short duration clinical testing, the patient is advised to lie on resting condition so as to minimize the noise. However, for long duration Holter monitoring, this EMG noise is unavoidable.

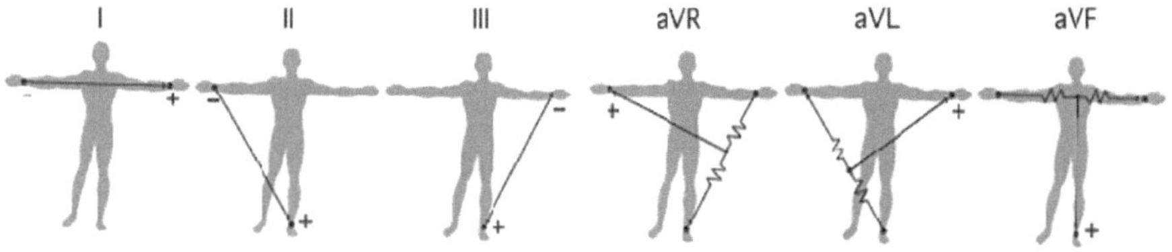

FIGURE 2.7
ECG lead conventions and body positions.

TABLE 2.1

ECG Lead Positions

Lead Name	Exploring/Measuring Electrode	Reference
I	Left arm	Right arm
II	Left leg	Right arm
III	Left leg	Right arm
aVL	Left arm	Wilson central terminal
aVR	Right arm	Wilson central terminal
aVF	Left leg	Wilson central terminal
V1	Right fourth intercostal space	Wilson central terminal
V2	Left fourth intercostal space	Wilson central terminal
V3	Halfway between V2 and V4	Wilson central terminal
V4	Left fifth intercostal space, midclavicular line	Wilson central terminal
V5	Horizontal to V4, anterior axillary line	Wilson central terminal
V6	Horizontal to V5, midaxillary line	Wilson central terminal

2. **Power line interference (PLI):** These are picked up on the lead wires of neighboring power cables, due to capacitive coupling with ECG lead wires [7]. So, a 50/60 Hz±0.2Hz current flows through the lead wires to the ground through the patient's body. Sometimes, the equivalent voltage drop, which appears as a common-mode signal to the input of the ECG amplifier can be as high as 20mV, which is itself many times greater than the maximum ECG amplitude. Driven right leg (DRL) circuit, as shown later in Figure 2.10, can minimize the PLI effect.
3. **Electrode pop or contact noise:** During the collection of ECG, loss of contact between the patient's skin and electrode may cause a temporary saturation of the amplifier output for certain period of time.
4. **Baseline wander:** The lung volume change due to the respiration of the patient changes the impedance between the heart muscle and electrode. This causes baseline (isoelectric segments) of the ECG to oscillate at a very low frequency drifting between 0.15 and 0.3Hz.
5. **Motion artifacts:** Due to improper "preparation" of the skin for electrode placement, or patient movement, a slow movement of the electrodes may occur in long-term recording using wearable sensors. Motion artifact has a significant overlap with ECG signal spectrum in the range 1–10Hz. This results in an abrupt baseline jump or complete saturation of amplifier output for 0.5s.
6. **Electrosurgical noise:** This is the noise generated by neighboring medical equipment in the clinical set-up at frequencies between 100kHz and 1MHz.
7. **Amplifier noise:** Noise and drift are two unwanted signals that are generated within the amplifier that contaminate a biopotential signal under measurement. 'Noise' generally refers to undesirable signals with spectral components above 0.1Hz, while 'drift' generally refers to slow changes in the baseline at frequencies below 0.1Hz. The noise and drift are measured either in microvolts peak to peak (μV_{p-p}) or microvolts root mean square (RMS) (μV_{RMS}) and applies as if they were applied as differential input voltage.

These artifacts can be minimized by suitable designs and clinical set-up; however, it is impossible to completely eliminate those altogether using hardware designs. An ECG amplifier must be able to minimize these external interferences and faithfully amplify the potentials due to cardiac systole and diastole only. Additionally, the design should also ensure patient safety arising out of accidental direct electric current flow through the cardiac muscle. The basic specifications of an ECG amplifier is given below:

1. **Gain:** The gain of an ECG amplifier is expressed in decibels as

$$\text{Gain (dB)} = 20\log_{10}(\text{linear gain}). \tag{2.2}$$

   The nominal linear gain of ECG amplifier is around 1,000.
2. **Bandwidth:** The frequency bandwidth of the amplifier should be designed so as to amplify, without attenuation, all frequency components of the ECG.
3. **Common-mode rejection (CMR):** Some of the artifacts picked up from external sources appear as 'common-mode signal' (i.e., at both input leads) at the input of the amplifier. Thus CMR of a biopotential amplifier (BPA) qualitatively describes its capability to reject these low-amplitude signals at the input stage and is measured by common-mode rejection ratio (CMRR), defined as

$$\text{CMRR (dB)} = 20\log_{10}\left(\frac{A_{\text{d}}}{A_{\text{cm}}}\right), \tag{2.3}$$

   where A_{cm} stands for gain for common-mode input and A_{d} stands for gain for differential input. Most biopotential amplifiers offer a typical CMRR of 120–140. A novel, balanced AC coupled differential amplifier with AC coupled input stage and a third stage providing ground path using a third (common) electrode is described in [8]. The CMRR of ECG amplifier greatly depends on matching of the resistance pairs in the instrumentation amplifier (INA) block, which, in practical circuits, is difficult to achieve. This results in an amplification of the DC offset voltage, mainly caused due to electrode–skin impedance and PLI. In [9], a design approach is proposed which does not rely on matching of resistors. It includes a fully differential DC suppression circuit in addition to AC coupling for full restoration of DC biopotential signals. The power line frequency (and its harmonics) induces stray capacitive coupling with the patient's body and lead wires. The order of these capacitances is 60 pfd, which corresponds to an impedance of 64 kΩ at 50 Hz. The leakage current produces a voltage drop w.r.t. ground, which appears as common-mode signal to the first stage of the ECG amplifier. To minimize this interference effect, the CMR plays a vital role in rejecting this noise.
4. **Recovery:** Sudden movement of the patient during ECG procedure can saturate the amplifier output due to high amplitude input transient pulses. After a small interval, called 'recovery time', the amplifier slowly comes back to the normal condition.
5. **Input impedance:** This should be sufficiently high so as to ensure that input signal is not attenuated. Most common ECG amplifiers offer an input impedance of 10 MΩ.

   The schematic of a typical ECG signal conditioning circuit to generate a unipolar voltage is shown in Figure 2.8. The lead selector circuit can be manually operated or controlled by a microcontroller in simultaneous or sequential mode. The objective of the isolation circuit is to provide galvanic isolation (1,000 MΩ or more) between the patient's body and the signal conditioning to prevent any accidental current flowing through the patient's body. The total gain of the amplifier,

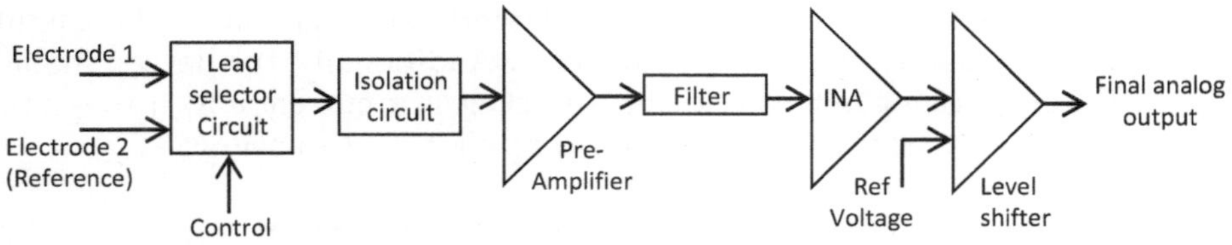

FIGURE 2.8
Block schematic shows components of a typical ECG amplifier. (Printed with permissions from "Biomedical sensors and their interfacing" in *Advanced Interfacing Techniques for Sensors*, B. George et al. (Eds.), Springer Nature, 2017.)

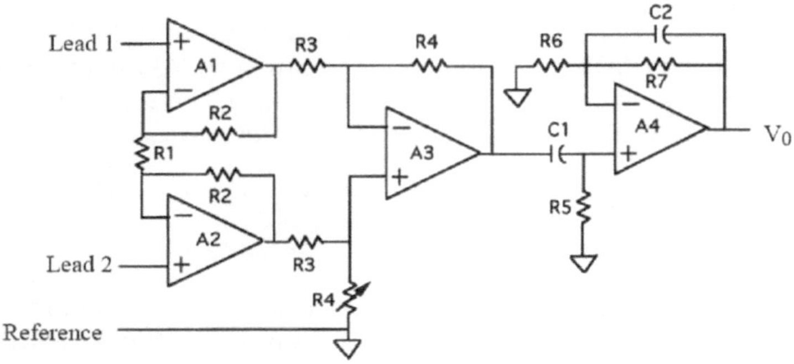

FIGURE 2.9
An ECG amplifier circuit using INA.

distributed over a preamplifier and an INA, is around 2,000. In the first stage, a preamplifier provides a low gain, typically 10–20. The filter circuit mainly discards high-frequency noise and passes the clinical bandwidth of 0.05–100 Hz to the INA. The INA provides a gain of 100–200. After INA, the final level shifter is used to get a unipolar voltage of 0–5 V.

The main component of the ECG signal component is the INA, which should provide a high CMR and amplify the weak ECG signal. The first stage (two op-amps A1 and A2) of the circuit shown in Figure 2.9 represents a standard INA, which provides high differential gain and unity common-mode gain without the requirement of close resistor matching. The differential output from A1 an A2 represents a signal with substantial relative reduction of the common-mode signal and is used to drive a standard differential amplifier (op-amp A3) which further reduces the common-mode signal. CMRR of the output op-amp as well as resistor matching in its circuit are less critical than in the another variant of INA, the follower type. Offset trimming for the whole circuit is done at one of the input op-amps. Complete INA integrated circuits based on this standard INA configuration are available from several manufacturers [10]. The third stage (op-amp A4) represents a band-pass filter with some gain. The total gain of the circuit as shown in Figure 2.9 is given as

$$G = \left(1 + \frac{2R_2}{R_1}\right)\left(\frac{R_4}{R_3}\right)\left(1 + \frac{R_7}{R_6}\right). \tag{2.4}$$

The filter passband is represented as

$$f = \frac{1}{2\pi}\left(\frac{1}{R_1C_1} - \frac{1}{R_2C_2}\right). \tag{2.5}$$

A good example of commercial ECG amplifier is AD620 from Analog Devices, a modification of the classic three OPAMP INA approach. The gain of the amplifier is determined by a single external resistance, and potentiometer for offset trimming are externally connected. In addition to the certain desirable characteristics, specific design enhancements of the ECG amplifier are required:

1. **PLI reduction**: Stray capacitances are formed between the ECG lead wires and neighbouring power lines. The displacement currents flow to ground through the patient's body and right leg, which is used as reference. Considering a small resistance r_0 of common lead wire (at right leg) and displacement current i_d, the small drop $V_c = r_0 \cdot i_d$ appears as common to the first stage of INA. A high-CMRR INA can minimize this effect. Additionally, a driven leg circuit (DRL) is used [3,11], as shown in Figure 2.10a. In the DRL circuit, two sensing registers (R_1) are used at the output of the first stage of the INA. They invert, amplify, and feedback the voltage V_c to the right leg of the patient. The modified common-mode voltage is given by

$$V_c = \frac{i_d r_0}{1 + 2R_2/R_1}. \tag{2.6}$$

   Hence, the interference signal is minimized at the amplifier input. A low-power VLSI implementation of DRL circuit is described in [12]. A second option to minimize the PLI is to implement a 50/60 Hz notch filter in the INA, as shown in Figure 2.10b.

2. **Patient isolation:** The objective of patient safety in a biosignal measurement application is to ensure that current from the acquisition circuit or its applied part can flow through the patient's body to ground, termed 'leakage current', and remain within safe levels in case of a fault. Thus, all biomedical amplifiers acquiring physiological signals from human body must satisfy some safety criteria for the worst-case voltage breakdown and maximum leakage currents through the electrodes attached to the human body. The accepted international standard is IEC-601 for Medical Electrical Equipment adopted by Europe as EN-60601. This is

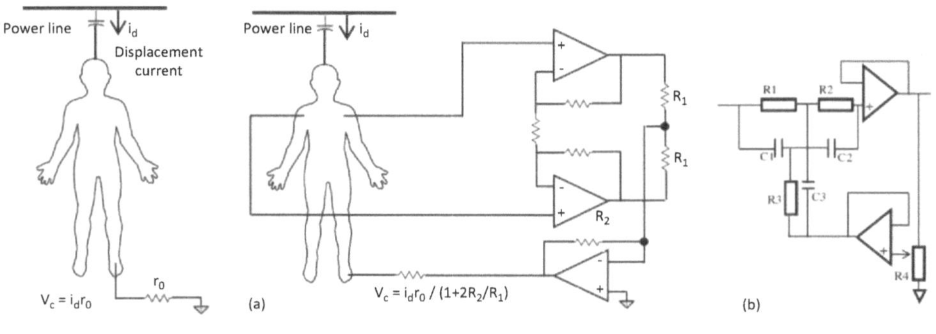

FIGURE 2.10
PLI reduction techniques in ECG measurements: (a) driven right leg; (b) notch filter. (Printed with permissions from "Biomedical sensors and their interfacing" in *Advanced Interfacing Techniques for Sensors*, B. George et al. (Eds.), Springer Nature, 2017.)

combined with UL (Underwriters Laboratories Inc.) Standard 2601-1 for the United States and endorsed by the Health Industries Manufacturers' Association (HIMA), National Electrical Manufacturers Association (NEMA), and US Food and Drug Administration (FDA) [3]. IEC-60601-1 standard allows a patient auxiliary current (current that can flow between two separate leads connected to the patient's body) up to 100 μA at not less than 0.1 Hz. The objective is to achieve complete galvanic isolation between the patient and acquisition circuit, and using surge protection arising from defibrillator or electrosurgical equipment.

Galvanic isolation is achieved by electrically separating the input stage of an isolation amplifier (IsoA) from the output stage. An IsoA is realized by any of the following techniques; viz., (a) isolation transformer through high-frequency magnetic coupling, (b) optical isolation using LED–phototransistor combination, and (c) capacitive coupling using a signal-modulated high-frequency digital carrier from the input stage (isolated) through a pair of capacitors to a demodulator at the output stage. Here, the input stage has a separate floating power supply and a 'ground' that are connected to the output side of the IsoA by a high impedance (around 1,000 MΩ) and a low parallel capacitance (around pfd range). The signal terminals of the input stage are isolated from the IsoA's output by similar high impedance. The schematic circuits of optical, magnetic, and capacitive isolation are shown in Figure 2.11. An optical isolation amplifier uses a linear opto-isolator using infrared LED and a phototransistor to separate the input and output, as shown in Figure 2.11a. Using optical isolation, a breakdown voltage up to 4–7 kV can be achieved. In magnetic isolation, a transformer is used to achieve nearly 7 kV of isolation between the primary and secondary. The low-frequency ECG signal modulates a high-frequency carrier (around 500 kHz) signal, and is transferred to the secondary through transformer action. At the output, a demodulator (low-pass filter) extracts the ECG. This is schematically shown in Figure 2.11b. In capacitive IsoAs, an internal oscillator is used to modulate the analog differential signal to a digital pulse train, which is transmitted across the isolation barrier, built around

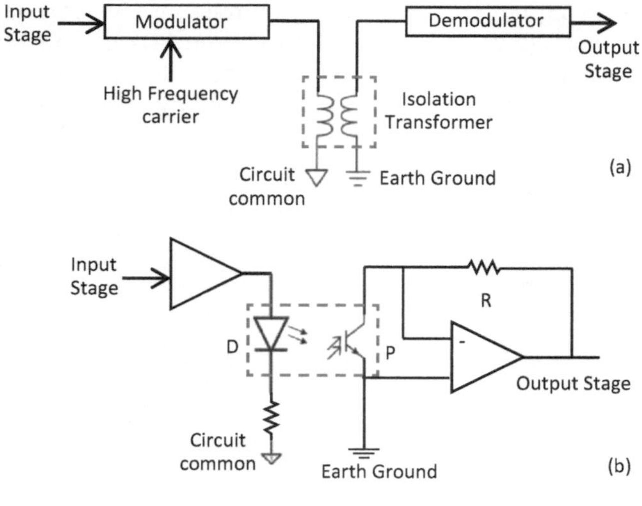

FIGURE 2.11
Schematics of patient isolation using isolation amplifier: (a) optical isolation; (b) magnetic isolation. (Printed with permissions from "Biomedical sensors and their interfacing" in *Advanced Interfacing Techniques for Sensors*, B. George et al. (Eds.), Springer Nature, 2017).

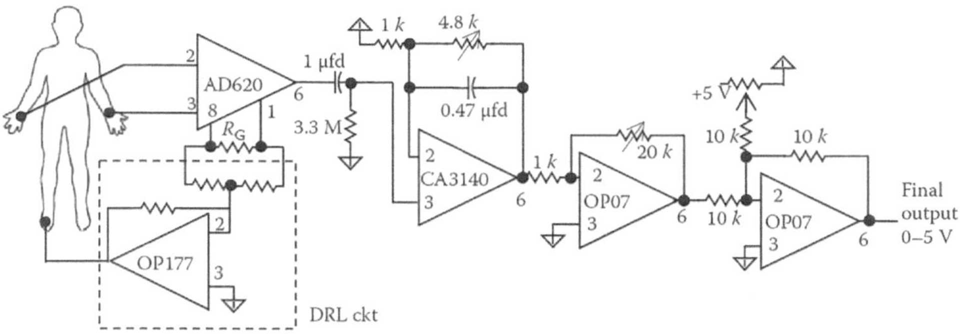

FIGURE 2.12
A practical ECG amplifier circuit. (Printed with permissions from "Telecardiology" in *Telemedicine and Electronic Medicine*, J.G. Webster, H. Eren (Eds.), CRC Press, 2017.)

a matched pair of capacitors (1–3 pfd range). At the output stage, the modulated signal is converted back to analog voltage by an averaging technique. The performance of an isolation amplifier is described by isolation-mode rejection ratio (IMRR), which refers to its ability to suppress the feed-through isolation-mode voltage that arises across the barrier and output stage, detailed in [28].

Accidental voltage surge protection from the biomedical equipment can be achieved by connecting a voltage limiting device (e.g., zener diode) between the connecting electrode and ground. The device absorbs the extra current inrush when the voltage across it raises a certain value, typically 300 mV or higher.

A practical single-channel ECG amplifier is shown in Figure 2.12. The first stage uses AD620 with a gain of 10–20 to avoid saturation of its output by electrode offset potential, which is around 300 mV. The gain setting resistance can be chosen as per the equation

$$R_G = \frac{49.4k}{G-1}. \tag{2.7}$$

The output of AD620 is used to implement a DRL circuit, as well as AC coupled to block the DC and low frequency (up to 0.05 Hz) using a low-pass filter (1 μF and 3.3 MΩ combination). The second stage uses a 70 Hz low-pass filter (CA3140) with a small gain. The last two stages (OP07s) provide a gain of 20 and a variable DC offset to obtain unipolar output. Nowadays, many soft computational techniques are available which are used for denoizing of digitized ECG.

2.3.2 Photoplethysmography (PPG) and Pulse Oximetry

Photoelectric plethysmography, commonly known as photoplethysmography (PPG), is an optical technique for measuring blood volume changes in peripheral body parts (fingertips, toe, earlobes, and forehead). The commonest use of PPG is in pulse oximetry, the measurement of blood oxygen saturation in ICUs. Over the last 20 years, research of PPG has shown its great potential toward various physiological functions like cardiac output, blood pressure, respiratory functions, heart rate, and many more [13]. There are two basic sensing techniques for PPG, viz., reflection mode and transmission mode [14]. Both use a matched pair or LED–photodiode (PD) combination attached to body part to illuminate a small portion of skin tissue and sense the blood volume information based on difference

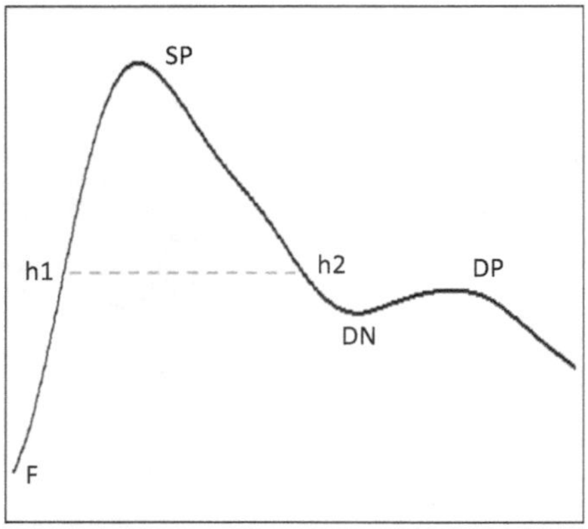

FIGURE 2.13
A typical transmission PPG waveform and clinical features. (Printed with permissions from "Biomedical sensors and their interfacing" in *Advanced Interfacing Techniques for Sensors*, B. George et al. (Eds.), Springer Nature, 2017.)

of absorbance of light radiation between blood-full and bloodless skin at near infrared (IR) wavelengths. The transmission mode, LED (operating in 0.8–1 μm wavelength), and PD are attached to opposite surface of skin, and the PD captures the light after passing through the bones, tissues, and blood capillaries. Under this wavelength, light is least absorbed by water content of the tissue. The PPG waveform shows a steady (DC) and a pulsatile component. The steady part represents thermoregulation and autonomic functions. The pulsatile component, as shown in Figure 2.13, is related to blood volume changes and consists of two phases, anacrotic (relates to ventricular systole) and catacrotic (relates to ventricular diastole), represented by four major fiducial points, viz., foot (F), systolic peak (SP), dicrotic notch (DN), and diastolic peak (DP). During ventricular systole, there is a momentary rise in blood volume in the peripheral capillaries, and in the mentioned wavelength, the received light intensity at the PD is less. The absorption of light by blood is quantitatively expressed by Beer–Lambert's law:

$$I = I_0 \mathrm{e}^{-d\alpha c}, \tag{2.8}$$

where I is the light intensity at a distance d, I_0 is the intensity at source, α is the absorption coefficient of blood, and c is the concentration of the medium. However, other factors like scattering, reflection, etc. from bones and skin tissues are not considered. By convention, the PPG waveform is inverted so that it correlates positively with blood volume. In catacrotic phase, more light is absorbed by the PD due to less blood flow. The anacrotic phase is characterized by SP. In the catacrotic phase, there is momentary rise of blood flow in peripheral capillary due to reflection from pulse wave pressure from the closed aortic valve. This results in the DN and DP. Like ECG, there are certain morphological features which can provide significant clinical information on cardiovascular parameters, as shown in Figure 2.13. Transmission-mode PPG sensing is feasible only for tissue thickness of up to a few centimeters. For thicker body sites, the detectable light intensity is too faint to produce a satisfactory signal to noise ratio (SNR). Transmission-mode measurements are typically performed on digits and the earlobes. In reflection-mode PPG, the LED and PD are attached to same surface of skin, and green wavelength in the range 0.5–0.6 μm has become popular in recent times. Unlike transmission mode, this configuration can

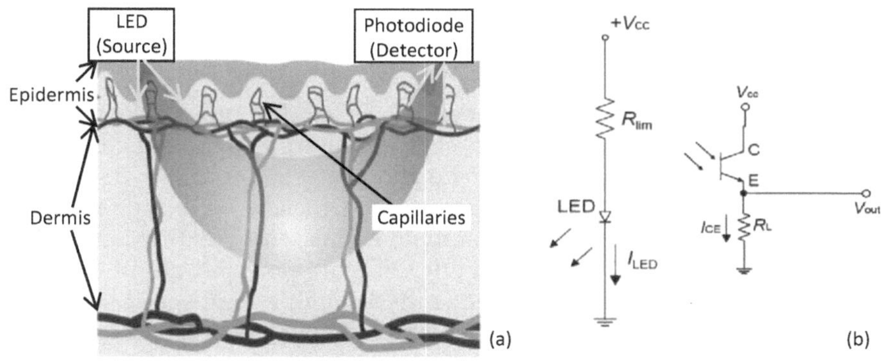

FIGURE 2.14
Reflective PPG: (a) tissue penetration and back scattering; (b) source and detector circuit. (Printed with permissions from "Biomedical sensors and their interfacing" in *Advanced Interfacing Techniques for Sensors*, B. George et al. (Eds.), Springer Nature, 2017.)

be applied to any site of the body. In this mode, the emitted photons that arrive at the PD follow a banana-shaped primary light path through the tissue, barely penetrating deeper than the skin, as shown in Figure 2.14a. This is due to the reason that with the increase in blood flow in a capillary, non-hemolyzed erythrocytes act as little mirrors which reflect the incident radiation to the PD [15].

The PPG sensor electronics shown in Figure 2.14b consists of a forward biases LED with a current limiting resistance R_{lim}, with the light intensity produced being directly proportional to the diode current (few mA I_{LED}) to produce a voltage drop of nearly 2 V across the diode. In the detection circuit (a phototransistor), a load resistor (R_L) converts the output current into a proportional voltage V_{out}:

$$V_{out} = I_{CE}R_L. \tag{2.9}$$

A typical PPG signal conditioning circuit is shown in Figure 2.15. It consists of two identical cascaded stages of passive high-pass filter (cutoff at 0.5 Hz) followed by active low-pass filter (cutoff at 3.4 Hz). The potentiometer *P* provides adjustable gain to the final output. The reference voltage is kept fixed at 2 V.

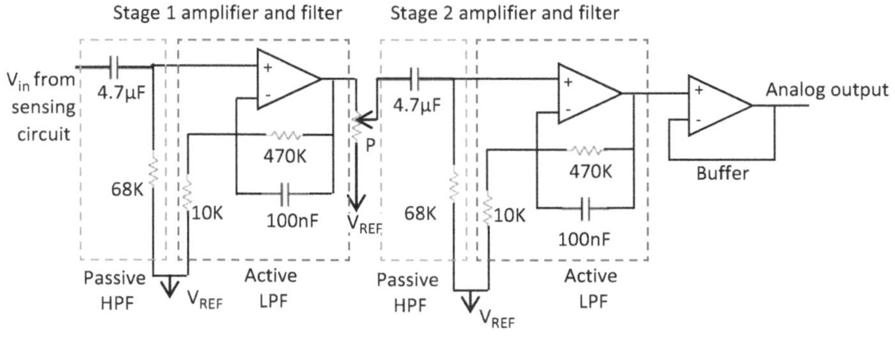

FIGURE 2.15
PPG signal conditioning circuit. (Printed with permissions from "Biomedical sensors and their interfacing" in *Advanced Interfacing Techniques for Sensors*, B. George et al. (Eds.), Springer Nature, 2017.)

In recent times, there has been resurgence over the use of PPG for cardiovascular function monitoring, and many improvised integrated sensors have been proposed which facilitate wearable and wireless monitoring [16–18].

A useful application of transmission-type PPG is found in pulse oximetry, which measures oxygen content in blood. Here, two LEDs are used as light sources, one at 660 nm (red) and the other at 940 nm (IR) with a single photodetector (normally, a silicon PD) as the sensor. In these two wavelengths, the absorption coefficients of oxygen-rich (HbO_2) and oxygen-less blood exhibit maximum separation, expressed by extinction coefficient [19]. In each cycle, the red LED is turned on, both LEDs are powered off, IR LED is powered on, and both LEDs are powered off in sequence through a high-speed switching circuit. The photodetector is mounted on the same plane but opposite side of the finger to receive radiation pulses from both the LEDs. The output of the photodetector is very small (order or few mV), and hence protection against electromagnetic interference (EMI) and radiofrequency interference (RFI) is essential. The most common practice is to use a high-CMRR INA to minimize the common-mode interference signal. The schematic block diagram of a standalone pulse oximeter is shown in Figure 2.16. The LEDs and PD are mounted in a flexible, light-resistant enclosure. A single screened cable powers the LEDs as well as carries the PD output to the electronics module. The switching of the LEDs is controlled by a microcontroller unit, which also calculates the SpO_2 after separating the signal components corresponding to oxygen-rich and oxygen-less blood from the PD output. The PD output is the representation of the absorption components due to tissue, venous blood, arterial blood, and variation due to pulsation in arterial blood (plethysmogram information). The amplified signal is filtered to get the representation of red, IR, and 'dark' light components, from which the DC or steady parts are extracted and scaled to equal values by the microcontroller unit. Then the pulsating components [AC] Red and [AC] IR corresponding to red and IR are extracted, and their amplitude ratio provides a measure of SpO_2 as

$$SpO_2 = \frac{[AC]_{Red}}{[AC]_{IR}}. \tag{2.10}$$

Improved accuracy in SpO_2 measurement is obtained by calculating the ratio described by Equation (2.10) multiple times instead of once per cardiac cycle.

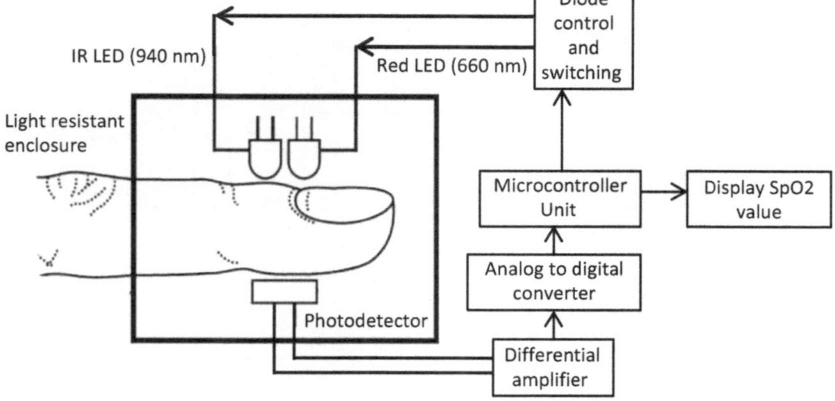

FIGURE 2.16
Schematic of pulse oximetry standalone instrument using PPG.

2.3.3 Respiration Signal Sensing

Two key functions in the respiratory system are the transport of the gas from the environment and alveoli via the branching airway tree and diffusion of the gases across the physical barrier separating the alveoli from the pulmonary capillaries. These key functions can be related to direct or indirect measurement of various parameters which can lead to the objective assessment of lungs, airways, and chest wall functionality. The various measurements under the respiratory system can be categorized under the following heads: (a) gas pressure, (b) gas flow and volume, (c) concentration and partial pressure of gases, (d) pulse oximetry and capnography, (e) lung mechanical function (spirometry and body plethysmography), and (f) gas exchange [1]. The current discussion is confined to airflow sensing and breathing rate using plethysmography.

Thermal convection flowmeters using electrically heated sensing elements like metal wires, metal films, and thermistors are used to measure gas flow rate through the nasal cavity. This provides indirect measurement of respiration in terms of lung volume or its changes [20]. These thermal flow sensors are based on the principle of convective heat exchange between the fluid and the active sensing element while the fluid passes over the hot body (sensor). The sensing element is connected in a full Wheatstone bridge and normally designed to operate either in constant temperature mode (Figure 2.17a) or in constant current mode (Figure 2.17b). The airflow due to breathing takes away the heat from the sensing element. In constant temperature mode, the sensing (heater) element is

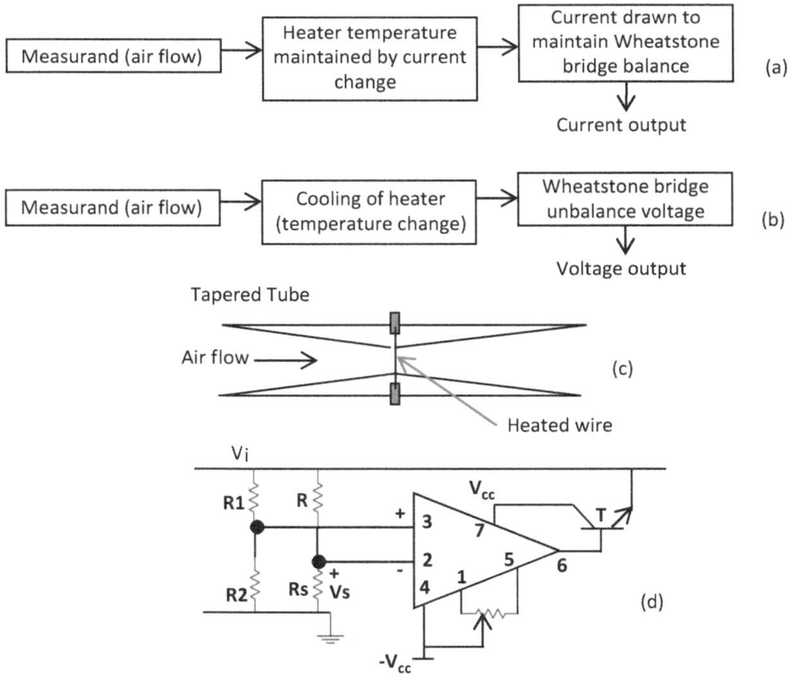

FIGURE 2.17
Basic scheme of thermal convection flowmeters for respiration measurement: (a) constant temperature mode; (b) constant current mode; (c) hot wire anemometer sensor configuration; (d) signal conditioning circuit of bridge output. (Printed with permissions from "Biomedical sensors and their interfacing" in *Advanced Interfacing Techniques for Sensors*, B. George et al. (Eds.), Springer Nature, 2017.)

maintained at constant temperature by the adjustment of current from the source, which is used as a measure of airflow rate. In constant current mode, the heater is supplied with a constant current, and bridge unbalance provides a measure of airflow rate. A typical hot wire anemometer configuration used in clinical setting is shown in Figure 2.17c, where a heated wire is mounted perpendicular to the flow line (nasal cavity or mouth) in a tapered tube. The bridge unbalance voltage is fed to a high-gain DC amplifier and an emitter follower stage, as shown in Figure 2.17d [21].

Microwave Doppler radar (MDR) [22,23] is another popular form of respiratory airflow sensing technique that is fully non-contact type measurement. A typical MDR system employs a pair of transmitting (Tx) and receiving (Rx) antennas mounted close to the body of the subject. Here, a radio wave, typically in the GHz range, is transmitted, and the modulated version of the signal due to quasi-periodic motion of the chest wall/abdomen due to inhalation, exhalation and the pause in between is received. The signal processing circuit captures the frequency shift of the reflected wave from the target (chest or abdomen). The theoretical analysis [24] shows that the inhalation, exhalation, and the pause in between them can be represented as a phase-modulated signal, with the phase shift directly proportional to the subject's chest movement. A typical test set-up of MDR respiration measurement system with interfacing schematic is shown in Figure 2.18. A pair of Tx–Rx antennas transmits and receives the frequency-modulated continuous wave (CW) microwave radiation in the GHz range, controlled by the MDR unit. The received signal is passed through a direct quadrature demodulator with RF and baseband automatic gain control, facilitating quadrature demodulation into direct baseband frequencies. The final output is passed through a low-pass filter and amplified to extract the respiration signal [25,26]. Additional advantages of MDR technique are robustness against environmental factors, interference from other sensor signals, etc.

Ultrasonic respiration sensors can use either transit time principle or the Doppler effect, i.e., change in frequency or phase shift of the reflected signal based on quasi-periodic motion of the chest due to respiration [27]. Direct time-of-flight measurements suffer from the inaccuracies that occur in measurement of the flight time (~μs or less) and inertial delay problem in piezo sensor. A piezo crystal operating at 40 kHz frequency transmits short bursts of ultrasonic wave to the subject's chest. The receiver calculates the time of travel from the reflected pulses, considering the speed of sound wave in air. The resolution in measurement can be improved with higher frequency (~240 kHz) activation of the sensor. Since the echo signal of an ultrasonic wave contains baseline wandering, high-frequency signals, and bursts; envelope detection principle from the received pulse waveform provides an indirect measurement of the time of flight of the ultrasonic pulses [28,29]. As shown in Figure 2.19, the transceiver is placed near the body, typically within 100 cm to avoid heavy attenuation of reflected ultrasonic wave

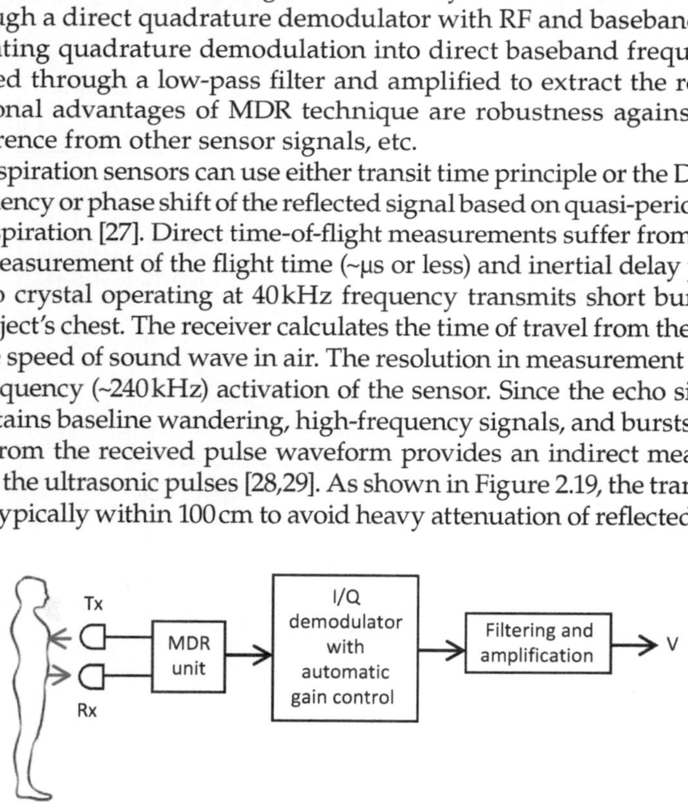

FIGURE 2.18
Schematic set-up of MDR for respiration rate measurement. (Printed with permissions from "Biomedical sensors and their interfacing" in *Advanced Interfacing Techniques for Sensors*, B. George et al. (Eds.), Springer Nature, 2017.)

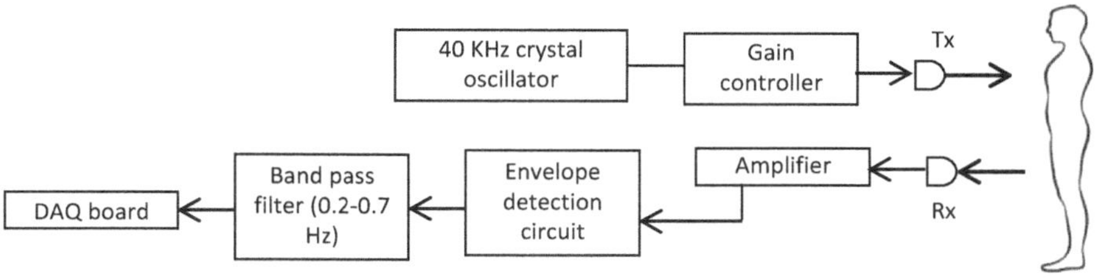

FIGURE 2.19
Ultrasonic respiration measurement schematic. (Printed with permissions from "Biomedical sensors and their interfacing" in *Advanced Interfacing Techniques for Sensors*, B. George et al. (Eds.), Springer Nature, 2017.)

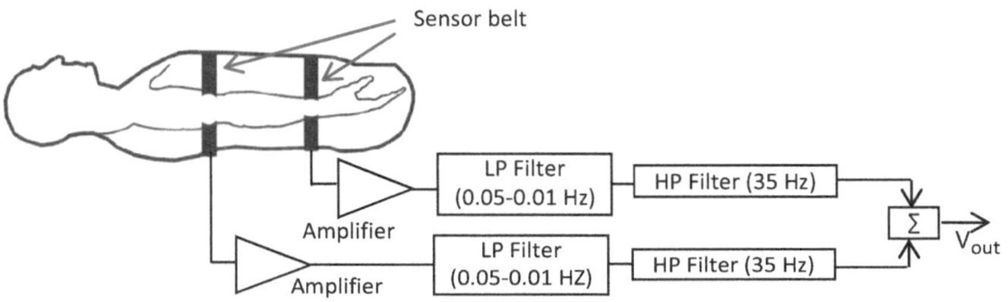

FIGURE 2.20
Basic RIP sensor configuration. (Printed with permissions from "Biomedical sensors and their interfacing" in *Advanced Interfacing Techniques for Sensors*, B. George et al. (Eds.), Springer Nature, 2017.)

energy. The reflected signal is passed through an amplifier, an envelope detection unit, and a band-pass filter to extract the low-frequency respiration information. An envelope detection circuit detects the peaks from the amplitude-modulated high-frequency signals employing a combination of moving average filter and adaptive threshold at the output of amplifier. Finally, the analog output is digitized and collected in a personal computer.

Respiratory inductive plethysmography (RIP) is a general technique used for measuring respiration or breathing effort by change in abdominal or chest sectional area. The configuration can utilize either elastic belt tension (piezo or strain gauge as sensing element) or electrical impedance (change in current between two electrode positions) or inductive-type plethysmography (frequency change in an RF signal looped in a chest belt) [30–32]. Among these, RIP sensors of the chest belt type are popular and often taken as standard to calibrate other respiration sensors. The common RIP sensor consists of an elastic chest and abdomen belt (around 1 in. wide) with a zigzagging wire sewn into it and worn around the rib cage under the armpits and abdomen. A small current (~20 mA) is passed through the coil from an oscillator. During inspiration, the abdomen (rib cage) area decreases (increases) with the lung volume. A typical RIP configuration is shown in Figure 2.20. The outputs of the sensing units are separately amplified and filtered to extract the respiration information before being added. Since the amplitudes of abdomen and chest sensors are unequal, the outputs of filters are normalized before addition.

2.3.4 Phonocardiogram (PCG) Signal

The electrophysiological action of the heart is manifested by sounds and vibrations due to its mechanical pumping actions. These vibrations are classified as sounds and murmurs. Generally, heart sounds are of short duration, whereas the murmurs have a noisy character and are of longer duration. The audible and most prominent heart sounds, generally known as 'lub-dub', are related to the closing of heart valves. There are two other heart sounds that are weaker in amplitude and originate due to blood flow through the heart chambers.

Phonocardiography is the recording of heart sound and murmurs resulting from the motion of blood through the different cardiac chambers and opening and closing of different valves to facilitate this blood flow. It was first recorded by Einthoven in 1907 using carbon microphone and string galvanometer. Phonocardiography is one of the primary auscultation techniques that can be used for cardiac function assessment and often synchronously recorded with standardized cardiovascular signals like ECG. A typical PCG recording consists of heart sounds and heart murmurs. While heart sounds are generated due to resonant phenomena of cardiac structures like heart valves, the heart murmurs are said to originate from blood flow turbulence through the cardiac chambers.

Figure 2.21 shows a typical PCG record with ECG and closure and opening of four cardiac valves – tricuspid, aortic, mitral (or bicuspid), and pulmonary – indicated by binary logic states (open: '0' and close: '1') [33].

Each cardiac cycle is considered to comprise of two periods, viz., systole, representing the contraction or depolarization of cardiac chambers, and diastole, representing the relaxation or repolarization of cardiac chambers. Generation of ECG has already been briefly stated in Section 2.3.1. The PCG consists of four major components I, II, III, and IV, also named as S1, S2, S3, and S4. Out of these, S1 and S2 are generated due to valve closure, have the largest intensity, and are audible as 'lub' and 'dub', respectively. The first heart sound S1 is observed between closing of mitral (and bicuspid) valve and opening of aortic (and pulmonary) valve. The second heart sound (S2) is audible between the exactly opposite events, i.e., opening of aortic (and pulmonary) valves and closing of tricuspid (and bicuspid) valves. S3 and S4 are with comparatively dull and weak in intensity, observed in children and certain adults, and not related to valve activity.

In Figure 2.21, the R-peak position is considered the beginning of ventricular systole, representing closure of AV valves (tricuspid and mitral), also known as isovolumetric

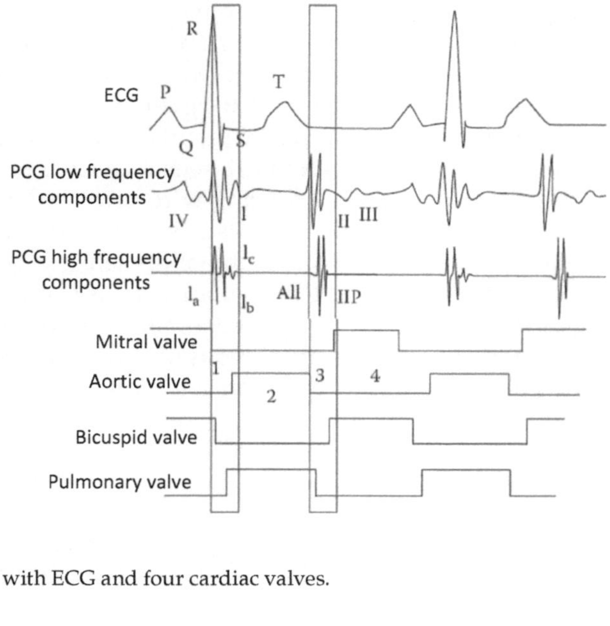

FIGURE 2.21
Representation of PCG with ECG and four cardiac valves.

ventricular contraction. This generates the first component S1 of the PCG, audible as 'lub'. This can be decomposed into four subcomponents. The initial component is due to shifting of ventricular blood toward the atria, closing both AV valves. The second component is due to abrupt tension in closed AV valves, slowing down the blood inside ventricles. The third component of S1 is due to oscillations of blood between root of aorta and ventricular walls. The fourth component is due to oscillations caused due to turbulent blood in the aorta and pulmonary valve. After the end of ventricular ejection, isovolumetric ventricular relaxation starts, represented by the T-wave of the ECG. This causes the fall of BP inside the ventricles below the large arterial pressure, resulting in the closure of aortic and pulmonary valves. The second PCG component, S2, audible as 'dub', starts after the end of T-wave. Its two subcomponents represent the closure of aortic valve (louder) and closure of pulmonary valve. The third heart sound component, S3, is due to rapid filling of blood from atria to ventricles, with both AV valves open. This causes the ventricular walls to vibrate. The final and fourth heart sound, S4, also known as atrial heart sound and occurs during atrial contraction, synchronous to the PQ segment of ECG. It is not audible [1].

Pathological patterns in the PCG signal can be observed in case of cardiac malfunction, especially in case of aortic and mitral regurgitation and stenosis, both causing heart murmurs. Wide splitting in S1 is caused by right bundle branch block (RBB), narrowing of tricuspid valve, and atrial septal defect. On the other hand, the splitting between two observable components of S1 disappears in case of left bundle branch block (LBB). Wide splitting of S2 is caused by delayed pulmonary or advanced aortic valve closing. Decreased intensity of S2 may be due to stiffened leaflets of aortic and pulmonary valve leaflets.

The heart sounds generated by all the valves travel in radial directions and can be audible by placing a stethoscope in chest surface. However, best auscultation sites are those where the intensity of the heart sounds is maximum. Right second intercostal space corresponds to aortic area; left second intercostal space corresponds to pulmonic area; midclavicular line located at fifth intercostal space corresponds to mitral area; and right fifth intercostal space corresponds to the tricuspid area.

2.3.4.1 PCG Signal Recording

Recording of PCG signal is a typical vibration measurement problem in soft tissue. It requires a vibration sensor, required amplification, filtering, recording in a computer, and visualization. The useful bandwidth of the signal is 0.2–1,000 Hz. However, for manual measurement using a stethoscope, the lower limit is bounded to 20 Hz for human hearing ability. The normal heart sounds are low-amplitude, low-frequency vibration signals with major frequency components in the range 10–400 Hz. However, murmurs have a higher frequency range (up to 600 Hz). The frequency analysis of the first heart sound S1 shows peaks in low-frequency range 10–50 Hz and in medium-frequency range 50–140 Hz. For the second heart sound S2, the analysis shows peaks in low-frequency range 10–80 Hz, medium-frequency range 80–120 Hz, and high-frequency range 220–400 Hz.

For manual diagnosis, acoustical stethoscope is the popular instrument that is used. Its most important part is an airtight (sealed) chamber from measurement site (patient's chest) to the expert's ear and good contact of the diaphragm surface, the primary sensing element, with the skin. These measurements suffer from the disadvantage of loose fitting at expert's ear, leading to poor sound quality and also creeping of noise from external sources. Due to the aforementioned problems and also to facilitate digitized recording and computerized analysis of PCG signals, the electronic stethoscope has been

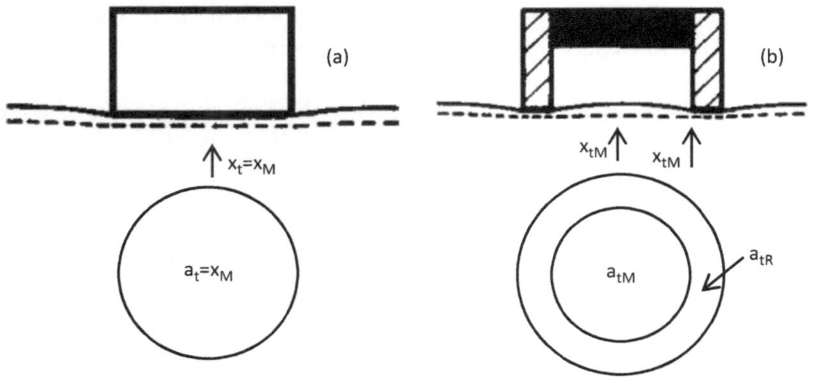

FIGURE 2.22
Schematic of PCG sensing: (a) absolute pickup; (b) relative pickup.

introduced. The principle of heart sound sensing is about converting the acoustic vibrations into electrical signals through a contact (piezoelectric/accelerometer) or non-contact (air-coupled) sensor. These are also known as absolute pickup and relative pickup sensors, respectively, as shown in Figure 2.22a,b, respectively. In contact type, the sensor is rigidly fixed on measurement site, and the vibration is averaged over the measurement site. Thus the measuring area (a_M) determines the contact area (a_t). In air-coupled sensor, an air-filled circular cavity is fixed (sealed) at the periphery of the chest wall. The vibration is measured as a difference over displacement under the cavity (measuring area, a_{tM}) and that under the edge (reference area, a_{tR}) of cavity. Thus, it is a kind of differential measurement. The air pressure within the cavity is measured. The electronic stethoscope is thus a relative pickup-type sensor.

The absolute and relative pickup PCG sensors need calibration with a standard accelerometer. The signal conditioning requirement of the raw sensor output specifies preamplification and filtering of respiration signals (in case of patients suffering from lung diseases), environmental vibrations, and air transmitted noises. A common practice is to use four high-pass filters, with gradually increasing cutoff frequency, to get rid of unwanted noises and artifact signals.

2.3.5 BP Measurement

Arterial blood pressure (ABP) is one of the prime physiological parameters used for determining the cardiovascular function and overall general healthiness. Arterial pressure is defined as the hydrostatic pressure exerted by the circulating blood over the arteries as a result of pumping action of the heart. Systolic pressure (SBP) is the highest pressure in a cardiac systole (ventricular contraction), while diastolic pressure (DBP) refers to the lowest (ventricular relaxation) one. Mean arterial pressure is the algebraic difference between the SBP and DBP and is determined by dividing the area under the BP curve of one cardiac cycle by its period. Pulse pressure (PP) is the algebraic difference between SBP and DBP. The techniques for ABP measurements are generally classified as direct and indirect methods. The direct technique, which measures the absolute BP at the measurement sites by direct sensing are considered most accurate, and their accuracy is determined by the physical characteristics of the sensor. However, it requires the insertion of the sensor inside the measurement site. Hence, the direct ABP measurements are limited to critical

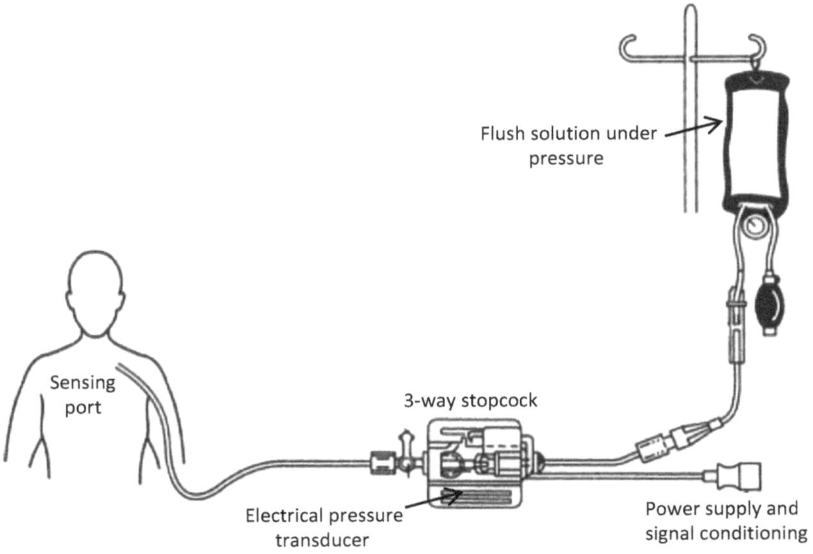

FIGURE 2.23
Schematic of an extravascular direct ABP measurement system.

clinical applications (anesthesia adjustments in surgery or in ICUs). In this subsection, the discussion will be confined to direct ABP measurements, which are of two types, extravascular sensing and intravascular sensing. In extravascular sensing, the vascular pressure is connected to a pressure transducer through a liquid-filled catheter. The block diagram of extravascular sensing is shown in Figure 2.23 [3]. A catheter tip (sensing port) is inserted into the measurement site through a surgical cut using a special needle. The catheter is flushed with heparin solution. The pressure is transmitted through this liquid to a transducer placed in a chamber which connects the catheter on one side and the supply solution and electrical connector on the other. Secondary transducers such as strain gage and capacitive type are used to generate an electrical output which is suitably conditioned and recorded. The disadvantages of this system are that the frequency response is limited by the hydraulic characteristics of the catheter liquid and also the time delay by the liquid. Intravascular sensors, also called catheter-tip sensors, eliminate the liquid-filled catheter and directly insert the transducer to the sensing point. Here, piezoresistive, capacitive, piezoelectric, and fiber-optic sensing principles are used to get an electrical output from the diaphragm element used as a primary sensing element. In piezoresistive sensing, a four-arm Wheatstone bridge can be used with four active resistive elements, each connected to an arm, with two in compression mode and the rest in expansion mode. Common piezoresistive transducers are ion-implanted into a thin silicon monocrystalline membrane [34,35]. Typical values are 100Ω–3 kΩ and powered by 3 V.

The output (V_0) of the four-arm Wheatstone bridge excited by supply V_s is obtained as

$$V_0 = V_s \frac{\Delta r}{r}, \tag{2.11}$$

where $\Delta r/r$ denotes the fractional change in resistance, which is nearly equal in all arms. The bridge output is fed to an INA. The schematic connection is shown in Figure 2.24.

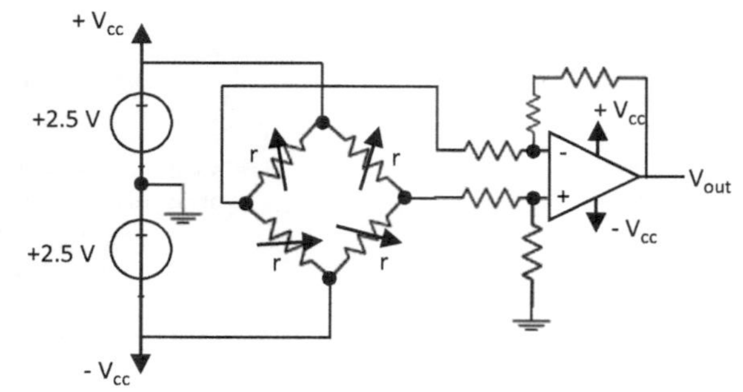

FIGURE 2.24
Typical signal conditioning circuit for piezoresistive ABP sensor. (Printed with permissions from "Biomedical sensors and their interfacing" in *Advanced Interfacing Techniques for Sensors*, B. George et al. (Eds.), Springer Nature, 2017.)

In case of a capacitive sensor, the sensing diaphragm is coated with a conductive material and used as the movable plate. The lead connectors and sensor can be micromachined in a small encapsulated structure [36].

The risk of inducing current to the patient's body can be avoided by using an optical sensor. Also, the decoupling of associated electronics at the sensor site can be got rid of. The basic principle of optical direct ABP sensing can be divided into the following categories: (a) intensity-reflective fiber-optic sensor, (b) light coupling between two fiber-optic sensors, (c) microbending fiber optic sensor, and (d) fiber-optic sensing based on interferometer principles. The most simple configuration is based on passing or reflecting a light beam through the deflection system (elastic member) [37,38] and measuring the modulation. A classical configuration of a cantilever fiber-optic catheter probe along with its signal conditioning schematic are shown in Figure 2.25. A cantilever arrangement accepts the pressure on a membrane coupled to a reflector, which faces the single fiber cable acting as

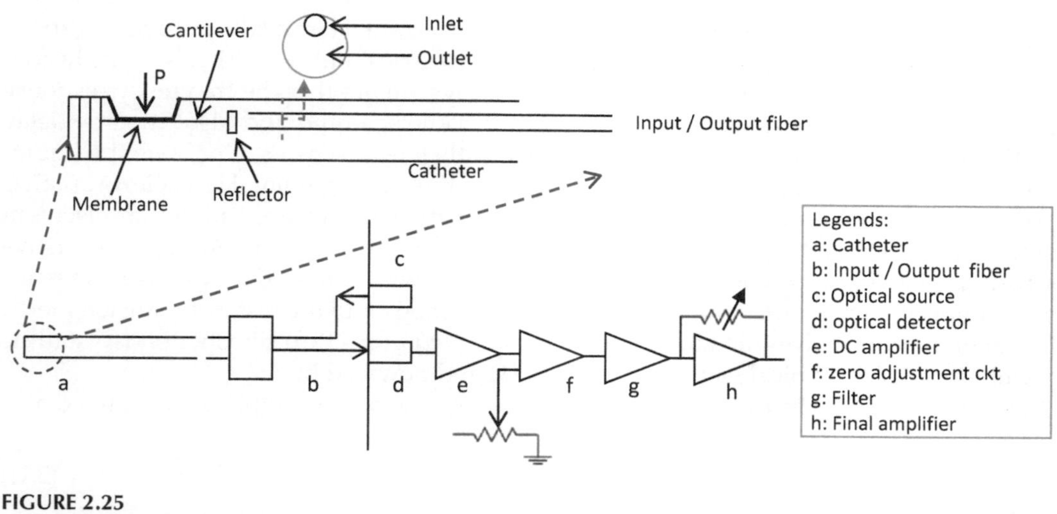

FIGURE 2.25
Schematic of a cantilever-type fiber optic (FO) ABP sensor and its signal conditioning. (Printed with permissions from "Biomedical sensors and their interfacing" in *Advanced Interfacing Techniques for Sensors*, B. George et al. (Eds.), Springer Nature, 2017.)

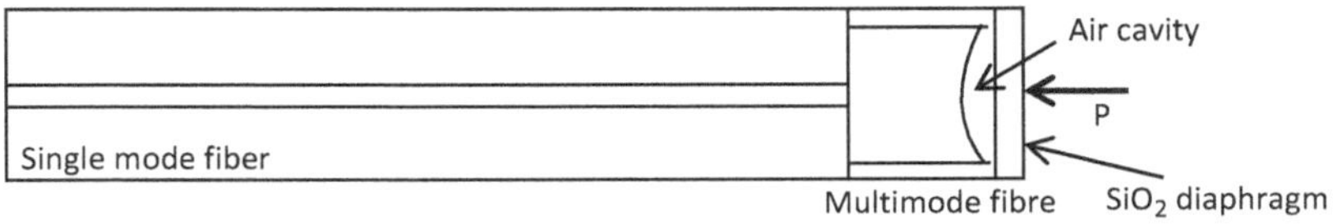

FIGURE 2.26
Schematic of ABP sensing using FP interferometry. (Printed with permissions from "Biomedical sensors and their interfacing" in *Advanced Interfacing Techniques for Sensors*, B. George et al. (Eds.), Springer Nature, 2017.)

both inlet and outlet, and guides the light pulses to the catheter tip. A change in pressure alters the amount of reflected light into the output fiber. The modulation in light intensity is sensed by a photoelectric detector and the final output is filtered and amplified to get a scaled output. Fabry–Perot (FP) interferometry is described in [39], to measure aortic arch and right coronary artery pressure. As shown schematically in Figure 2.26, it consists of a single-mode fiber, a multimode fiber, and a SiO_2 diaphragm. The SiO_2 diaphragm couples the pressure at the catheter tip. An FP cavity is formed by multimode fiber–air cavity interface and air cavity–diaphragm interface. The single-mode fiber guides the light into the FP cavity and also collects the reflected light. The multiple reflections of light in the FP cavity form an interference pattern. The change in pressure on the diaphragm alters the FP cavity length, and this is reflected in the phase change of the interference pattern. This in turn brings sinusoidal changes to the intensity of the final reflected optical beam. The reflected light can be split into two fibers using an optical splitter, one directly connected to an optical detector (broadband channel) and the other via a tunable filter (narrowband channel). The difference in pattern between the two detectors can be mathematically correlated to the blood pressure.

The schematic of an extrinsic fiber-optic interferometer ABP sensor is shown in Figure 2.27. It contains an integrated fiber Bragg grating (FBG) core with periodic change in the refractive index inside a single-mode fiber. The fiber is drawn into a glass capillary, at the end of which a diaphragm is sealed by splicing a multimode fiber, which accepts the blood pressure. An FP cavity is formed between the diaphragm and the FBG which facilitates the reflection of light. The outside pressure can change the cavity length (ΔL) through deformation of the diaphragm [39]. The measurement set-up consists of an optical

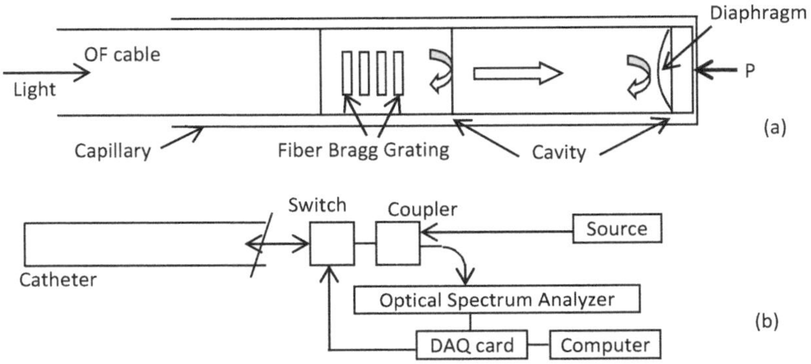

FIGURE 2.27
FBG-type FP interferometric ABP sensor: (a) schematic; (b) measurement set-up. (Printed with permissions from "Biomedical sensors and their interfacing" in *Advanced Interfacing Techniques for Sensors*, B. George et al. (Eds.), Springer Nature, 2017.)

source, coupler with switch (operating in time multiplexing mode and controlled by data acquisition (DAQ) card), an optical spectrum analyzer (OSA), DAQ card, and computer. The coupler directs the light pulses from sources and redirects the reflected light into the receiver fiber. The OSA converts the received radiation into intensity-modulated electrical pulses, which are finally analyzed using a computer.

2.4 Muscle and Nervous System Activity

2.4.1 Electroencephalogram (EEG)

EEG reflects the global electrophysiological activity of human brain and can be measured from scalp surface using specially designed electrodes. Since 1960s, when digital computers were introduced in medical signal processing, EEG has been widely used as a popular tool for brain imaging in basic and clinical neurosciences. Advanced applications of EEG include sleep staging and brain function monitoring in the ICU.

The origin of EEG signal is attributed to equivalent current diploes (ECDs), which result from synchronous depolarization (excitatory input) of a group (around 1,000,000) of pyramidal cell neurons, which are the major constituents of human brain [40]. EEG is a record of spatio-temporal summation of ECDs, their orientations, locations, strength over time, and impedance characteristics of the head tissues as picked up by electrodes placed on scalp. Typically, the EEG voltages are very small, in the range of 50 μV and having 0–100 Hz as clinical bandwidth. However, most of the brain-related studies involve a frequency range of 0–30 Hz, where the principal signatures related to brain activity are reflected. These waves are detailed in Table 2.2.

EEG must be recorded from multiple measurement positions distributed over the scalp, resulting in a number of different measured signals. To measure EEG signal, at least three electrodes are required, two active and the other reference. There are few accepted international standards or protocols which provide the electrode placement locations for EEG measurements. The most common and popular is the 10–20 electrode system, which was adopted in 1958 at the International Federation in Electroencephalography and Clinical Neurophysiology for electrode placement, as shown in Figure 2.28 [41]. The nomenclature of 10–20 system divides the cerebral cortex into four regions, viz., frontal (F), parietal (P), occipital (O), and temporal (T) and uses four anatomical landmarks as references, viz., Nasion: N (at the top of nose), Inion: I (extrusion at the back of the head), and pre-auricular: A1 and A2 (in front of the ears), as shown in Figure 2.28. Gaps between adjacent electrodes are either 10% or 20% of the distances between these anatomical landmarks. The electrode positions are further specified as Z: midline, odd numbers: left hemisphere, and

TABLE 2.2

Common EEG Waves and Frequency Bands

Name	Frequency (in Hz)	Characteristics
δ	0.5–4	Present in deep sleep
θ	4–8	Present when drowsy
α	8–13	Present with eyes closed
β	13–30	Present with eyes closed or open, sensitive to drugs/medication

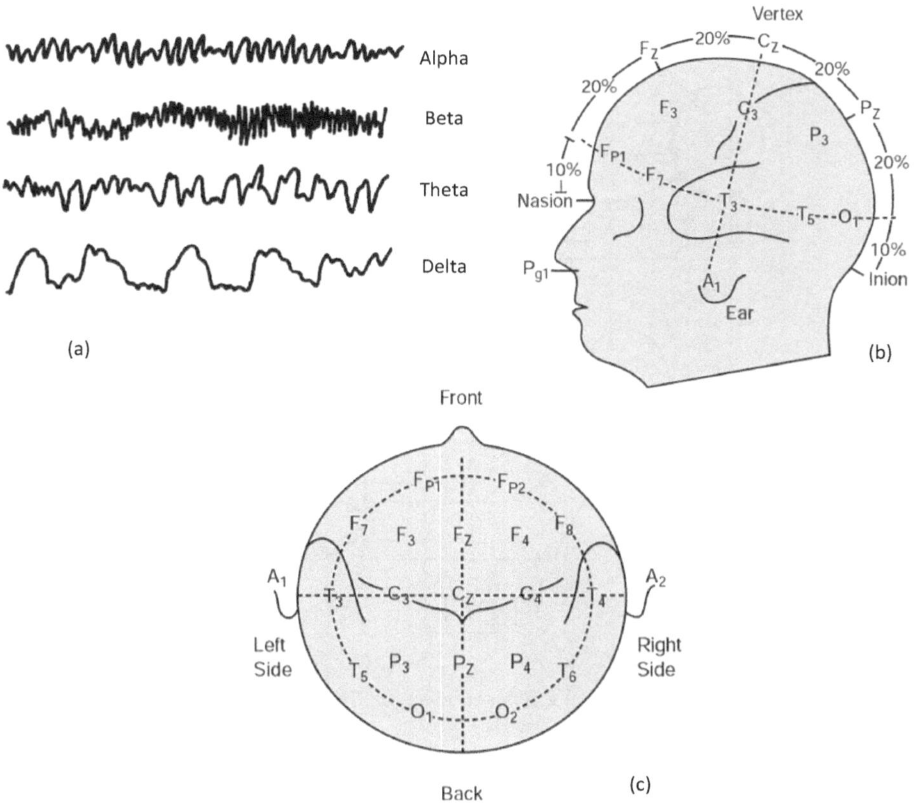

FIGURE 2.28
(a) EEG waveforms; (b) and (c) lead placement convention as per 10–20 system.

even numbers: right hemisphere. The 10–20 electrode system provides totally 21 electrodes (19 placed in the scalp, 2 at the ears). Accordingly, modern EEG measurement systems provide 21 different signals, each referred to as a 'channel'. For advanced applications, 32–256 channel systems are available commercially. Since an EEG voltage represents a difference between the voltages at two electrodes, the display of the EEG may be set up in one of several ways.

2.4.1.1 EEG Signal Conditioning

The representation of the EEG channels is referred to as a 'montage'. There are three different montages for EEG recording, viz., bipolar montage, unipolar montage (common reference), and average reference montage. The schematic connections of electrodes are shown in Figure 2.29. In the bipolar montage, shown in Figure 2.29a, the channels are connected in chains, in the sense that the second lead in first connection becomes the first lead in the second connection. Here the external interferences can be easily cancelled out since differences in electrode potentials are being measured in contiguous layers, thus amplifying local potentials only. In unipolar (or referential) montage, as shown in Figure 2.29b, the channel potentials are measured with respect to one common reference. Thus, it is susceptible to external noise but captures both local and far field potentials. The average

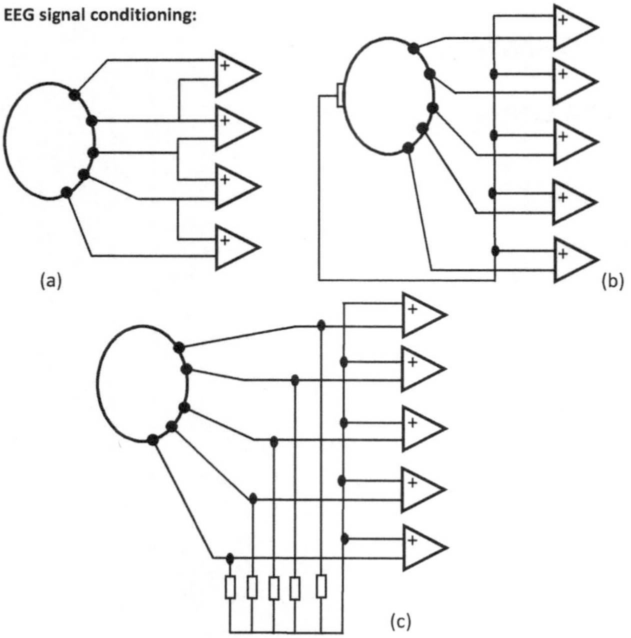

FIGURE 2.29
Schematic of EEG montage connections: (a) bipolar connection; (b) unipolar connection; (c) average electrode connection (lead positions are only representatives).

electrode connection, shown in Figure 2.29c, is a variant of the monopolar connection and similar to WCT in ECG measurements.

Electrical contact between the skin and the amplifier input is an important issue for faithful acquisition of the small EEG signal. This contact impedance (typically below 20 kΩ) should be negligible with respect to the amplifier input resistance (typically above 10 MΩ) and almost constant over the low frequencies. To meet these requirements, electrically conducting electrodes (usually made of metal) are fixed on the skin by various techniques. The electrode double layer generates a small potential, called half-cell potential at the electrode–electrolyte interface. This differs from one metal to the other. Hence, heterogeneous materials cannot be used for EEG measurements as the common-mode voltage at the differential amplifier inputs will not be identical and may not cancel each other. The small potentials measured at the EEG electrodes are noisy signals, containing internal and external artifacts. Among the major artifacts are muscle noise (electromyogram), residual ECG, and PLI. In addition, electrode potentials of several magnitudes of EEG are also sensed at the input of the EEG amplifier. Hence, to cancel out these effects, differential amplifiers are employed. Use of electrolyte gel keeps the contact resistance between the skin and electrodes low. Sometimes, notch filters are used to cut off the PLI at 50 Hz or 60 Hz.

The basic structure of an EEG amplifier is similar to that of an ECG amplifier, with the total amplification factor (10,000–20,000) split into 2–3 stages. The first stage provides a low gain 10–20. After the first stage, there is a high-pass filter (0.05 Hz) to block the DC offset. This helps to interpret EEG signals which are superimposed on low-frequency components arising out of artifacts or pathological activity. Also, the electrical safety of the patients due to accidental current flowing through the brain due to failure of the electronic components is incorporated in the design as per IEC601 standard, which specifies the maximum voltage flowing through the skin tissue at 100 μV. This can be achieved in two ways, viz., connecting high resistances between electrode and input of amplifier and using optical isolation. The CMRR of common EEG amplifier is around 80 dB or higher.

The structure of an EEG measurement system may vary among applications and types of post-acquisition processing [41]. For a standalone system, a preamplifier and a main amplifier transmit the ECG analog data to a local computer for storage. For a hospital environment, this computer may be connected to a remote-end server for data sharing and clinical evaluation by multiple experts. A third configuration uses an integrated digital front end (signal conditioning plus digital transmission unit) to wirelessly transfer the EEG data to a local server. This can be used for monitoring multiple patients in activity [42]. Over the last two decades, there has been some improvisations on EEG front-end signal conditioning circuit development [43]. Among them, the initial designs aimed for low-noise, low-offset ECG acquisition amplifier design [44–47]. Later on, the designs incorporated non-contact sensing, portability, and wireless connectivity for ambulation analysis [48]. In [49], an application-specific integrated circuit (ASIC) for an eight-channel EEG recording system is described which uses AC coupled chopper stabilized INA to achieve 120 dB CMRR, and the overall consumption is 66 μA at 3 V power. The digitized EEG is transferred at 11 bit resolution at high sampling rate (25 ksps) using a wireless radio interface. A capacitive-type EEG electrode design using polydimethylsiloxane (PDMS) and adhesive PDMS for EEG monitoring is proposed in [50]. A report on advanced applications discussed the integration of EEG and near infrared spectroscopy (NIRS) for enhanced brain imaging [49].

2.4.2 EMG Signal

The electrical signal generated by muscle activity is referred to as the EMG. It is a complicated signal affected by the anatomical and physiological properties of muscles and the control scheme of the central and peripheral nervous system, as well as the characteristics of the measurement system used to record it [33]. Recording of small electrical signals from muscle cells was reported in 1849 by Raynold, in 1922 by Gasser and Erlanger, and in 1944 by Inman. Traditionally, EMG has been primarily used for diagnostic purposes. However, in the last two decades, it has been extensively used for assistive technologies like artificial limb movement.

Unlike the myocardium, the skeletal muscles do not have any natural excitation source. Rather, electrical excitation in skeletal muscle is initiated and regulated by the central and peripheral nervous systems. The exact origin of electrochemical activity is beyond the scope of this chapter. The muscle cells are activated by the central nervous system through electric signals transmitted by motoneurons. A motoneuron innervates a group of muscle fibers which constitute the smallest functional unit of the muscle, called a motor unit (MU), the junction being called the end plate (Figure 2.30). The muscle fibers have variable length and, if properly excited, can shorten their length [51]. The innervation ratio determines how many muscle fibers are regulated by a single MU and varies among different body parts, smallest in the eye and largest in the leg.

The motoneurons carry nerve impulses from the brain through the spinal cord to the nerve endings, where they connect to the muscle fibers. In brief, the origin of EMG is attributed to spatial and time-averaged summation of APs that are generated as the end-plate potentials due to the release of acetylcholine (Ach), a neurotransmitter, at the MU to excite a muscle fiber. The electrical signal that emanates from the activation of the muscle fibers of an MU that are in the detectable vicinity of an electrode is called the motor unit action potential (MUAP). This consists of a fundamental unit of EMG signal. The shape of

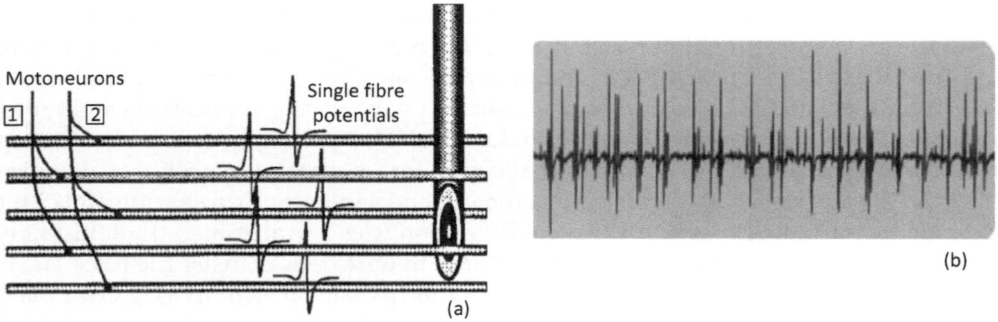

FIGURE 2.30
EMG signal generation: (a) innervation of muscle fibers showing two MUs with intermingled muscle fibers; (b) EMG signal collected using surface electrodes on the hand skin.

the MUAP is influenced by many factors, viz., relative position of the electrodes on muscles, relative geometrical relationship between the electrode surfaces and muscle fibers on the MU, size of the muscle fibers, and number of muscle fibers on the MU in the vicinity of the electrode. The electrical manifestation of the MUAP is accompanied by a contractile twitch of the muscle fibers. The MU must be activated continuously to sustain a muscle contraction. This results in a MUAP train whose morphology remains constant if the geometric relation between the electrode and the active muscle fiber remains constant. The MUAP train can be represented as interpulse intervals; and the waveform of the MUAP, as a sequence of delta functions convoluted over a filter $h(t)$ that represents the MUAP:

$$p_i(t) = \sum_{k=1}^{n} h_i(t - t_k), \tag{2.12}$$

where t_k represents the time locations of the MUAPs, n is the total number of inter-pulse intervals in a MUAPT, and i and k are integers that denote specific events.

The EMG signal over a surface area can be represented as a summation of such MUAP trains [33]:

$$m(t, F) = \sum_{k} p_k(t, F), \tag{2.13}$$

where F is the force generated by the muscle.

EMG signals can be acquired by placing an electrode either on the skin surface or inside the muscle whose activity is to be monitored. In monopolar configuration, as shown in Figure 2.31a, there is one sensing or measuring electrode placed on the muscle fiber, and a reference electrode, normally placed on an electrically uncorrelated body surface. The common-mode signals picked up can be eliminated using a high CMRR INA and appropriate filtering to select the usable frequency band of EMG. However, the monopolar sensing electrode also picks up other signals (like ECG, electrooculogram, or, EOG) as noise from other physiological activity if the measurement site is in the vicinity. In bipolar configuration (Figure 2.31b), this problem is avoided by placing two sensing electrodes from same muscle tissue (nominally 1 cm apart) and one reference electrode on an electrically uncorrelated body surface. The two sensing electrodes are connected in 'differential' mode to the biomedical amplifier which cancels out common physiological signals and other external artifacts. An important characteristic of the EMG signal is that it rapidly decays (amplitude becomes 25% of original over a distance of 100 μm) while propagating through

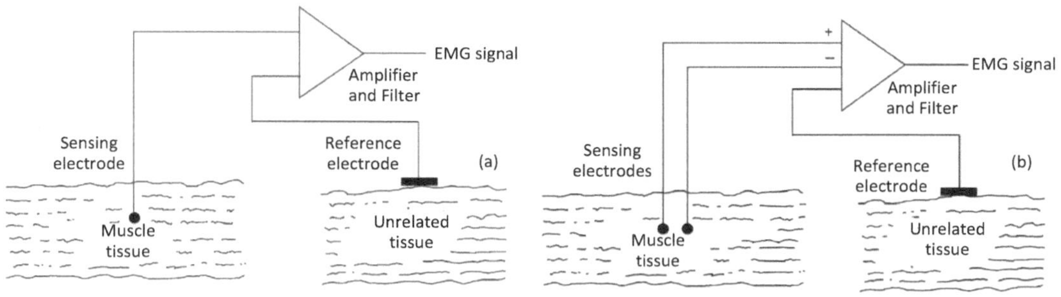

FIGURE 2.31
EMG acquisition using (a) monopolar configuration; (b) bipolar configuration.

the muscle tissue. Thus the electrodes can pick up signals from only nearby tissues. The other factors like distance between the sensing muscle fiber and detection surface and orientation of the detection surface of the electrode with respect to length of muscle fiber play important roles in EMG measurements.

As a special design consideration of EMG amplifiers, the electrode cables (leads) to the input to the amplifier should be as short as possible and should not be susceptible to movement. This is sometimes accomplished by placing the preamplifier near the electrode housing, within 10 cm for surface EMG measurements. The recommended amplifier characteristics of EMG measurements are as follows [33]:

CMRR: 85 dB or above

Input impedance: $10^{15}\ \Omega$ in parallel with 7 pfd

Input bias current: less than 15 fA

Bandwidth: 20–500 (surface electrodes), 20–2,000 (wire electrodes), 20–5,000 (Needle electrodes).

Most common EMG measurement configurations are available to collect 8–32 channel data.

2.5 Galvanic Skin Response

2.5.1 Measurement of Bioimpedance

The application of bioimpedance measurement is widespread, starting from body composition analysis (muscle mass, fat mass, etc.), impedance plethysmography, electrical impedance tomography to the performance analysis of implantable devices like pacemakers and defibrillators. Bioimpedance is the basic characteristic of a body tissue (both resistive and capacitive) to oppose the flow of electric current at the stimulation frequency through it. The capacitive reactive components of the measured tissue impedance change at higher frequencies due to the relative conductive properties of tissue fluids and cellular membranes. Bioimpedance methods can be categorized into two areas, viz., impedance plethysmography and impedance cardiography (ICG), both using a low-frequency stimulus current (<100 Hz) where most of the element paths in current are primarily resistive. In impedance plethysmography, the changing impedance waveform (ΔZ) is used to determine the volume change in a heterogeneous tissue segment (cardiac, respiration and

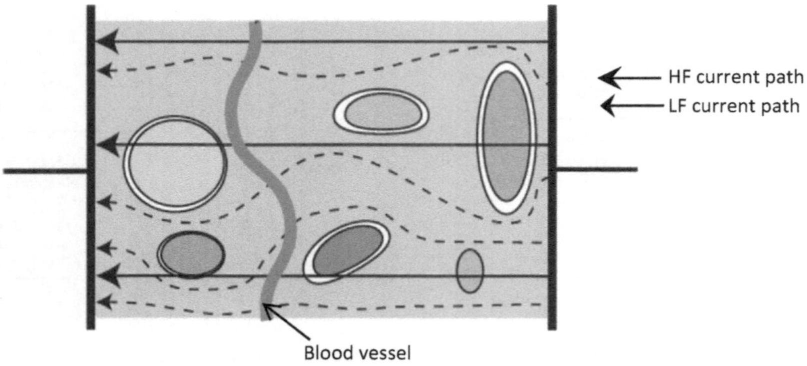

FIGURE 2.32
Current path through body tissue containing cells, membranes, and blood vessels at HF and LF.

peripheral tissue) as a function of time. In ICG, a subdivision of impedance plethysmography, certain cardiac events (opening and closing of heart valves) are detected by the first derivative of the ΔZ to measure cardiac stroke volume [33].

Living tissue is neither an ideal capacitor nor an ideal conductor. Many types of membranes with special electrical properties are found, like the dead, horny layer at the surface of the human skin and cell membranes and the large peritoneum membrane in the abdominal cavity. Figure 2.32 shows current pathways though blood vessels, tissues, and cell membrane at lower (<1 kHz) frequencies (LFs) and higher (>100 kHz) frequencies (HFs). The HF current passes right through the intracellular volume, whereas the LF passes around the membrane obstacles. This leads to a frequency-dependent phase shift between measured voltage and current, and by measuring a frequency spectrum, the tissue character can be analyzed [52].

There can be two modes of bioimpedance measurement: first, placing two current carrying electrodes (CC) on body tissue (of interest) and measuring the current passing through them, and second, measuring the small potential between two points due to living tissue activity using pickup electrodes (PUs). Figure 2.33 shows different monopolar and bipolar configurations. In monopolar configuration (a,c), one electrode contributes more than the other. In (a), the CC and M have different electrode areas, with different current densities. As a result, with DC current flow, they will act as anode and cathode. The larger

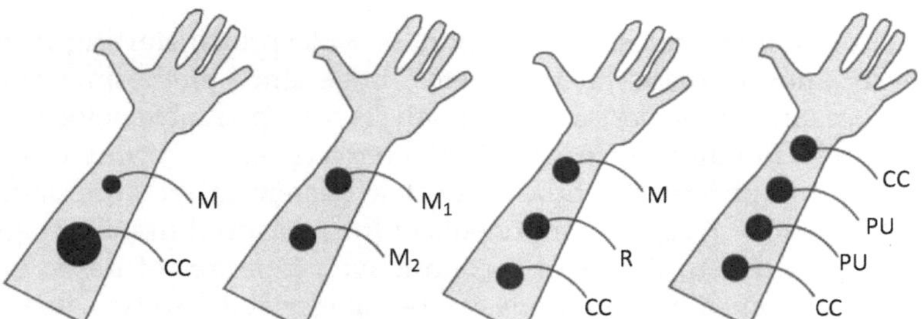

FIGURE 2.33
Different surface electrode configurations for bioimpedance measurement (M: measuring, R: reference; CC: current carrying; PU: pickup).

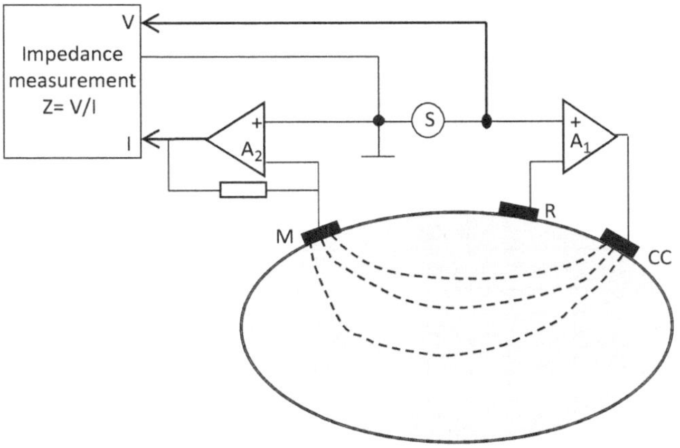

FIGURE 2.34
(a–d) Typical 3-electrode measurement circuit for skin bioimpedance (M: measuring electrode; CC: current carrying electrode; R: reference electrode).

electrode CC, having much lower current density, will have negligible effect on measurement, and may be regarded as neural electrode. A potential difference will be obtained if the larger electrode (CC) is placed at an indifferent skin position and the smaller one (M) at a nerve-activated skin position. If both the electrodes are placed on an active position, the recorded voltage difference may be very low. A second monopolar configuration, a three-electrode system (c), has one reference electrode and is used in high-resistivity skin tissue. The measuring electrode (M) contributes more to the measurement result than the other two electrodes. A bipolar electrode (configuration b) has equal contribution from both the electrodes M_1 and M_2. Under PU configuration, these may be used to measure potential difference due to remote organ activity resulting in a volume current flow spreading out from the organ. Configuration (d) shows four-electrode tetrapolar system with two CC and two PU electrodes, placed close to each other.

Figure 2.34 shows a quasi-monopolar three-electrode configuration for skin bioimpedance measurement [52]. Current is injected through CC at a body surface point using the AC source (*S*) and amplifier A1. Considering the body surface in isopotential condition (M and R at the same potential), A1 brings the internal body potential equal to excitation potential (~30 mV). A2 is the current reading amplifier. The impedance between the CC and M are measured as $Z = V/I$, where V is the constant PD between CC and R and I is the sensed current. Due to the capacitive behavior of the skin, the measured current is phase-shifted with respect to the excitation voltage. By using a synchronous rectifier circuit with the excitation voltage (*S*) as reference, the impedance can be decomposed so that only the in-phase conductance component G is measured.

2.6 Smart Biomedical Sensors and Data Acquisition Technology

Over the last few decades, technological innovations in very large scale integration (VLSI) circuits, low-power communication, and computing technology have revolutionized biomedical sensors, specially built for monitoring purpose. Driven by various societal

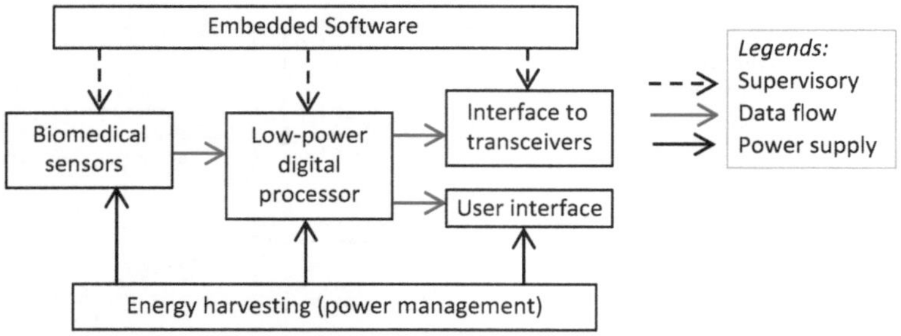

FIGURE 2.35
Smart biomedical sensor block diagram.

demands, to achieve flexibility and remote access, these biomedical sensors are smaller, compact, and intelligent for self-monitoring, enabled with energy harvesting for continuous operation and embedded with powerful data analysis algorithms to determine a patient's condition. Thus, it is possible to develop autonomous systems which can be used for the standalone (single patient monitoring) or networked (multiple patient monitoring) mode [53]. The block diagram of a typical smart biomedical sensor is shown in Figure 2.35.

The list of sensors includes ECG; EEG; pulse oximeter; and sensors for position, motion (accelerometers), blood pressure (non-invasive), glucose, airflow, etc. Various types of microelectromechanical systems (MEMS) sensors find wide application in smart biomedical sensors. A low-power digital processor (DSP) is the heart of the unit and controls the operation of the sensors (acquisition rate of the AD converter), wireless transceivers (data delivery rate, error control, synchronization within a group, sleep mode, etc.), energy harvesting unit (battery condition monitoring), and the user interface (AV alarm generation). In addition to this, the overall healthiness and functioning of the unit is supervised by an embedded software. The energy harvesting unit uses one of the principles like thermoelectricity, piezoelectricity, electromagnetism, and RF to recharge a battery unit which powers the different modules. Connectivity to local/remote relaying node/base station can be achieved by using low-power industrial, scientific and medical (ISM) band standards like Bluetooth, WLAN, ZigBee, RFID (radio frequency identification), etc. [54]. These modules exchange data through one of the serial protocols like I^2C, SPI, UART. The low-power DSP can extract useful features/events from a short span of biomedical signals to communicate any abnormal situation/condition of the patent.

In the context of smart biomedical sensor applications in patient monitoring, there are three types of applications, viz., (a) constant monitoring of vital signs of ambulatory patients [55], (b) persons (non-patient) looking for improvement of their health condition (like calories or, diabetes control), and (c) healthy persons (sportsperson, military personnel, rescue operators) tracking their activity during exercise or operations. Some essential features to suit these demands are wearability [56,57], extreme low-power operation and wireless connectivity and extremely small (chip area 2–4 cm^2) board size. Three different technologies have significantly contributed to achieve the mentioned features: (a) incorporation of mixed circuits (analog and digital) in the same circuit board using a technology named systems-in-package (SiP) [58], where the individual dies are interconnected through wires. (b) Three-dimensional stacking of circuit boards. Here, the circuits consist of a layer of heterogeneous stack in a three-dimensional fashion. Thus, sensors can be placed directly on top of processing and communication modules, leading to further reduction of overall board area. (c) Adoption of flexible integrated circuits, especially for

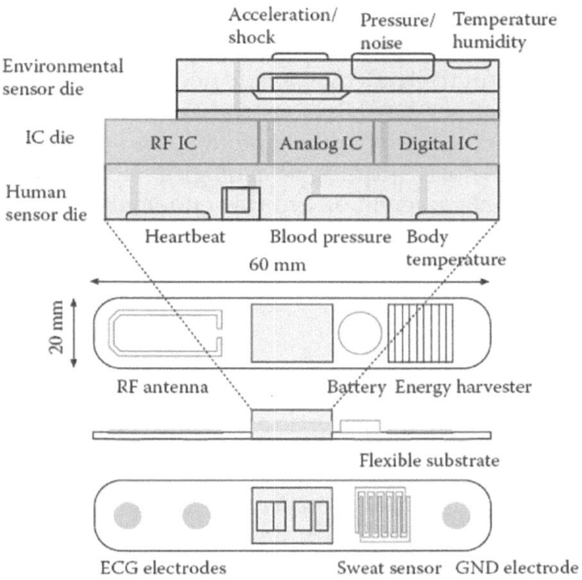

FIGURE 2.36
Conceptual diagram of wearable sensing device with adhesive tape. (Printed with permissions from *Telemedicine and Electronic Medicine*, J.G. Webster, H. Eren (Eds.), CRC Press, 2017.)

ambulatory monitoring. These circuits employ special types of fabrication techniques and materials to fix them on flexible or even bendable parts (of the human body). An excellent application of these flexible circuits is realized in the disposable adhesive tape wearable biomedical sensor, a schematic of which is shown in Figure 2.36. The device is intended to measure ECG, HR, skin temperature, blood pressure, body movement along with some surrounding environmental condition (e.g., humidity). The most advantageous feature is that it fits well on the skin surface (size of 20 mm × 60 mm) and allows normal activity of the human subject. As shown in the figure, the components are stacked in a three-tier structure, from human sensor die in the bottom (touching the skin) to analog and digital ICs in the middle and environment sensing ICs on top. This smart integration and performance is not possible in conventional electronics ICs mounted on printed circuit boards (PCBs), due to enhanced cost of manufacturing and more bulky circuit. This ASIC uses a T8051 core based on reversible logic as the central controller with 4 kB data memory, CMOS-based sensor interface and ADC, interconnecting digital configuration bus-on-chip, 1.2 GHz RF interface, and an in-built energy harvesting module using solar energy. The details of development are available in [59].

2.7 Conclusion

From the era of dedicated, single patient monitoring, we are now in the age of distributed, ubiquitous patient monitoring. Over the last few decades, microelectronics, materials science, and low-power computing technology have been the dominant players in development of biomedical sensors used for monitoring applications. Wireless connectivity, self-monitoring, plug and play, and energy harvesting are some of the features of today's biomedical sensors. In some distributed monitoring applications, determining the patient's condition using vital sign analysis at the sensor node is another feature that has been introduced. This has reduced the burden of centralized data analysis as well as data traffic volume in sensor network systems. Many autonomous systems are now commercially available which can continuously acquire physiological signals, analyze them, and provide advice for lifestyle management. Toward this end, mobile devices have played a key role. With the advent of technological domains like *m*-health and *e*-health in the current century, there is a broader scope for applications and R&D initiatives in the area of new biomedical sensor development.

References

1. J. G. Webster, *The Physiological Measurement Handbook*, CRC Press, Boca Raton, USA, 2015.
2. R. Plonsey and R. Barr, *Bioelectricity: A Quantitative Approach*, vol. 136, no. 1. Springer, New York, 2007.
3. J. G. Webster, *Medical Instrumentation: Application and Design*, John Wiley & Sons, Hoboken, USA, 1997.
4. M. R. Neuman, Biopotential electrodes, in *The Biomedical Engineering Handbook*, J.D. Bronzino (Ed.), 2nd ed., CRC Press, Boca Raton, USA, 2000.
5. R. Gupta, M. Mitra, and J. Bera, *ECG Acquisition and Automated Remote Processing*. New Delhi: Springer India, 2014. ISBN: 9788132215578.
6. S. Chaudhuri, T. D. Pawar, and S. Duttagupta, *Ambulation Analysis in Wearable ECG*, Springer, New York, 2009.
7. J. C. Huhta and J. G. Webster, 60-Hz interference in electrocardiography, *IEEE Trans. Biomed. Eng.*, vol. BME-20, no. 2, 91–101, 1973.
8. E. M. Spinelli, R. Pallàs-Areny, and M. A. Mayosky, AC-coupled front-end for biopotential measurements, *IEEE Trans. Biomed. Eng.*, vol. 50, no. 3, 391–395, 2003.
9. E. M. Spinelli, N. Martínez, M. A. Mayosky, and R. Pallàs-Areny, A novel fully differential biopotential amplifier with DC suppression, *IEEE Trans. Biomed. Eng.*, vol. 51, no. 8, 1444–1448, 2004.
10. R. B. Northorp, *Application and Analysis of Analog Electronic Circuits to Biomedical Instrumentation*, 2nd ed. Boca Raton, FL: CRC Press, 2012.
11. J. G. Webster, *The Measurements, Instrumentation and Sensors Handbook*. Boca Raton, FL: CRC Press, 1999.
12. A. Wong, K. P. Pun, Y. T. Zhang, and C. S. Choy, An EGG measurement IC using driven-right-leg circuit, in *IEEE International Symposium on Circuits & Systems*, Greece, 2006, pp. 345–348.
13. J. Allen, Photoplethysmography and its application in clinical physiological measurement, *Physiol. Meas.*, vol. 28, no. 3, 1–39, 2007.
14. J. A. Nijboer, J. C. Dorlas, and H. F. Mahieu, Photoelectric plethysmography-some fundamental aspects of the reflection and transmission methods, *Clin. Phys. Physiol. Meas.*, vol. 2, no. 3, 205–215, 1981.
15. A. Reisner, P. A. Shaltis, D. McCombie, and H. H. Asada, Utility of the photoplethysmogram in circulatory monitoring, *Anesthesiology*, vol. 108, no. 5, 950–958, 2008.
16. K. M. Warren, J. R. Harvey, K. H. Chon, and Y. Mendelson, Improving pulse rate measurements during random motion using a wearable multichannel reflectance photoplethysmograph, *Sensors (Switzerland)*, vol. 16, no. 3, 2016.
17. I. Borik, S. Cap, Implementation of wireless data transfer to photoplethysmographic measurements, in *ELEKTRO*, Slovakia, 2012, pp. 407–410.

18. K. Davoudi, M. Shayegannia, and B. Kaminska, Vital signs monitoring using a new flexible polymer integrated PPG sensor, in *Computing in Cardiology Conference (CinC)*, Spain, 2013, vol. 510, pp. 265–268.
19. J. G. Webster, *Design Of Pulse Oximeters*. Bristol: IOP Publishing, 1997.
20. S. Silvestri and E. Schena, Micromachined flow sensors in biomedical applications, *Micromachines*, vol. 3, no. 2, 225–243, 2012.
21. G. A. L. Araujo, R. C. S. Freire, J. F. Silva, A. Oliveira, and E. F. Jaguaribe, Breathing flow measurement with constant temperature hot-wire anemometer for forced oscillations technique, in *Conference Record - IEEE Instrumentation and Measurement Technology Conference*, Italy, 2004, vol. 1, pp. 730–733.
22. Y. S. Lee, P. N. Pathirana, C. L. Steinfort, and T. Caelli, Monitoring and analysis of respiratory patterns using microwave Doppler radar, *IEEE J. Transl. Eng. Heal. Med.*, vol. 2, 2014.
23. C. Gu and C. Li, Assessment of human respiration patterns via noncontact sensing using Doppler multi-radar system, *Sensors (Switzerland)*, vol. 15, no. 3, 6383–6398, 2015.
24. Y. S. Lee, P. N. Pathirana, R. J. Evans, and C. L. Steinfort, Noncontact detection and analysis of respiratory function using microwave Doppler radar, *J. Sens.*, vol. 2015, 2015.
25. M. Zakrzewski, A. Vehkaoja, A. S. Joutsen, K. T. Palovuori, and J. J. Vanhala, Noncontact respiration monitoring during sleep with microwave Doppler radar, *IEEE Sens. J.*, vol. 15, no. 10, 5683–5693, 2015.
26. O. Postolache, P. S. Girao, E. Lunca, P. Bicleaknu, and M. Andrusca, Unobtrusive cardio-respiratory monitoring based on microwave Doppler radar, in *EPE 2012 - Proceedings of the International Conference and Exposition on Electrical and Power Engineering*, Romania, 2012, pp. 597–600.
27. Y. Qi, C. B. Soh, E. Gunawan, K. S. Low, and R. Thomas, An ultrasonic contactless sensor for breathing monitoring, *Sensors (Switzerland)*, vol. 14, no. 8, 15434–15457, 2014.
28. Y. Yamana, S. Tsukamoto, K. Mukai, H. Maki, H. Ogawa, and Y. Yonezawa, A sensor for monitoring pulse rate, respiration rhythm, and body movement in bed, in *Proceedings of the Annual International Conference of the IEEE Engineering in Medicine and Biology Society, EMBS*, Boston, USA, 2011, pp. 5323–5326.
29. S. D. Min, J. K. Kim, H. S. Shin, Y. H. Yun, C. K. Lee, and M. Lee, Noncontact respiration rate measurement system using an ultrasonic proximity sensor, *IEEE Sens. J.*, vol. 10, no. 11, 1732–1739, 2010.
30. G. D. Gargiulo, U. Gunawardana, A. O'Loughlin, M. Sadozai, E. S. Varaki, and P. P. Breen, A wearable contactless sensor suitable for continuous simultaneous monitoring of respiration and cardiac activity, *J. Sens.*, vol. 2015, 2015.
31. S. T. A. Hamdani and A. Fernando, The application of a piezo-resistive cardiorespiratory sensor system in an automobile safety belt, *Sensors (Switzerland)*, vol. 15, no. 4, 7742–7753, 2015.
32. Z. Zhang, J. Zheng, H. Wu, W. Wang, B. Wang, and H. Liu, Development of a respiratory inductive plethysmography module supporting multiple sensors for wearable systems, *Sensors (Switzerland)*, vol. 12, no. 10, 13167–13184, 2012.
33. J. G. Webster, *Encyclopedia of Medical Devices and Instrumentation*, vol. 3, Wiley-Interscience, Hoboken, USA, 2006.
34. K. D. Wise and J. B. Angell, An IC piezoresistive pressure sensor for biomedical instrumentation, *IEEE Trans. Biomed. Eng.*, vol. BME-20, no. 2, 101–109, 1973.
35. M. Esashi, H. Komatsu, T. Matsuo, M. Takahashi, T. Takishima, K. Imabayashi, and H. Ozawa, Fabrication of catheter-tip and sidewall miniature pressure sensors, *IEEE Trans. Electron Devices*, vol. 29, no. 1, 57–63, 1982.
36. K. E. Babbitt, L. Fuller, and B. Keller, A surface micromachined capacitive pressure sensor for biomedical applications, in *Proceedings of the UGIM Symposium, Microelectronics Education for the Future. Twelfth Biennial University/Government/Industry Microelectronics Symposium (Cat. No. 97CH36030)*, USA, 1997, pp. 150–153.
37. A. Lekholm and L. Lindström, Optoelectronic transducer for intravascular measurements of pressure variations, *Med. Biol. Eng.*, vol. 7, no. 3, 333–335, 1969.

38. H. Matsumoto, M. Saegusa, K. Saito, and K. Mizoi, The development of a fibre optic catheter tip pressure transducer, *J. Med. Eng. Technol.*, vol. 2, no. 5, 239–242, 1978.
39. N. Wu, Y. Tian, X. Zou, Y. Zhai, K. Barringhaus, and X. Wang, A miniature fiber optic blood pressure sensor and its application in in vivo blood pressure measurements of a swine model, *Sens. Actuators, B Chem.*, vol. 181, 172–178, 2013.
40. J. R. Gawryluk and R. C. N. D'Arcy, Electroencephalography: Basic concepts and brain applications, *Handb. Phys. Med. Biol.*, vol. c, 548, 2010.
41. H. Hinrichs, Electroencephalography, in *Biomedical Technology and Devices Handbook*, J. Moore, H. Hinrichs (Eds.), CRC Press, Boca Raton, USA, 2004.
42. M. Teplan, Fundamentals of EEG measurement, *Meas. Sci. Rev.*, vol. 2, no. 2, 1–11, 2002.
43. A. B. Usakli, Improvement of EEG signal acquisition: An electrical aspect for state of the art of front end, *Comput. Intell. Neurosci.*, vol. 2010, 2010.
44. R. Martins, S. Selberherr, and F. A. Vaz, A CMOS IC for portable EEG acquisition systems, *IEEE Trans. Instrum. Meas.*, vol. 47, no. 5, 1191–1196, 1998.
45. K. A. Ng and P. K. Chan, A CMOS analog front-end IC for portable EEG/ECG monitoring applications, *IEEE Trans. Circuits Syst. I Regul. Pap.*, vol. 52, no. 11, 2335–2347, 2005.
46. T. J. Sullivan, S. R. Deiss, and G. Cauwenberghs, A low-noise, non-contact EEG/ECG sensor, in *IEEE Biomedical Circuits and Systems Conference*, Canada, 2007, pp. 154–157.
47. A. K. Whitchurch, J. K. Abraham, M. A. Lonkar, and V. K. Varadan, Design of a compact amplifier and signal conditioning module for wireless EEG monitoring, in *IEEE Region 5 Technical Conference, TPS*, USA, 2007, pp. 153–156.
48. J. Safaie, R. Grebe, H. A. Moghaddam, and F. Wallois, Toward a fully integrated wireless wearable EEG-NIRS bimodal acquisition system, *J. Neural Eng.*, vol. 10, no. 5, 1–12, 2013.
49. R. F. Yazicioglu, P. Merken, R. Puers, and C. Van Hoof, A 200 µW eight-channel EEG acquisition ASIC for ambulatory EEG systems, *IEEE Journal of Solid-State Circuits*, vol. 43, no. 12, 164–166, 2008.
50. S. M. Lee, J. H. Kim, H. J. Byeon, Y. Y. Choi, K. S. Park, and S. H. Lee, A capacitive, biocompatible and adhesive electrode for long-term and cap-free monitoring of EEG signals, *J. Neural Eng.*, vol. 10, no. 3, 1–10, 2013.
51. K.-Å. Henneberg, Principles of electromyography, in *The Biomedical Engineering HandBook*, J.D. Bronzino (Ed.), CRC Press, Boca Raton, USA, 2000.
52. S. Schreiner, J. D. Bronzino, and D. R. Peterson, *Medical Instruments and Devices Principles and Practices*. Boca Raton, FL: CRC Press, 2016.
53. M. F. Abbod, D. A. Linkens, M. Mahfouf, and G. Dounias, Survey on the use of smart and adaptive engineering systems in medicine, *Artif. Intell. Med.*, vol. 26, no. 3, 179–209, 2002.
54. V. Potdar, A. Sharif, and E. Chang, Wireless sensor networks: A survey, in *Proceedings - International Conference on Advanced Information Networking and Applications, AINA*, UK, 2009, pp. 636–641.
55. H. Alemdar and C. Ersoy, Wireless sensor networks for healthcare: A survey, *Comput. Networks*, vol. 54, 2688–2710, 2010.
56. D. N. Mathias, S. Il Kim, J. S. Park, and Y. H. Joung, Real time ECG monitoring through a wearable smart T-shirt, *Trans. Electr. Electron. Mater.*, vol. 16, no. 1, 16–19, 2015.
57. U. Anliker, J. A. Ward, P. Lukowicz, G. Troster, F. Dolveck, M. Baer, F. Keita, E. B. Schenker, F. Catarsi, L. Coluccini, and A. Belardinelli, AMON: A wearable multiparameter medical monitoring and alert system, *IEEE Trans. Inf. Technol. Biomed.*, vol. 8, no. 4, 415–427, 2004.
58. A. Fontanelli, System-in-package technology: Opportunities and challenges, in *9th International Symposium on Quality Electronic Design (ISQED 2008)*, USA, 2008.
59. A. Chan, K. Higuchi, and M. Kazusuke, Development of disposable adhesive wearable human-monitoring system, in *Tememedicine and Electronic Medicine*, H. Eren and J. Webster, Eds. Boca Raton, FL: CRC Press, 2016.

3

Data Compression in Health Monitoring

Sourav Kumar Mukhopadhyay
Ryerson University

Rajarshi Gupta
University of Calcutta

CONTENTS

3.1 Introduction

Continuous monitoring of patients under treatment in a biomedical instrumentation set-up generates large volume of data every day. In a hospital environment, the data volume assumes an enormous amount due to monitoring of multiple critical patients and demands high-capacity memory devices for data archiving. On the other hand, for remote monitoring applications, patients' physiological data are required to be transmitted to a distance for analysis and interpretation by medical experts. Volume reduction of biomedical data before transmission can save transmission time. For monitoring multiple patients using wireless communication technology, this also enhances the transmission link efficiency.

Data compression for telemedical applications and continuous monitoring of patients have been an active area of research from 1970s, especially after the introduction of microprocessor-based systems in medical applications. A primary and essential condition in biosignal compression is that the clinical features should be minimally distorted so that reconstructed data is properly diagnosed. There are two specific characteristics prevalent in most of the physiological signals. First, the most important clinical information is represented in a limited frequency band. Second, many of these signals, being quasi-periodic in nature, have ample redundancy between successive cycles. Based on these, the two general principles adopted for biomedical data compression are time-domain and transform-domain methods. The former processes the raw samples based on redundancy between small groups of data samples. These methods have low complexity and are easy to implement using low-end microprocessors. Most of the early algorithms for biomedical signal compression developed in 1970s–1980s followed time-domain methods. However, the compression achieved using these methods is generally low. Moreover, implementing lossy compression using these methods may result in unacceptable distortion of the clinical features. On the other hand, the transform-domain methods can achieve higher compression with low distortion by mapping the time series data to a different domain, thus leveraging more scope to discard non-important information. Over the years, many parameters or performance indices have been developed for estimating the compression performance. This chapter will primarily discuss the major compression techniques, their advantages and limitations, and their performance on popular cardiovascular signals like those of electrocardiogram (ECG) and photoplethysmography (PPG).

3.2 Redundancy: An Essential Property for Compression

Redundancy in a biomedical signal can be realized in different forms. For example, in electrocardiogram (ECG), redundancy is observed in intra-cycle, inter-cycle, and inter-lead data. This is shown in Figure 3.1.

In Figure 3.1a, a short duration ECG lead II record of data from a healthy human is shown, with consecutive cardiac cycle (or beat) boundary demarcations marked on waveforms with small circles. Figure 3.1b shows the plot of individual beats by aligning R-peaks of 60 s duration from same record, clearly illustrating similarity between cardiac cycles. The equipotential segments in PQ, ST, and TP regions in panels (a) and (b) show intra-beat redundancy. This redundancy is illustrated in transform-domain in Figure 3.2. In Figure 3.2a, the R-peak–aligned beat data from a single lead is considered as a 2-D matrix and decomposed using principal component analysis (PCA), by eigenvalue decomposition of the covariance matrix. Considering p-measurements (beats), out of the p number of principal component (PC) scores, only first few PC scores show some variance, as evident from y-axis scales. Thus, the signal can be reconstructed discarding lower-order components. This redundancy can also be mathematically established by computing the energy contribution from different PC scores.

In Figure 3.2b, the intra-cycle redundancy of the same signal (at 1 kHz sampling) is illustrated by discrete wavelet transform (DWT), using ten-level decomposition with Daubechies6 wavelet as the basis function. Among the detail (D) and approximation (A) coefficients, the $D4$, $D5$ show maximum energy contribution, whereas $D10$ represents nearly the DC level. Here also, the energy from the respective A's and D's can be calculated for discarding low-contributing components.

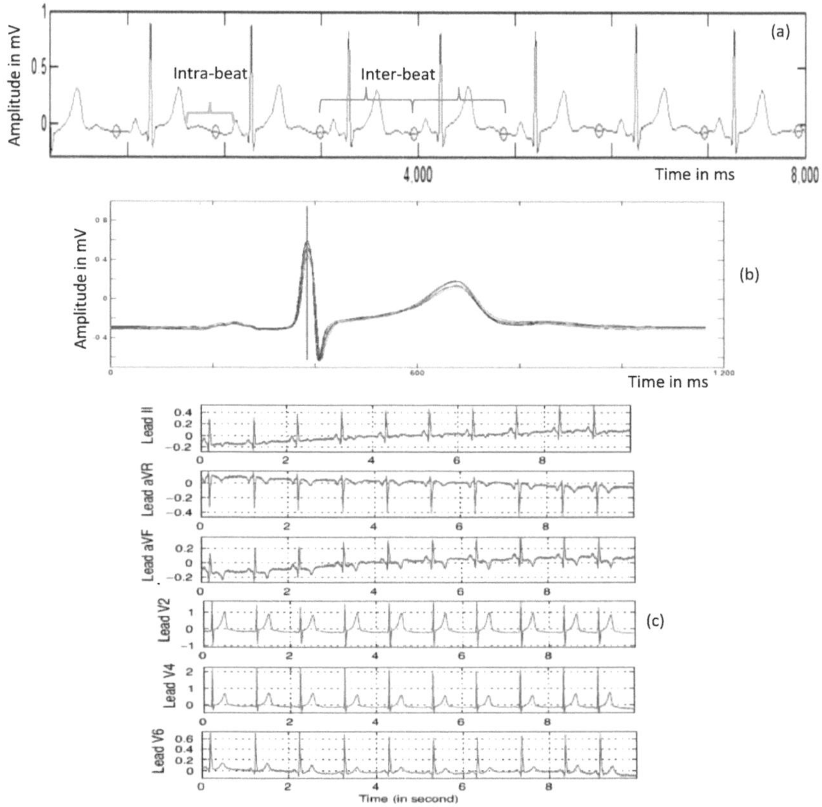

FIGURE 3.1
ECG patterns showing data redundancy among (a) intrabeat, (b) interbeat, and (c) interleads.

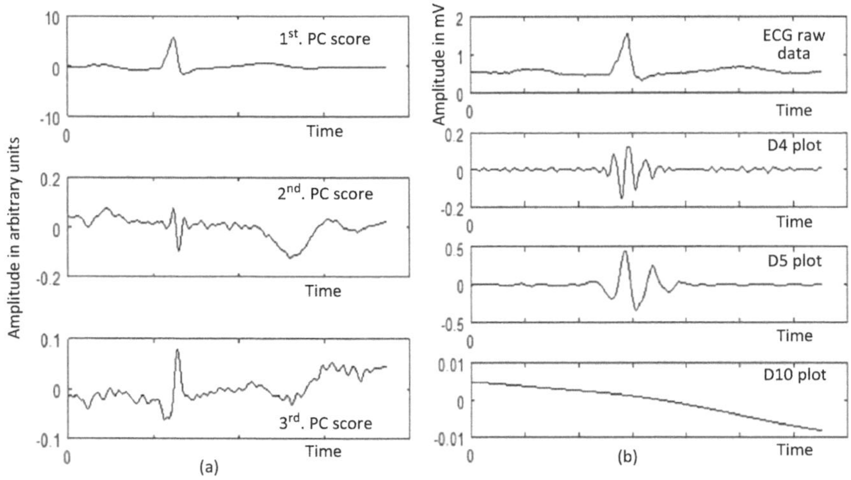

FIGURE 3.2
Showing redundancy in time series ECG using (a) principal component decomposition and (b) DWT using Daubechies6.

The same principle follows when a biomedical signal is mapped in another domain using a linear transform. The extent of redundancy may vary depending upon the frequency spectrum, more specifically, the spread of clinical information in the signal.

3.3 Distortion Measures

Acceptability of biosignals upon reconstruction could be verified by only well-trained medical practitioners or doctors. A few numerical measures are also there to gauge the similarity between the original and reconstructed biosignals. The fidelity assessment techniques of reconstructed biosignals can broadly be categorized into two types: objective assessment and subjective assessment techniques.

3.3.1 Objective Assessment of Reconstructed Biosignals

Popular techniques, which are used for the objective assessment of reconstructed biosignals include (a) percent root mean square difference (PRD), (b) normalized percent root mean square difference (PRDN), (c) root mean square error (RMSE), (d) cross correlation (CC), and (e) quality score (QS).

PRD is calculated as

$$\text{PRD}(\%) = \sqrt{\frac{\sum_{i=1}^{N}\left(x_i - \overline{x_i}\right)^2}{\sum_{i=1}^{N} x_i^2}} \times 100, \tag{3.1}$$

where x_i = original biosignal, $\overline{x_i}$ = reconstructed biosignal, and N = total number of samples.

As per the globally considered standard [1], the quality of a reconstructed biomedical signal is considered to be 'very good' if the PRD value lies in between 0% and 2%; it is considered to be 'good' if the value lies in between 2% and 9%; a PRD value in between 9% and 19% is considered to be 'not good'; and any PRD value above 19% is considered as 'bad'. But, there is a problem associated in the denominator part of the expression of PRD (Equation 3.1). In the presence of low-frequency noises such as baseline wander, the denominator part of Equation (3.1) becomes very high due to the added DC level and provides a low PRD value, which is truly misguiding. Hence, the original signal must be filtered to remove the low-frequency noises beforehand.

The above-mentioned problem could be diminished to some extent by revamping the denominator part of Equation (3.1), and this modified equation is termed "normalized percent root mean square difference" (PRDN):

$$\text{PRDN}(\%) = \sqrt{\frac{\sum_{i=1}^{N}\left(x_i - \overline{x_i}\right)^2}{\sum_{i=1}^{N} (x_i - m)^2}} \times 100, \tag{3.2}$$

where m = mean value of the original signal.

PRDN provides the actual measure of signal distortion, if the low-frequency noise present in the original signal is a constant DC steady-state. On the other hand, it is very interesting to note that (a) the values of PRD and PRDN become equal if the low-frequency noises are efficiently filtered out from the original signal prior to compression and (b) in the case of a strict lossless compression, both the values of PRD and PRDN should be zero.

RMSE is another means of measuring the signal distortion and is expressed as

$$\text{RMSE} = \sqrt{\frac{\sum_{i=1}^{N}\left(x_i - \overline{x_i}\right)^2}{N}}. \tag{3.3}$$

Cross correlation (CC) is the measure of similarity between two signals and is often used to evaluate the closeness of the reconstructed biosignal to its original counterpart. CC is expressed as

$$\text{CC} = \frac{\frac{1}{N}\sum_{i=1}^{N}(x_i - m)\left(\overline{x_i} - \overline{m}\right)}{\sqrt{\frac{1}{N}\sum_{i=1}^{N}(x_i - m)^2}\sqrt{\frac{1}{N}\sum_{i=1}^{N}\left(\overline{x_i} - \overline{m}\right)^2}}, \tag{3.4}$$

where $\overline{m}$ = mean value of the reconstructed signal. Only in the case of a strict lossless compression, CC becomes 1.

QS is the ratio of compression ratio (CR) and PRD and is often used to quantify the overall performance of a biomedical signal compression algorithm using a single metric. A high QS value is always desirable:

$$\text{QS} = \frac{\text{CR}}{\text{PRD}}. \tag{3.5}$$

Again, it is required to mention that, if the original biosignal contains low-frequency noises, PRD becomes low, and as a consequence, QS becomes high, which is misguiding too.

Maximum amplitude error (MAX) is a comparatively less popular yet efficient biosignal distortion measure technique, which is expressed as

$$\text{MAX} = \max\left(\left|x - \overline{x}\right|\right). \tag{3.6}$$

The metric MAX enumerates the maximum amplitude of the reconstruction error. The smaller the value of PRD, PRDN, RMSE, and MAX, and the higher the value of CC and QS, the better the reliability of the algorithm. However, except PRD, other objective assessment techniques are not standardized yet.

3.3.2 Subjective Assessment of Reconstructed Biosignals

Perhaps the result of subjective assessment could provide better insight into the quality of the reconstructed biosignal, as the test is conducted by experts. There are two types of subjective assessment tests, semi-blind and blind mean opinion score (MOS) tests. In the case of a blind MOS test, first, the original signal is provided to the clinicians along with a few diagnostic queries. At a later time, the reconstructed biosignal is also provided to those clinicians along with those same queries. Now, answers of those queries are compared to evaluate the MOS score. On the other hand, both the original and reconstructed signals are provided to the clinicians in the case of semi-blind MOS test. Finally, average MOS error values are calculated from the scores given by the experts [2]. Steps involved in conducting a semi-blind MOS test are explained below.

STEP I: Print both the original and reconstructed biosignals in proper scale at which doctors are habituated to see.

STEP II: Prepare and print a table as follows, and provide it to the doctors along with the printed signals.

The Semi-Blind MOS Test

Doctor's Name: Date: Place: Signature:

Detail of the signal provided to the doctor:

Q1. How would you evaluate these two signals? (mark with ✓)

1 2 3 4 5

where 1→ completely different and 5→ identical

Q2. Would your diagnosis be different if you were asked to make a decision based on only the reconstructed signal? Answer: YES/NO

Q3. Similarities between signal features: Detail interpretation (mark with ✓)

Features	Quality ratings				
	1	2	3	4	5
Feature #1					
Feature #2					
...					
Feature # fe					

Q4. Please provide your clinical insight/notion. This will help us to improve the performance of the algorithm.

STEP III: Calculate MOS of the reconstructed signal as

$$\text{MOS} = \frac{1}{N_e N_{fe}} \sum_{e=1}^{N_e} \sum_{fe=1}^{N_{fe}} Q_r(e, f_e), \tag{3.7}$$

where

N_e = Total number of evaluators (doctors)

N_{fe} = Total number of considered features

Q_r = Quality rating of the f_e^{th} feature given by the eth evaluator.

The MOS value of a particular feature is expressed as

$$\text{MOS}(f_e) = \frac{1}{N_e} \sum_{e=1}^{N_e} Q_r(e, f_e). \tag{3.8}$$

STEP IV: Calculate Gold standard MOS error of each feature as well as of the overall reconstructed signal using Equations (3.9) and (3.10), respectively:

$$\text{MOS}_{fe} = \left(1 - \frac{\text{MOS}(f_e)}{5}\right), \tag{3.9}$$

$$\text{MOS}_e = \left(1 - \frac{\text{MOS}}{5}\right). \tag{3.10}$$

According to the Gold standard MOS error criteria, the reconstructed signal and also each feature fall under the category 'very good' if the error values lie in the range 0–15; they fall under the category 'good' if the error values lie in the range 15–35; if the MOS error values lie in between 35 and 50, it is considered as 'not good', and 'bad' beyond 50.

It is not imperative to have all the evaluators be clinicians. Researchers who are working in biosignal processing domains and have good understanding about the chosen signal could also take part in the evaluation process [3]. Nevertheless it is good to have more clinicians in the evaluation panel.

3.3.3 Diagnostic Distortion Measure for ECG

All the metrics discussed in Section 3.3.1, for measuring the amount of distortion present in the reconstructed biomedical signal, are non-diagnostic in nature. In many cases, it is seen [4] that the compression algorithm itself reduces some noises, which were present in the original signal beforehand, and reconstructs the signal without jeopardizing its clinical morphology. However, the values of PRD, PRDN, and other metrics become high.

Fortunately, there are a few metrics, namely, weighted diagnostic distortion (WDD) [1], wavelet-energy–based diagnostic distortion (WEDD) [2] for ECG signal, which only enumerate the amount of diagnostic information present in the reconstructed ECG signal and then compare it to that of its original counterpart. These are called objective diagnostic-distortion measurement techniques. Steps involved in calculating WEDD as well as its MATLAB® code are given below.

STEP I: Mean values of the original and reconstructed ECG signals are subtracted from the respective signals.

STEP II: Decompose both the original and reconstructed ECG signals into $L + 1$ number of bands using DWT selecting Daubechies or biorthogonal wavelet filters as the mother wavelet function. The value of L is strongly related to the sampling frequency of the ECG signal. $L = 10$ for an ECG signal sampled at 1 kHz rate [5]. In this example, we assume that the sampling rate of the ECG signal is 1 kHz.

STEP III: Calculate the dynamic weight of each of the bands of the original ECG:

$$W(m) = \frac{\sum_{j=1}^{n} \left(B_m(j)\right)^2}{\sum_{i=1}^{L+1} \sum_{j=1}^{n} \left(B_i(j)\right)^2}, \quad m = 1, 2, \ldots, L+1, \tag{3.11}$$

where $W(m)$ is the weight of the mth band, n is the number of samples in each band, and $B_m(j)$ is the jth wavelet coefficient of the mth band of the original ECG. In the above expression, m varies from 1 to 11. $m = 1$ signifies $D1$ wavelet band, $m = 2$ signifies $D2$ wavelet band, and likewise $m = 10$ signifies $D10$ wavelet band. $m = 11$ signifies $A10$ wavelet band.

STEP IV: Calculate the error present in the wavelet coefficients of each band of the original and reconstructed signals.

$$\mathrm{WPRD}_m = \sqrt{\frac{\sum_{j=1}^{n}\left[B_m(j) - \overline{B_m}(j)\right]^2}{\sum_{j=1}^{n}\left(B_m(j)\right)^2}}, \quad m = 1, 2, \ldots, L+1, \tag{3.12}$$

where $\overline{B_m}(j)$ is the jth wavelet coefficient of the mth band of the reconstructed ECG.

STEP V: Finally, WEDD is calculated as

$$\mathrm{WEDD}\,(\%) = \left(\sum_{m=1}^{L+1} W(m) \times \mathrm{WPRD}_m\right) \times 100. \tag{3.13}$$

MATLAB® code

```
original = original - mean(original);     % original→ original ECG signal
rec = rec - mean(rec);                     % rec→ reconstructed ECG signal
w='bior6.8' or 'db6';                      % w→ mother wavelet function
n = 10;                                    % n→ number of decomposition level
[C,L] = wavedec(original,n,w);
for i = 1:n
          A(i,:) = wrcoef('a',C,L,w,i); % A→ approximation coefficients
          D(i,:) = wrcoef('d',C,L,w,i); % D→ detail coefficients
End
deno = 0;
for i = 1:n
          deno = deno + sum(D(i,:).*D(i,:));
end
deno = deno + sum(A(10,:).*A(10,:));
for i = 1:n
        weight(i)= sum(D(i,:).*D(i,:))/deno;
end
weight(11)= sum(A(10,:).*A(10,:))/deno;
[CC,LL] = wavedec(rec,n,w);
for i = 1:n
    AA(i,:) = wrcoef('a',CC,LL,w,i);
    DD(i,:) = wrcoef('d',CC,LL,w,i);
end
for i = 1:n
          a = D(i,:); b = DD(i,:); c = (a-b).*(a-b); d = sum(c); e = a.*a;
f = sum(e); wprd(i) = sqrt(d/f);
end
```

```
a = A(10,:); b = AA(10,:); c = (a-b).*(a-b); d = sum(c); e = a.*a; f =
  sum(e); wprd(11) = sqrt(d/f);
WEDD=0;
for i = 1:11
          WEDD = WEDD + weight(i).*wprd(i);
end
WEDD = WEDD *100;
fprintf('\n The WEDD value = %f', WEDD);
```

3.4 Compression Techniques for Biomedical Signals

Implementing data compression in biosignals requires a deep understanding of the signal properties, specially the diagnostic features and their correlation with the signal morphology [6]. From this aspect, compression can be carried out by reducing data redundancy and information redundancy. Data redundancy can be realized in the form of inter-sample correlation, as already shown in Figure 3.1. Information redundancy can be reduced by discarding redundant information, as shown in Figure 3.2, e.g., discarding (or applying some thresholding) lower-order coefficients in the transform-domain representation of the data. The generalized structure of a data compression technique involves two components, viz., a model and a coder. The model can exploit any of the following: mapping a larger region (window) of data into a smaller region, transforming the input into a different domain where the redundancy can be visualized or realized better, and characterization of fitting the input into an underlying mathematical model. The model provides an opportunity to discard non-important clinical information from the biomedical signal. The model is followed by a lossless encoder (Arithmetic, Huffman, etc.) which basically exploits probability distribution of the retained information (coefficients) and assigns a minimum number of bits for compressed data representation. In the following subsections, the different time- and transform-domain methods for ECG and PPG signal compression are discussed.

3.4.1 Time-Domain ECG Data Compression

ECG compression techniques/algorithms are broadly categorized into three types: (a) direct, (b) transformation-based, and (c) prediction-based. Time-domain ECG compression techniques usually fall under the direct category, and they are often found to be computationally less complex as well as less power consuming [4,7–9]. In recent times, a number of lossless [10,11], lossy [12], and partially lossless–partially lossy (PLPL) [13–16] ECG compression algorithms based on ASCII (American standard code for information interchange) character encoding have been proposed in the literature. In [13,14], first, the first-difference of the ECG signal is computed and squared. Now, those indices are noted where the square of the first-difference becomes maximum within a window size 0.25 s. Finally, maximum amplitude is searched in the original time-domain ECG signal within 160 ms about the maximum squared position, which gives a true R-peak. Then, the detected QRS-complex regions are compressed using a strict-lossless compression algorithm based on ASCII character encoding [10], and the non-QRS-complex regions such as T waves, P waves, ST segment regions are compressed using a tolerable lossy compression algorithm based on ASCII character encoding [12]. Similar types of PLPL ECG compression algorithms are also proposed in [15,16], where the QRS-complex regions are identified through a heuristically determined standard-deviation–based approach. First, the standard deviation (SD) of a group of samples is calculated, and if the SD value is found to be higher than that of a predefined threshold, then those samples are compressed using the lossless compression algorithm [10]. Otherwise, the lossy compression algorithm [12] is used to compress those samples.

Besides providing a high compression performance, these compression algorithms based on ASCII character encoding also offer another benefit. The compressed ECG data files contain only ASCII characters, and therefore the compressed data can easily be transmitted to remote clinics or hospitals using low-cost telenetwork systems. The direct compression technique based on ASCII character encoding is proposed in [17] for a lossless compression of ECG signal is discussed elaborately hereunder. The technique is completely reversible, i.e., the original ECG signal can be fully recovered using an algorithm, which is the reverse of the compression algorithm.

Block schematic of the lossless ECG compression algorithm based on ASCII character encoding is shown in Figure 3.3.

Let us consider an array, X, containing the original ECG signal of length N. Now, the first-difference of the ECG signal is computed:

$$Y(i) = X(i) - X(i-1), \quad i = 2,\ 3,\ 4,\ldots,N. \tag{3.14}$$

Next, from the first-difference data (Y), at a time, eight consecutive samples are taken and stored in another array (say Z). Let us take an example.

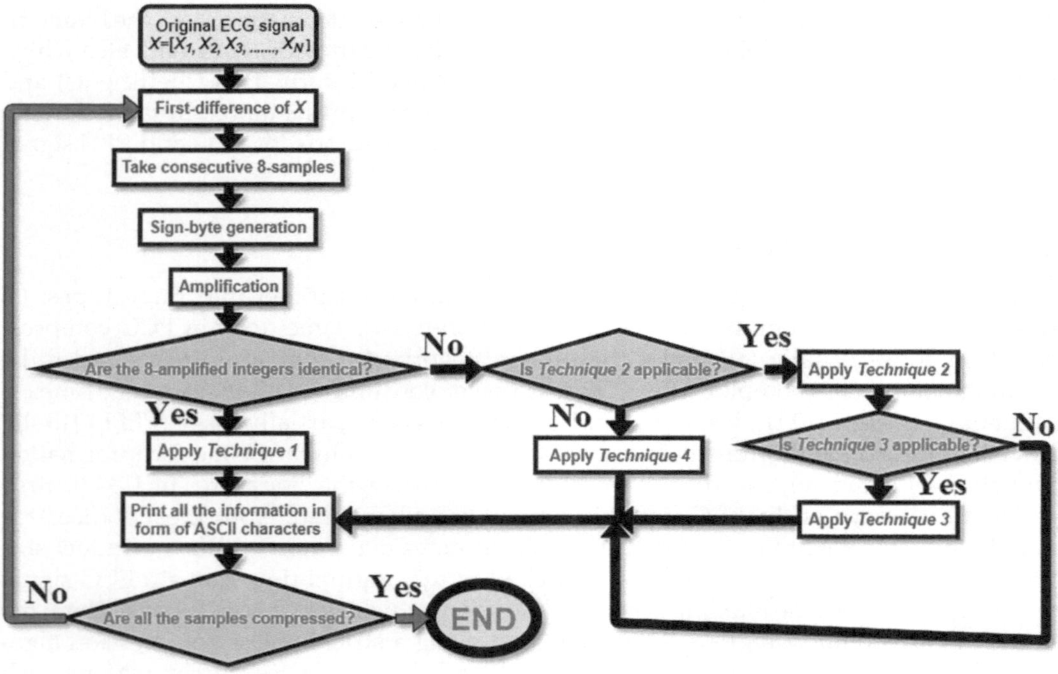

FIGURE 3.3
Block schematic of the lossless ECG compression algorithm based on ASCII character encoding.

0.001	0.005	–0.001	–0.002	0.000	0.000	0.001	0.003
Z(1)	Z(2)	Z(3)	Z(4)	Z(5)	Z(6)	Z(7)	Z(8)

Now, all the samples of the *Z* array are checked as to whether they are positive or negative. Positive samples of the array are marked by binary '0's, and the negative samples are marked by binary '1's. Therefore, a binary string of length 8 is created with these 0s and 1s. For the above example, the binary string looks like that given below.

0	0	1	1	0	0	0	0

The decimal equivalent of this binary string is calculated. Here, for this particular example, the decimal equivalent of $(00110000)_2$ is $(48)_{10}$, and the decimal equivalent is considered as the sign-byte of these eight samples. After calculating the sign byte, all the samples in the array *Z* are made positive and amplified by a factor of 1000 thereafter and rounded off to the nearest integer. For the above-given example, after sign modification and amplification, the content of the *Z* array, which is shown above, will be as shown below.

1	5	1	2	0	0	1	3
Z(1)	Z(2)	Z(3)	Z(4)	Z(5)	Z(6)	Z(7)	Z(8)

Now, these eight amplified integers of *Z* array are merged using various grouping techniques to represent them using as few numbers of integers as possible. The grouping techniques are described below.

Technique 1: If all those eight amplified integers are found identical, they are represented using only one integer. Let us take an example.

5	5	5	5	5	5	5	5
Z(1)	Z(2)	Z(3)	Z(4)	Z(5)	Z(6)	Z(7)	Z(8)

In this case, all the eight amplified integers are represented using only one integer, i.e., 5.

Technique 2: If it is found that all the integers of *Z* are not identical, i.e., if the implementation of Technique 1 is not possible, then it is checked as to whether any two neighbor integers are less than 10 or not. If it is found that two neighbor integers are less than 10, then the first integer is multiplied by 10 and added to the second. Let us take an example.

1	5	1	2	0	0	1	3
Z(1)	Z(2)	Z(3)	Z(4)	Z(5)	Z(6)	Z(7)	Z(8)

For this example, all the neighbor integers are less than 10, and therefore Z(1) is multiplied by 10 and added with Z(2); Z(3) is multiplied by 10 and added with Z(4); Z(5) is multiplied by 10 and added with Z(6) , and Z(7) is multiplied by 10 and added with Z(8). Hence, eight amplified integers are now reduced to four grouped integers as shown below.

15	12	0	13
$(Z(1)\times 10)+Z(2)$	$(Z(3)\times 10)+Z(4)$	$(Z(5)\times 10)+Z(6)$	$(Z(7)\times 10)+Z(8)$

Technique 3: If all those eight integers are grouped into four integers using Technique 2, and if it is seen that all those four grouped integers are less than 15, then these four grouped integers are regrouped in their binary domain. Each of these four grouped integers is converted into 4-bit binary number; two neighbor nibbles (1 nibble = 4 bits) are concatenated to form a byte (1 byte = 8 bits) and converted into decimal. Let us take an example.

Amplified integers	1	5	1	2	0	0	1	3
Technique 1 grouping	15		12		0		13	
4-bit binary equivalent	1111		1100		0000		1101	
Byte formation	11111100				00001101			
Decimal equivalent	252				13			

Therefore, eight amplified integers are reduced to two integers.

Technique 4: Implementation of Technique 3 is possible only when all the eight amplified integers are grouped using Technique 2. But, what happens if the implementation of Technique 2 itself is not possible for all the amplified neighbor integers? In such cases, the non-grouped integers are kept unchanged. An extra array (say *K*) is taken to mark those corresponding indices of non-grouped integers, and finally the decimal equivalent of *K* is calculated. Let us take an example.

Amplified integers	1	5	14	9	12	14	1	3
Reduced dataset	15		14	9	12	14	13	
	Technique 2 grouping		Non-grouped integers		Non-grouped integers		Technique 2 grouping	
Content of *K*	0	0	1	1	1	1	0	0
Decimal equivalent of *K*	60							

Finally, the sign byte and grouped, regrouped, and non-grouped integers are printed in the output file in the form of their corresponding ASCII characters. The decimal equivalent of *K* is printed in the form of an ASCII character only when there are non-groped integers, i.e., Technique 4 is used. Let us take an example.

48	15	14	9	12	14	13	60
Sign byte		Grouped/regrouped/non-grouped integers					*K*
0	☼	♫	○	☺	♫	♪	<
ASCII characters							

The above-described lossless ECG compression algorithm is executed iteratively on each set of eight samples taken from the first-difference data. The 'ENTER' character is used to separate each set of characters from its previous and next in the output file. Figure 3.4 shows a snapshot of the compressed ECG data file. It is very important to remember that

1. in the case of Technique 1 grouping, the corresponding character set contains only two characters (sign byte and one character, which represent all the eight integers);
2. in the case of Technique 2 grouping, the corresponding character set contains five characters (sign byte and four characters for four grouped integers);

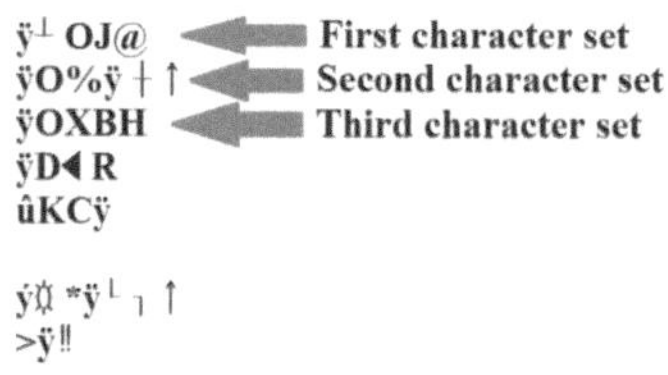

FIGURE 3.4
Snapshot of the compressed ECG data file.

3. in the case of Technique 3 grouping, the corresponding character set contains three characters (sign byte and two characters for two regrouped integers); and
4. in the case of Technique 4 grouping, the number of characters in the corresponding set may vary from 7 to 10 (sign byte, grouped or non-grouped integers, and *K*).

3.4.1.1 The Reconstruction Algorithm

The ECG signal is reconstructed from the compressed data using an algorithm, which is the reverse sequence of steps of the compression algorithm. One set of characters is taken at a time from the compressed ECG data for reconstruction. In every character set, the first character represents the sign byte, and the rest are the grouped, regrouped, and non-grouped integers and *K*.

If the number of characters in a set is 2, it signifies that Technique 1 was executed during compression. Therefore all the eight integers are regenerated easily.

If the number of characters in a set is 5, it signifies that Technique 2 was executed during compression, and therefore, grouped integers are separated using the reverse method of Technique 2.

If the number of characters in a set is found to be 3, it signifies that Technique 3 was executed. In this case, first, each of these two regrouped integers is converted into 8-bit binary equivalent. Next, these 8 bits are separated into 2 nibbles. Finally, each nibble is converted into decimal equivalent. Therefore, four integers are regenerated. Now, applying the reverse approach of Technique 2, these four integers are converted into eight. Let us take an example.

	Sign Byte	Regrouped Integers							
ASCII characters	0	3				♪			
Decimal equivalent	48	252				13			
8-bit binary equivalent		11111100				00001101			
Nibbles		1111		1100		0000		1101	
Decimal		15		12		0		13	
Reverse approach of Technique 2		1	5	1	2	0	0	1	3

If the number of characters in a set varies from 7 to 10, it indicates that the set contains either a combination of grouped and non-grouped integers or only non-grouped integers. In this case, the last integer of the set, i.e., *K*, is converted into binary equivalent. In the binary string, if any bit is found to be '0', the corresponding integer is ungrouped using the reverse approach of Technique 2. Otherwise, it is left unchanged. Let us take an example.

	Sign Byte	Grouped and Non-grouped Integers								*K*
ASCII characters	0	☼		♫	○	♀	♫	♪		<
Decimal equivalent	48	15		14	9	12	14	13		60
Binary equivalent of *K*		00		1	1	1	1	00		
Expanded dataset		1	5	14	9	12	14	1	3	

The above-described techniques decode grouped, regrouped, and non-grouped integers into eight amplified integers. Now, these decoded integers are divided by 1000. Next, the sign byte, i.e., the first character of every set is converted into 8-bit binary equivalent, and in the binary sequence if any bit is found to be '1', the corresponding deamplified number is made negative. Let us take an example.

	Decimal Equivalent of Sign Byte	Decoded Integers							
Expanded dataset	48	1	5	14	9	12	14	1	3
Deamplified numbers		0.001	0.005	0.014	0.009	0.012	0.014	0.001	0.003
Binary equivalent of sign byte		0	0	1	1	0	0	0	0
Deamplified numbers after sign modification		0.001	0.005	−0.014	−0.009	0.012	0.014	0.001	0.003

Thereafter, every number is added with its previous (reverse operation of first difference) to regenerate the original ECG data. The above-described decoding technique is executed iteratively on all the character sets present in the compressed ECG data file. Figures 3.5a–c show original ECGs, reconstructed ECGs, and the reconstruction errors

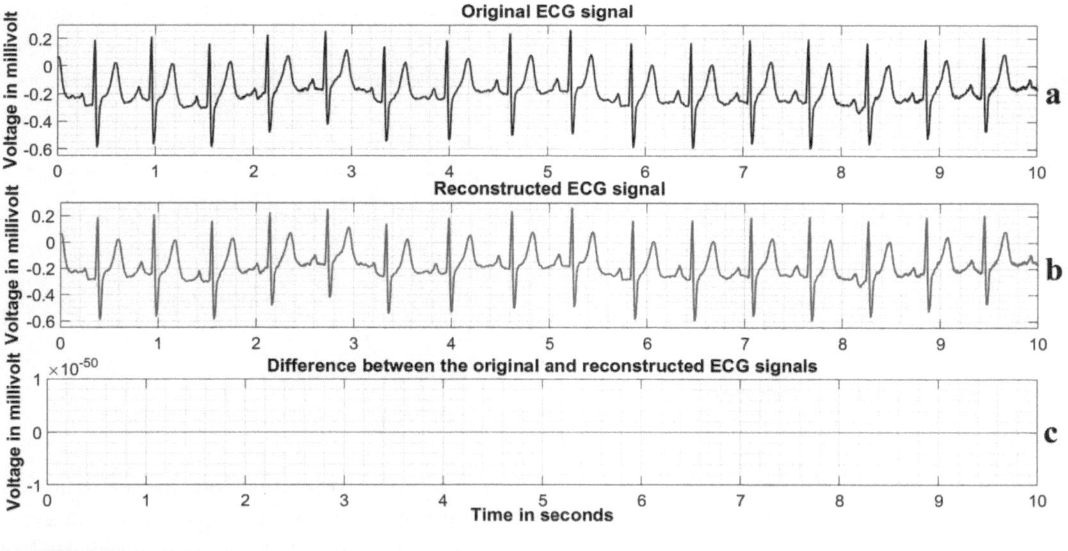

FIGURE 3.5
(a) Original ECG (MIT-BIH arrhythmia database, record no. 107), (b) reconstructed ECG, and (c) difference between the signals shown in (a) and (b).

using the above-described compression algorithm. From this figure, it can be seen that the compression algorithm is indeed lossless (Figures 3.6 and 3.7).

A simple and low-complexity lossless algorithm to encode the successive sample differences (SSD) is presented in [18]. The principle is based on the fact that in terms of variability, normally QRS complex denotes the maximum information and is clinically

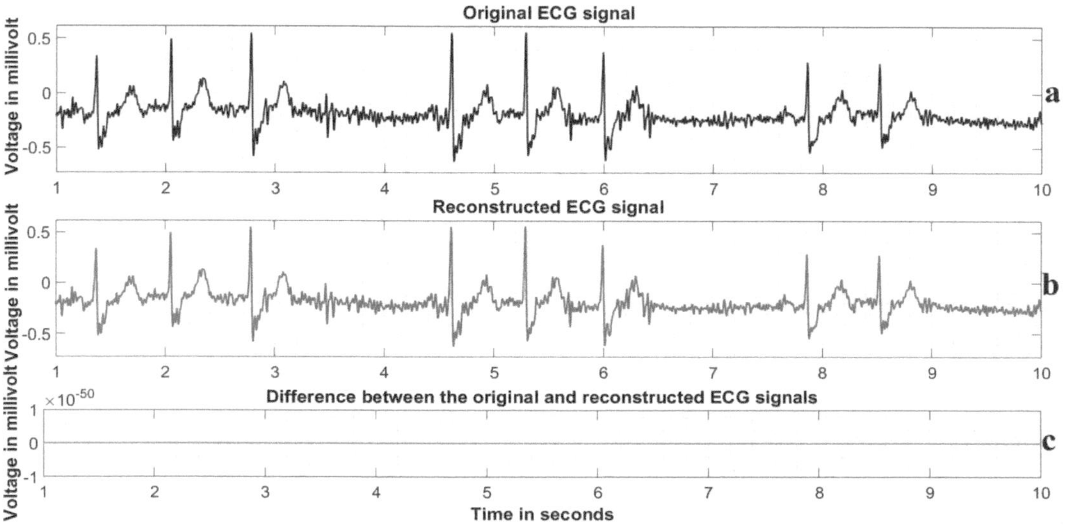

FIGURE 3.6
(a) Original ECG (MIT-BIH arrhythmia database, record no. 232), (b) reconstructed ECG, and (c) difference between the signals shown in (a) and (b).

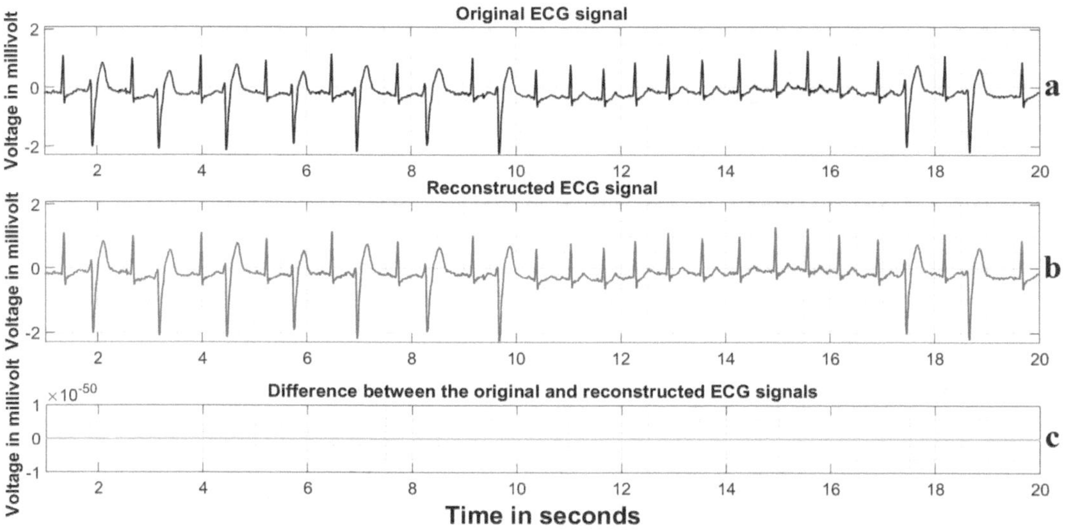

FIGURE 3.7
(a) Original ECG (MIT-BIH arrhythmia database, record no. 200), (b) reconstructed ECG, and (c) difference between the signals shown in (a) and (b).

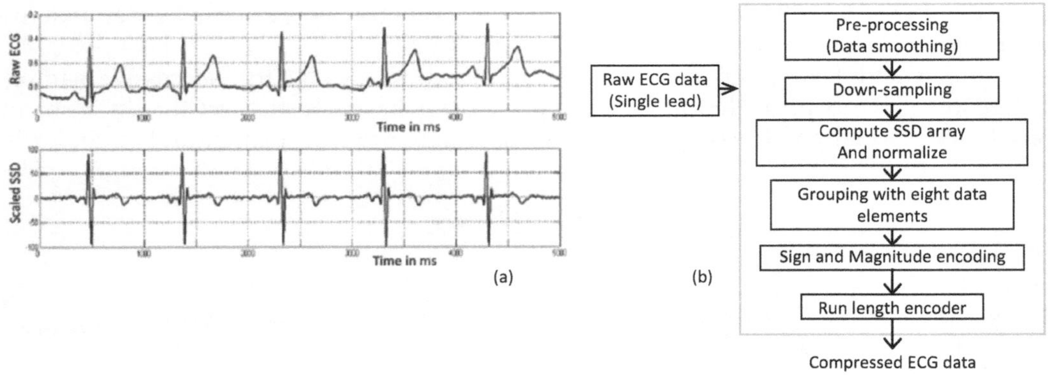

FIGURE 3.8
SSD encoder: (a) SSD plot and (b) signal processing stages.

most significant in an ECG. The other regions have either low variability (P, T wave) or almost no variability (equipotential segments PQ, ST, and TP). The normalized SSD plot along with the raw ECG plot is shown in Figure 3.8a. It is observed that fluctuations in QRS regions are only considerable, and the data in non-QRS regions can be represented by 4 bits or lower. Hence, there is scope for the direct encoding of SSD array using simple time-domain methods. The processing stages are shown in Figure 3.8b.

Although the process can be applied in real time using a short number of samples, the CR can be enhanced by taking around 1,000 samples together, using the run-length encoder (RLE). The objective of the different compression stages are briefed below:

Data smoothing: To reduce abrupt fluctuations and high-frequency noise in the dataset, data smoothing can be performed either by a low-pass filter, or moving average filter, or a cubic-spline interpolation operation.

Down sampling: In case the ECG is sampled at high frequency, say 500 Hz or more, then this step may select optimum number of samples keeping the peak amplitudes intact.

SSD array generation and normalization: Keeping the maximum amplitude of SSD elements at 99 by applying a normalization factor to convert them into signed integers.

Data grouping: This selects a group of eight consecutive elements from normalized elements.

Sign and magnitude encoder: Using a binary sign representation of eight elements (resulting a byte) and nibble combination of offset rule of encoder.

RLE: Consecutive zero elements at equipotential segments could be encoded using a two-byte encoded data.

The advantageous feature of the encoder is that it can be implemented using low-end microcontrollers [20] and requires practically no header information. Another successful implementation to apply this encoder using a GSM-based telemonitoring service is reported in [21].

A similar version of real-time ECG compression using zonal complexity measure is reported in [22]. Here, the window is attributed as 'complex', 'semi complex', or 'plain' based on SD, and accordingly, different binary encoding rules are applied on the first- and second-order SSD.

A cycle-by-cycle time-domain ECG compression technique based on the principle of delta encoder of SSD is reported in [19]. A simplified compression flow logic diagram is shown in Figure 3.9.

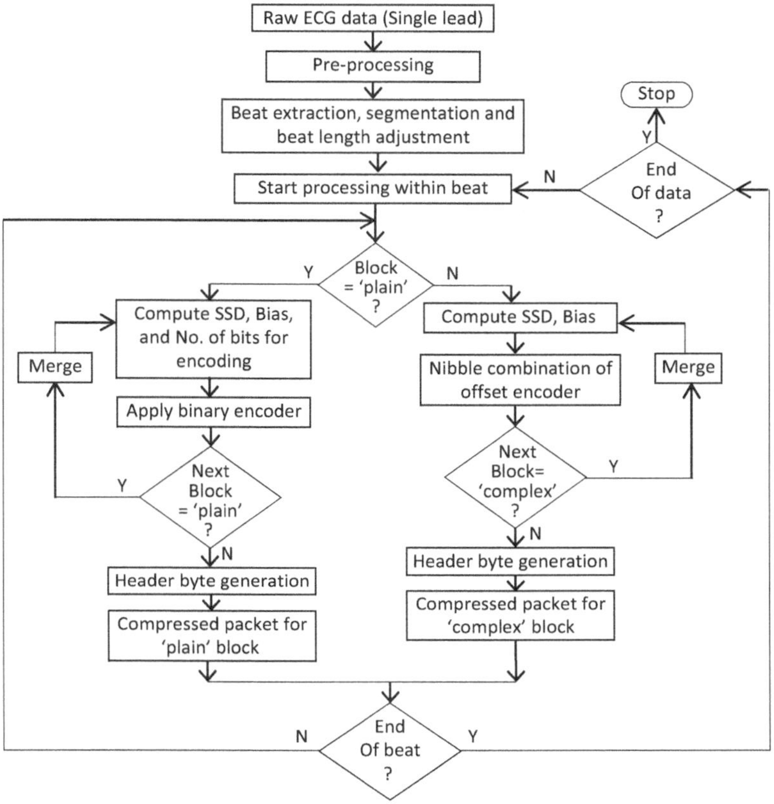

FIGURE 3.9
Logic flow diagram of the hybrid encoder [19].

Each beat or cardiac cycle is fragmented into blocks, which can be 'plain' or 'complex', depending on its complexity decided by the SD. Accordingly, for plain 'block', a flexible bit allocation is adopted where the SSD array is compressed by a binary encoder. Successive 'plain' blocks are merged to enhance the compression. For 'complex' blocks, fixed length nibble combination or offset rule encoder is adopted. In this sense, the compressor is hybrid. An important finding of the work is low bits/sample requirement using digitized samples at 10 bit resolution. This enhances the link efficiency in telemonitoring application.

3.4.2 Transform-Domain Compression Methods: Wavelet and PCA

In the recent years, transform-domain methods have become very popular due to their implementabilty using VSLI circuits. In general, transform-domain methods employ an orthogonal linear transform to map the time series samples to a different domain where the variance of the signal, representing the clinical information, can be represented by a small number of expansion coefficients [23]. Based on the signal representation, the ECG compression methods are categorized into one-dimensional (1-D) and two-dimensional (2-D) compression. In 2-D compression method, the time series 1-D ECG is represented as a 2-D matrix, by slicing out the ECG beats from the time-series data and aligning them with reference to some fiducial point [24]. This representation, shown in Figure 3.10 maximally exploits intrabeat as well as interbeat correlations in the ECG data to achieve higher compression and low distortion. In regard to lossy compression techniques, there are two basic signal processing steps. First, the basic mathematical model based on which the compression is done. This step provides the maximum data volume reduction. This includes the transform function, the rule based on which the coefficients are selected (or discarded), their resizing (truncation if any), and the quantization strategy for selected coefficients. The other step is an information coding which is usually lossless. The steps for transform-domain compression and their significance are summarized:

a. Preprocessing of data to remove unwanted noise and artifacts. Most of the physiological signals (ECG, EEG, EOG) are of low frequency (below 100 Hz). Hence, implementing a low-pass filter of appropriate frequency can significantly discard the radiofrequency and electromagnetic interference. However, some of the artifacts lie within the clinical spectra and require special digital signal processing technique to get a cleaner signal.
b. Delineation of the cardiac cycle is required if 2-D compression is implemented. For ECG, this requires the detection of the R-peaks accurately in each cardiac cycle. Till date, many successful implementation of R-peak detection algorithms are available in published literature [25–28].
c. For 2-D compression, the extracted cardiac cycles are aligned with respect to a major fiducial point (like R-peak in ECG), and beat lengths are equalized by 'period normalization' or 'padding' to form a 2-D matrix.
d. Orthogonal expansion is applied to get a representation of the 1-D or 2-D data in different orthogonal dimensions, represented by expansion coefficients. These coefficients have no redundancy among them.
e. The compression is carried out by two steps, viz., redundancy reduction and information reduction. For physiological signals like ECG, it is observed that most of the signal energy is represented in limited orthogonal dimensions (or scales) and in fewer number of expansion coefficients in those dimensions. In other words, the non-contributing dimensions (scales) and coefficients are considered redundant. So, identifying the useful dimensions (scales) and selection of 'contributing' coefficients based on a rule constitutes the redundancy reduction step. Further truncation of the selected coefficients using a thresholding rule (described later in this chapter) constitutes the information reduction stage.
f. For digitized representation of the selected expansion coefficients for further use, quantization is applied. The quantization can be fixed or adaptive based on their zonal significance toward data reconstruction. This involves some loss of information associated with rounding off the data.
g. Lossless ending of the coefficients is applied based on probability distribution. Most common are delta, Huffman, and Arithmetic coders.
h. To include essential side information for decoding, data packetization is done for each group of raw data samples considered for compression. The header bytes containing this extra information are normally prefixed before each data packet.

Among the different mathematical tools, wavelet transform, discrete cosine transform (DCT), Walsh transform, Karhunen Louve transform (KLT) (popularly known as

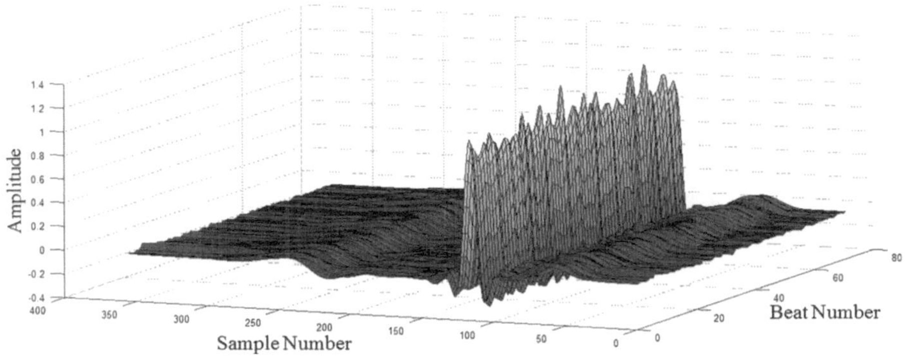

FIGURE 3.10
3D plot of beat matrix, with aligned R-peaks, showing interbeat redundancy.

principal component analysis or PCA), discrete Legendre transform, discrete Hermite transform, and discrete Fourier transform (DFT) have been used in ECG compression applications due to their excellent ability to compact energy in fewer sub-bands or coefficients. The general prerequisites for such techniques are preprocessing (filtering), R-peak detection, beat boundary estimation, beat alignment for the formation of 2-D matrix before the transform is applied. Since the heart rates vary among consecutive cardiac cycles, an approach based on padding or spline interpolation is used to make beat lengths equal. Following this, the 2-D transform (DWT, KLT, DCT, etc.) is applied to obtain the expansion coefficients. Some rule is framed to discard the insignificant coefficients. Finally, the selected coefficient is quantized, and a lossless coding is applied for further compression.

However, the inaccurate beat detection and low interbeat correlation as well as small and abnormal QRS morphology [27] pose a great challenge before the successful implementation of 2-D compression methods. Other concerns are additional time complexity for beat detection and space requirement for buffering 2-D ECG beat.

Among the transform-domain ECG compression methods, wavelet transform (WT) has received maximum attention by the research community due to its excellent energy compaction ability in a small number of decomposed coefficients. A wavelet is a waveform of limited duration that has an average value of zero. A wavelet family is the set of basis functions $\psi_{a,b}$ generated by translations and dilations of a mother wavelet $\psi(t)$, given as

$$\psi_{a,b}(t) = \frac{1}{\sqrt{a}}\psi\left(\frac{t-b}{a}\right), \tag{3.15}$$

where a and b are scaling (dilation) and shifting (translation) parameters, respectively, and t is the time. The mother wavelet $\psi_{a,b}$ is compressed ($a < 1$) and dilated ($a > 1$) and shifted over the waveform segment (for different b values) to realize the different frequency components occurring in different time by a convolution operation. Lower a values provide good frequency resolution but poor time information, whereas higher a values provide good time resolution but poor frequency resolution [29]. This, in effect, can be understood as a series of band-pass filters with varying band-pass and center frequencies being imposed on the investigated waveform. The utility of WT for ECG compression can be realized from the fact that most of the QRS energy is represented in the lower-order sub-bands, while the higher sub-bands mostly include noise representations [30].

The continuous wavelet transform (CWT) generates substantial redundant information as the parameters a and b can assume any values. A good alternative is to implement DWT where these parameters can be varied in discrete steps: $a = 2^j$ and $b = k2^j$. The DWT can be represented as a correlation of the mother wavelet $\psi(t)$ and the ECG $f(t)$ for scale and location parameters j and k, respectively:

$$X_{j,k} = \int_{-\infty}^{+\infty} f(t)\Psi_{j,k}(t)dt. \tag{3.16}$$

In dyadic wavelet transform, the parameters a and b vary in a geometric scale of ratio 2. This facilitates multiresolution analysis (MRA) decomposition of the ECG successively using two basic functions, viz., the wavelet $\psi(t)$ and the scaling $\phi(t)$, into two scales called coarse and finer components [5]. These functions can be obtained as a weighted sum of the shifted and dilated versions of the functions itself, represented as

$$\begin{aligned} \phi(t) &= \sum_n h[n]\phi(2t-n), \\ \psi(t) &= \sum_n g[n]\psi(2t-n), \end{aligned} \tag{3.17}$$

where $h[n]$ and $g[n]$ are the half-band low-pass filter and high-pass filter, respectively. Again the scaling and the wavelet functions $\psi(t)$ and $\phi(t)$ discretized in scale j and translation k can be obtained from their original versions, represented as

$$\begin{aligned} \psi_{j,k}(t) &= \frac{1}{\sqrt{2^j}}\psi\left(\frac{t}{2^j}-k\right), \\ \phi_{j,k}(t) &= \frac{1}{\sqrt{2^j}}\phi\left(\frac{t}{2^j}-k\right), \end{aligned} \tag{3.18}$$

where j and k control the dilation and translation, respectively. Filtering and down sampling of the original discretized signal $x[n]$ lead to an approximation A (lower frequency) and detail D (high frequency) coefficients, given as

$$\begin{aligned} A[k] &= \sum_n x[n]\cdot h[2k-n], \\ D[k] &= \sum_n x[n]\cdot g[2k-n]. \end{aligned} \tag{3.19}$$

A wavelet pyramidal structure can be obtained by further decomposition of A, where the original signal energy is distributed into various successive levels, represented as

$$\begin{aligned} x(t) &= A_1(t) + D_1(t) \\ &= A_2(t) + D_2(t) + D_1(t) \\ &= A_3(t) + D_3(t) + D_2(t) + D_1(t) \\ &= \ldots\ldots \end{aligned} \tag{3.20}$$

Among the various compression techniques that are based on DWT decomposition of ECG, the tree-based methods and threshold-based methods are most popular. Among the tree-based methods, embedded zero tree wavelet (EZW) [31] and its extended version, set partitioning in hierarchical trees (SPIHT) [32], have demonstrated good performance for ECG compression. Here, the ECG dataset is successively decomposed to form a pyramidal structure. Some of the decomposed coefficients are considered as 'significant' when they are greater than a threshold and otherwise considered as non-significant. EZW is based on following principles: (a) the position and sign of the significant coefficients are considered; (b) non-significant coefficients are compactly encoded based on self-similarity across sub-bands; (c) successive approximation of the significant coefficients. The SPIHT algorithm sorts the coefficients transformed by wavelet-based decomposition in order of their significance or magnitude. This partial ordering is done by comparing them with an octavely decreasing threshold. For transmission applications, coefficients with low bit rate are ordered first, with gradually decreasing significant ones.

Algorithms based on wavelet threshold are computationally most simple among transform-domain compression techniques. The wavelet coefficients are ordered based on their significance and thresholded based on a fixed energy packing efficiency (EPE) [33,34], either targeting a bit rate or the criteria based on reconstruction error. *A* binary significance map (BSM) is created to encode the positions of selected wavelet coefficients. The selected set of DWT coefficients are then quantized and compressed using a lossless encoder like delta or Huffman encoder. There are two types of thresholding applied, 'hard' and 'soft'. The hard thresholding is based on 'keep or kill', i.e., the coefficients greater than the threshold are retained, and those with lesser values are 'killed'. The soft thresholding is based on 'shrink or kill', i.e., the coefficients greater than the threshold are decreased by the amount of threshold, and those with lesser values are 'killed'. Mathematically, the thresholding rules are represented as

$$\begin{aligned} \hat{d}_{i,l} &= d_{i,l} \quad \text{if} \quad d_{i,l} \geq \text{th} \quad \text{if} \left|d_{i,l}\right| \geq \text{th} : \text{'hard' thresholding} \\ &= 0 \quad \text{otherwise,} \\ \hat{d}_{i,l} &= \text{sign}\left(\left|d_{i,l}\right| - \text{th}\right) \qquad \text{if} \left|d_{i,l}\right| \geq \text{th} : \text{'soft' thresholding} \\ &= 0 \quad \text{otherwise,} \end{aligned} \tag{3.21}$$

where $d_{i,l}$ represents the lth coefficient in ith decomposition level.

Since most of the ECG energy is found in the lower sub-bands, and with lower number of coefficients, the energy contribution from the each sub-band may be calculated as

$$\text{ECE} = \frac{\sum_{l} d_{i,l}^2}{\sum_{i}\left(\sum_{l} d_{i,l}^2\right)}, \tag{3.22}$$

where the inner (outer) summation in denominator indicates summation within (across) a sub-band. The selected coefficients are quantized using a linear quantization (LQ) strategy. The final data packet consists of encoded BSM followed by the lossless encoded amplitudes of the selected coefficients. In [35], an iterative threshold adjustment was carried out until a fixed percentage of DWT coefficients assumed zero value. Some applications of quality control over reconstructed data and/or bit rate control is available using DWT-based compression. In [36], a predefined PRD could be achieved by the DWT-based compression based on adaptive quantization strategy of the orthonormal DWT coefficients, followed by an entropy encoder utilizing Lempel–Ziv–Welch (LZW) encoder. In [37], noise estimation from each block of ECG data, consisting of 512 samples, was done and used as the threshold to compress the DWT coefficient in the coding stage. The noise level was estimated by adding different known noises to the clean ECG data and then computing the RMSE for each block between the noisy signal and clean signal. It was observed that the distortion level is minimum when the applied threshold equals the noise level. In another approach [38], the authors decomposed the ECG data up to five levels, packed them in three frames, and concatenated them. The less significant DWT coefficients were thresholded. A uniform scalar zero zone quantizer (USZZQ) was used for quantizing significant wavelet coefficients. Finally, Huffman coding of the difference between the significant coefficients was done.

For real-time applications like telecardiology, modification of the traditional, recursive wavelet-based decomposition of the ECG data was suggested to meet the desired latency as well as low distortion. In [39], a 1-D round-off non-recursive DWT scheme is presented to achieve a linear quality control over the reconstruction error. A bit allocation scheme for quantization of DWT coefficients based on zonal complexity of the ECG data is presented in [40] to achieve low delay in the compression process. The payload after DWT decomposition can be reduced by an approach described in [41], which preserves the bit depths to secure bits of interest after DWT decomposition, followed by RLE.

Major parameters which determine the performance of the DWT-based compression method in terms of CR, PRDN, latency are the choice of mother wavelet, levels of decomposition and their individual significance toward representing the total energy and clinical information, type of thresholding rule, quantizing principle (linear or non-linear), and finally, the lossless compression of the BSM and quantized coefficients. In the published literature on ECG compression, Coiflet, biorthogonal spline, and Daubechies wavelets have been the most popular use, due to their morphological similarity with the QRS zone of the ECG. This ensures that most of the ECG energy is captured in limited sub-bands with fewer number of DWT coefficients. In [42], the wavelet that best matches the shape of the ECG signal has been addressed. The threshold estimation is based on two principles, viz., retaining the coefficients with value higher than the threshold and compromising between the number of retained coefficients and target bit rate reconstruction error. In reported works, the following thresholding rules have been applied: (a) energy packing efficiency [33]; (b) total percentage of DWT coefficients to be zeroed [35]; (c) fixed number of coefficients from sorted DWT decomposed array, sorted in descending order of amplitude; and (d) recursive selection of coefficients for achieving distortion value [43]. The length of the window plays an important role for latency and buffer requirement for hardware implementation, the most common choices being 512, 1,024, and 2,048 number of samples.

There are a few wavelet-based 2-D compression works available in published literature [24,44,45]. The motivation behind all of these techniques is to exploit the interbeat correlation in a single-lead ECG data. After R-peak detection, the 1-D ECG data is cut and aligned with reference to the R-peak to form a 2-D image matrix. Thereafter, compression based on either wavelet packet or vector quantization (VQ) has been applied. In VQ, the compression is performed by quantizing wavelet coefficients into a reduced set of vectors, defined in the codebook, C, defined as

$$C = \left[c_1\, c_2 \ldots c_m\right], \tag{3.23}$$

where $c_k = [c_{k1}\ c_{k2}\ \ldots\ c_{kn}]$ is the kth codebook.

For a set of DWT coefficients X_k, the error measure with one code vector c_j is given as

$$d\left(X_k, c_j\right) = \sum_{j=1}^{n}\left[X_k - c_{j,k}\right]^2, \tag{3.24}$$

where n is the number of subsets where the X_k belongs to. The minimum error (d) provides the best match of X_k. In VQ, instead of the whole X_k, the index of c, i.e., j, is stored. In [45], the DWT coefficients are arranged in a tree structure to form a codebook vector. The energy concentration occurs in low-frequency sub-bands. These coefficients are lossless encoded to form compressed data packets.

A quality-controlled lossy compression using average beat subtraction and lossless encoder was proposed [46]. Since the consecutive ECG beats show adequate similarity (unless an abnormal beat occurs), a block of 10 ECG beats were considered at a time, and one progressive average beat (PAB) and 10 residues were computed using the equations

$$\begin{aligned} b_a(i) &= \frac{1}{2}\left[b_{av}(i-1) + b(i)\right], \\ \text{res}(i) &= b_{av}(10) - b(i), \end{aligned} \tag{3.25}$$

where $b(i)$ denotes the ith beat vector, formed by period equalization from the group; $b_a(1) = b1$ and i starts from 1; $b_a(10)$ is the final PAB; and res(i) is ith residue. The compression flow consists of preprocessing, beat extraction, formation of 2-D beat matrix, computing the PAB and residuals, applying DWT on them, and finally selection of coefficients and their lossless encoding, as shown in Figure 3.11. Coiflet 4 was used as mother wavelet to decompose the PAB and residuals.

The DWT coefficients from b_a and residual vectors were selected on 5% PRDN criteria and 99.9% ECE criteria, respectively, by ordering them in decreasing order of significance. Fixed quantization levels of 8 and 5 bits were used for selected PAB and residuals. Figure 3.12 shows the reconstruction quality for MIT-BIH arrhythmia 117 data, which is used by many researchers. Here, the ECE is used for reconstruction with a PRDN control limit of 5%, yielding a CR of 10.59 and PRDN of 4.91. As shown in the figure, the reconstructed plot closely follows the original signal, with minor deviations in the QRS regions.

Diagnostic ECG, specially arrhythmia records, often contain abnormal patterns like premature ventricular contraction (PVC), atrial premature beat (APB), left bundle branch block (LBB), etc. Here, it is essential that the pathological information should be

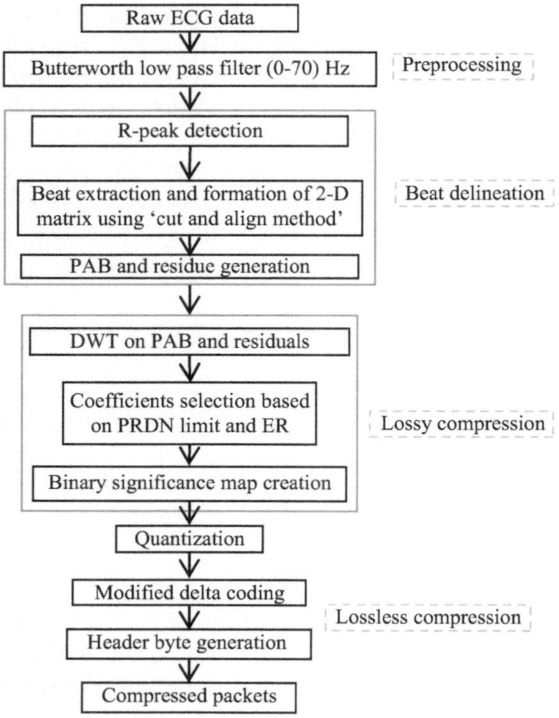

FIGURE 3.11
Process flow diagram of residual encoding using PAB.

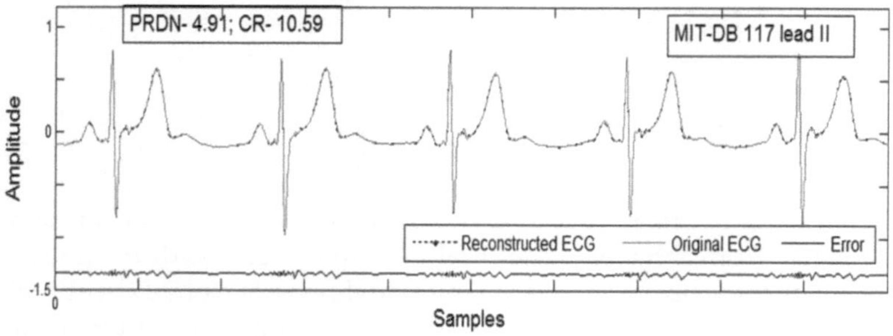

FIGURE 3.12
Original reconstructed ECG data and error plot for MIT-BIH arrhythmia data 117 using residual encoder compression.

preserved in the reconstructed record for clinical interpretation. However, adopting a uniform, beat-by-beat compression strategy may not be always helpful to provide the best compression results in such cases. Figure 3.13 shows the distortion in R-peak and QS region in the ECG abnormal beat (marked 'A', ventricular premature beat) using uniform compression strategy with DWT-based approach (Coiflet 4, Level 5 decomposition) for different levels of target CR. Here, the normal beats are also distorted to achieve higher compression (CR 31).

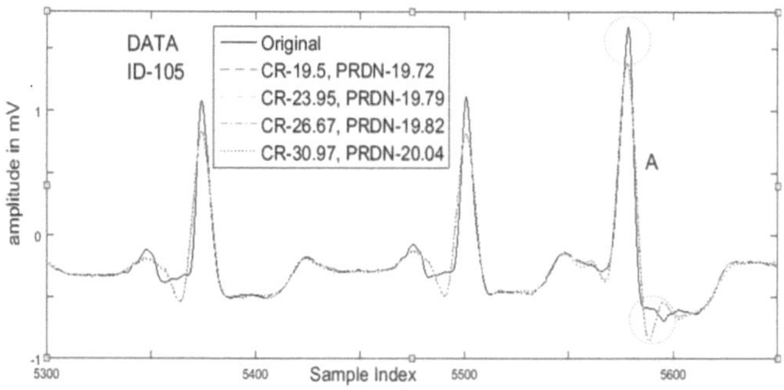

FIGURE 3.13
Abnormal beat morphology distortion in uniform compression for achieving higher compression.

PCA is the other ECG compression tool that received significant attention. Mathematically, it can be defined as a linear transform that projects the multiple ECG measurements in few orthogonal dimensions, called the basis vectors or eigenvectors, based on a hidden data pattern within the dataset. The projections, called scores, signify the variability within a dataset in the transformed space.

In the context of ECG compression, this variability can be explored by forming a 2-D matrix using 'cut and align method' to form a beat matrix $\boldsymbol{B}$ and then executing the PCA using the eigenvalue decomposition of the covariance matrix (also called variance–covariance matrix) of $\boldsymbol{B}$. The $[\boldsymbol{B}]_{m\times n}$ consists of m beat vectors, each consisting of n samples after period equalization, and is represented as

$$\begin{aligned} B &= \left[B_1\, B_2 \ldots B_n\right], \\ \text{where } B_k &= \left[b_{k1}\, b_{k2} \ldots b_{km}\right]^{\mathrm{T}}, \end{aligned} \tag{3.26}$$

where b_{ki} represents ith sample of kth beat vector, with $i = 1, \ldots, m$.

The orthogonal transformation is defined as

$$\begin{aligned} \mathrm{pc} &= \Psi^{\mathrm{T}} \times B, \\ \Psi &= \left[\psi_1\, \psi_2 \ldots \psi_m\right], \\ \mathrm{pc} &= \left[\mathrm{pc}_1\, \mathrm{pc}_2 \ldots \mathrm{pc}_m\right]^T. \end{aligned} \tag{3.27}$$

where ψ_k defines the eigenvector and the pc_k, the principal component score. It is observed that the ECG signal energy in the transform-domain is represented in fewer scores, called significant scores, as shown in Figure 3.2a. Thus, the m-dimensional data can be represented in the transform-domain by first few (say t, where $t < m$) scores, providing a scope of dimensionality reduction [47]. In PCA-based ECG compression techniques, the transformed scores and eigenvectors are used for further processing. Selection of the number of scores depends on the target reconstruction quality or bit rate. In [48], a PCA-based single-ECG compression is described where the PCs and their quantization levels are recursively selected based on predefined distortion measures and/or CR. The signal processing flow is shown in Figure 3.14, which consists of independent control over distortion and bit rate.

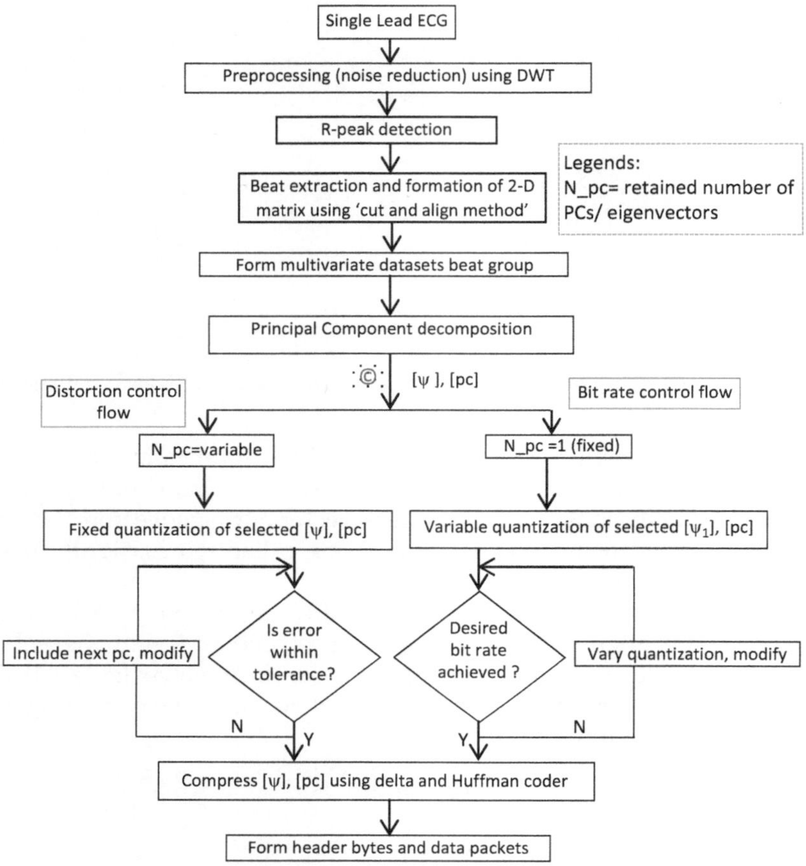

FIGURE 3.14
Signal processing flow for PCA-based compression with independent distortion and bit rate control strategy.

For distortion control flow, two parameters, viz., PRDN and MAE, were checked after each iteration, and the number of scores were recursively increased until the desired target distortion level was met with. However, quantization levels of eigenvectors and scores were kept fixed. For bit rate control, however, a single score was considered, and the quantization level of the score was varied until the CR per group of data was achieved. Two types of myocardial infarction (MI) abnormality, viz., anterior and interior, along with the normal pattern were evaluated from Physionet ptbdb data [49]. Figure 3.15 shows the distortion control (panel a) and bit rate control (panel b). Although both plots show acceptable reconstruction, the CR for bit rate control is 45.75, at the cost of higher PRDN at 11.16, whereas the same for distortion control is 12 but with PRDN of 4.57. For MIT-BIH arrhythmia data with normal rhythm (MIT-BIH arrhythmia 117), the reconstruction pattern with error control is shown in Figure 3.16a. The research [48] showed that there is a limit (CR of 68.96 MIT-BIH arrhythmia 117 from Physionet) up to which the bit rate control could be achieved keeping the reconstruction quality clinically acceptable. The variation of PRDN and MAE with target bit rate (or CR) is shown in Figure 3.16b,c, respectively.

Instead of the fixed quantization level, a joint optimal selection of for the PCs and eigenvectors along with their quantization level can provide better control over the

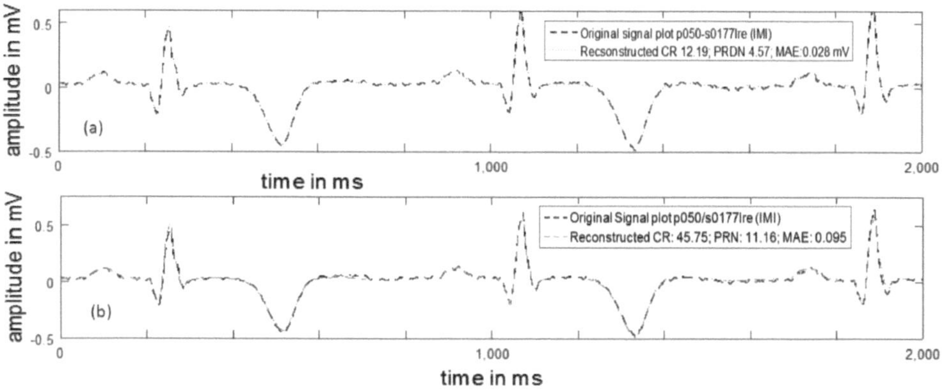

FIGURE 3.15
Compression performance with IMI abnormality from ptbdb: (a) reconstruction control and (b) bit rate control.

reconstruction quality. A stochastic optimization technique like PSO can be used to achieve this optimality at the point © in Figure 3.14. The objective function of the PSO can be formulated as

$$\begin{aligned} &\min_{q\in A} f(q), \text{ subject to} \left|\text{prd}(q) \sim 0.05\right| \approx 0, \\ &\text{where } q = Z\left(b_{\text{pc}}, b_{\text{eig}}\right), \end{aligned} \tag{3.28}$$

where $f(q)$ is the objective function; $(b_{\text{pc}})_k$ and $(b_{\text{eig}})_k$ are bits allocated for quantizing the kth PC and eigenvector, respectively; prd is the PRD value; and A denotes the search space

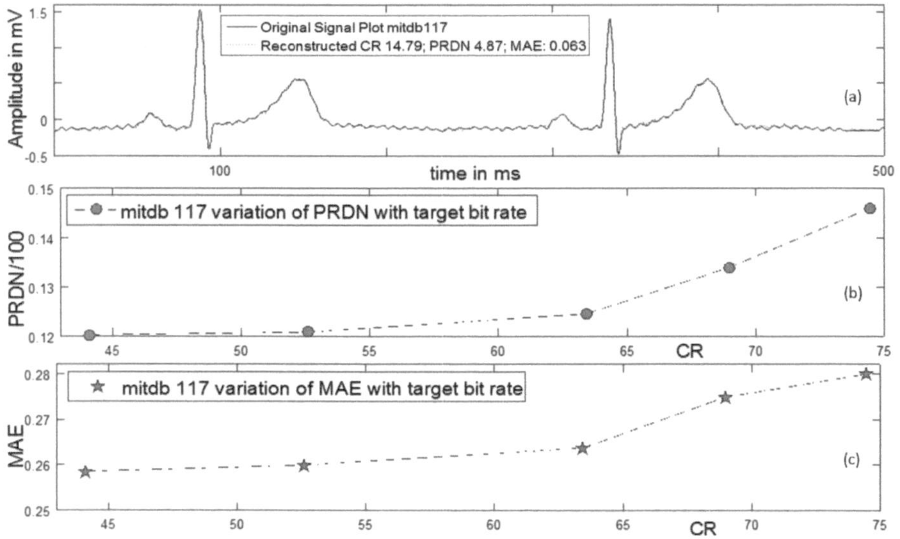

FIGURE 3.16
Compression performance with MIT-BIH arrhythmia data record 117: (a) using reconstruction control, (b) PRDN variation with CR, and (c) MAE variation with CR.

over which the $(b_{pc})_k$ and $(b_{eig})_k$ can traverse. The velocity and position of the participles are updated as

$$v_i(t+1) = wv_i(t) + c_1\varphi_1\left(p_{\text{best}} - x_i\right) + c_2\varphi_2\left(g_{\text{best}} - x_i\right),$$
$$x_i(t+1) = x_i(t) + v_i(t+1), \tag{3.29}$$

where c_1 and c_2 are positive constants (cognitive variables), φ_1 and φ_2 are two random variables (social variables) with uniform distribution between 0 and 1, w is the inertia weight which shows the effect of previous velocity vector $[v_i(t)]$ on the new vector $[v_i(t+1)]$. $x_i(t+1)$ is the updated position vector of $x_i(t)$. Best visited position for the particular particle is p_{best} and best position explored for all particles is g_{best}.

The results with optimal bit allocation are shown in Figure 3.17, with a limiting PRDN of 5%, using MIT-BIH arrhythmia data 117. Panel (a) shows the bit allocation for different PC scores and eigenvectors under different PRD constraints. An interesting point to note is that the bits are not always allocated as per the general notion, i.e., allocating a higher number of bits to PCs contributing more energy, e.g., for 2% PRDN limit, p_3 is allocated with 8 bits against p_2 which is allocated with 7 bits. Panel (b) shows the deviation in few clinical signatures: QRS width, QRS amplitude, *T*-height, *P*-amplitude, etc. against fixed bit and variable (optimal) quantization levels. It is clearly observed that except the *P*-amplitude, the optimal bit allocation provides least distortion in clinical signatures. The panel (c) shows the original, reconstructed plot along with sample to sample error. Compared to fixed quantization (shown in Figure 3.15), the optimal quantization technique provides better results in terms of PRDN, at the cost of lower CR.

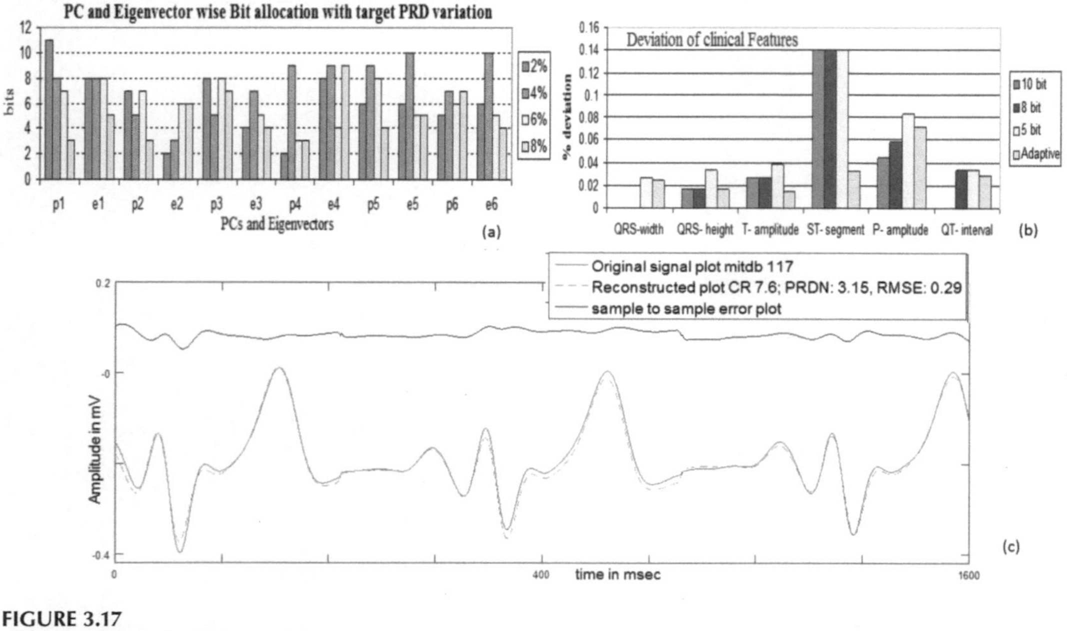

FIGURE 3.17
Performance analysis of optimal bit allocation for eigenvectors and PCs using MIT-BIH arrhythmia data record 117: (a) the bit allocation, (b) distortion measures, (c) original and reconstructed ECG data with error plot (target PRDN 5%). For ease of interpretation, the error signal is plotted with a bias.

3.4.3 Compression of PPG

Compression of PPG has received attention from the signal processing research community in the last decade only, due to gradual expansion in the use of pulse oximetry in clinical heath monitoring. Compared to ECG, a PPG signal has less complicated morphology, containing an anacrotic phase (ventricular systole) and a catacrotic phase (ventricular diastole). Till date, only limited published literature is available on PPG compression and can be divided into two broad categories, viz., time-domain methods [50–54] and transform-domain methods [55,56]. The time-domain methods majorly use delta-type encoder which are lossless or lossy.

Application of double derivative and Huffman coding for lossless PPG compression is illustrated in [51]. The signal processing flow diagram is shown in Figure 3.18. At first, the raw PPG samples are scaled in the range of 10–1,000. Then two successive first order derivatives (d_1 and d_2) are applied; each time a biasing was applied to make each data element positive. Huffman encoding is applied on second-derivative array. Its justification lies in the fact that the PPG morphology has slow slope changes in the anacrotic and catacrotic phases, as compared to ECG, where the slope changes in QRS complex are sharp. Thus, the number of symbols with larger occurrence will be higher, compared to those with smaller number of occurrence. Although the coding is simple, based on probability distribution of the unique symbols, the header bytes require a lot of information to be included for proper decoding. The work [51] shows that using 10 bit resolution data sampled at 128 Hz, major clinical features like systolic upstroke time and systolic amplitude in the reconstructed signal are within 1% from their respective original values. A salient feature of the work is that it can be implemented by a low-end microcontroller (like 8051) in real-time monitoring. The work reported in [50] shows another simple technique which is useful for real-time PPG compression. Here the morphological complexity of the current window is determined by SD of the first derivative and tagged as either 'simple' or 'complex'. Based on zonal complexity, RLE is implemented on 'simple' zones with a thresholding rule, followed by a selective biasing or nibble combination. The work achieves a CR of 3.84 with a PRDN of 7.57. The signal processing flow diagram is shown in Figure 3.19.

Till date, the work reported in [53] has achieved highest CR of 122 and low distortion (less than 1%) with PPG data sampled at 500 Hz and 24 bit resolution. The processing includes computing the second derivative from the raw samples, their scaling, and five types of grouping scheme to reduce the intra-zone redundancy. Additionally, the technique applied an encryption technique based on choosing an encryption key and XOR

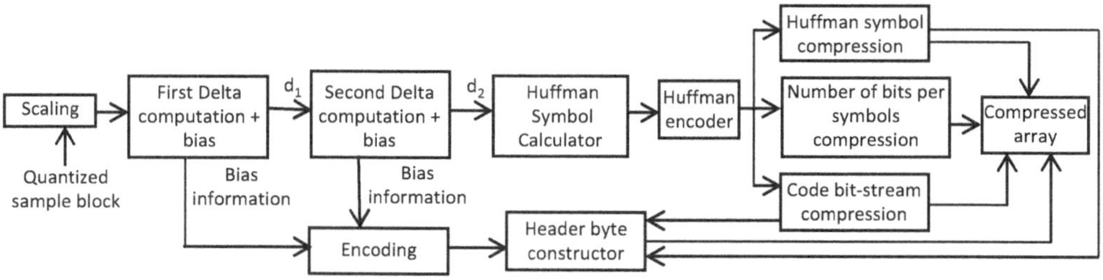

FIGURE 3.18
Compression flow for lossless PPG compression using Huffman coder.

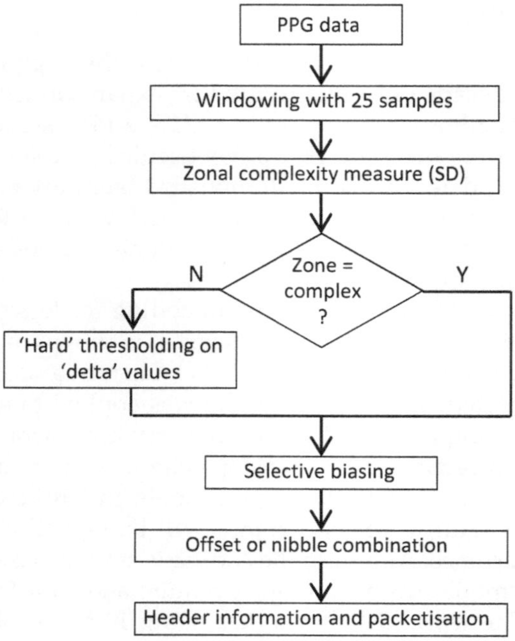

FIGURE 3.19
Compression flow of zonal-complexity–based PPG encoder [50].

operation. A real-time hardware implementation of PPG compression using adaptive delta modulation at 24 bit resolution data and 8 bit microcontroller (18LF4520) is reported in [54]. The raw signal, x, is acquired at 24 bit resolution and a 2 bit delta encoder was implemented to encode the error, e, represented as

$$e(n) = x(k+1) - x(k). \tag{3.30}$$

To mitigate the slope overload effect, the step size (resolution, or Δ_1) of the analog to digital converter (ADC) was adapted using adjustment of the reference voltage (V_r), represented as

$$\Delta_1 = \frac{V_r}{2^n}, \tag{3.31}$$

where n (the number of bits) was reduced from 16 to 10. The compressed data was acquired in a PC using Bluetooth communication, achieving a CR of 41 with a low PRD value.

Among the transform-domain methods, the first work described the use of Fourier series analysis for PPG compression [55]. At first, the cardiac cycles are extracted from single-channel PPG record, and then Fourier series expansion is applied as follows:

$$x(t) = a_0 + \sum_{k=1}^{\infty} \left[a_k \cos(k\omega t) + b_k \sin(k\omega t) \right]. \tag{3.32}$$

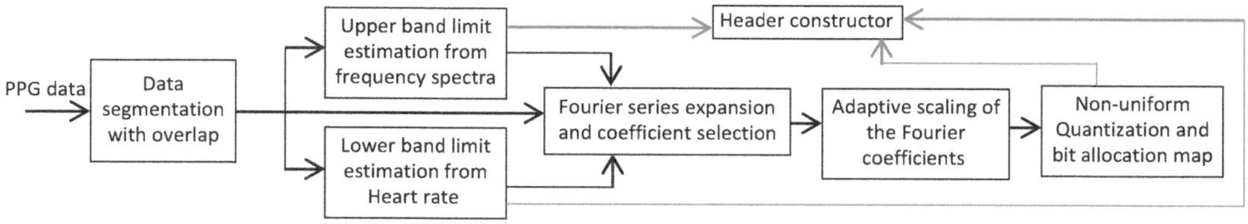

FIGURE 3.20
Compression flow diagram for band-limited estimation-based PPG compression.

The signal is reconstructed with reduced number of coefficients (nominally 50% of total number of coefficients), allowing some reconstruction error. This method reports a data compression factor on 12 and also simultaneously reports a motion artifact reduction by a factor of 35 dB.

An improved technique employing PPG compression based on Fourier series is reported recently [57]. The signal flow diagram is shown in Figure 3.20. The technique is based on the key concept that the PPG signal frequency content is highly person specific and computing the upper and lower limits of frequency band can significantly reduce the redundancy as well as the computation time for Fourier series expansion. It adaptively band-limits the signal components based on frequency content of contiguous overlapping PPG segments to determine the upper and lower cutoff frequencies. The systolic peak indices provide an estimation of average heart rate, which determines the lower cutoff. The upper cutoff is decided on choosing the 3rd to 4th harmonics of the fundamental. Each block is then decomposed to obtain the expansion coefficients, which are selected based on band limit estimation. The coefficients are scaled based on energy retention criteria, and bits are assigned to quantize them to form a binary significance map. This work reports a CR of 36 with a PRD below 4% with PPG data collected at 16 bit resolution.

Quality-controlled PPG compression is reported in [56]. The technique adopts cycleby-cycle compression by segmenting the PPG data array into the constituent beats and forms a 2-D matrix, similar to [48]. PCA decomposition of the covariance matrix of beat-segmented data is done to get PC scores and eigenvectors. The error control is performed by recursive selection of PCs and their quantization level to limit the PRD and MAX at certain predetermined values while reconstructing the data. The compression of PC scores and eigenvectors were performed using lossless delta encoder. The work also reports effect of HR variability as well as noise due to motion artifacts (MAs) on the compression parameters.

3.4.4 Simultaneous Compression of Multiple Biomedical Signals

Recent advancements in electronics and miniaturized device fabrication technologies have enabled simultaneous acquisition as well as processing of multiple biosignals (MBioSigs) [58]. Presence of cardiac disorder is normally reflected in the shape of ECG waveform, and cardiologists often prefer to examine multilead ECG (MECG) to infer the anomaly. In recent years, the PPG signal has become a very promising tool for non-invasive assessment of blood pressure, blood oxygen saturation (SpO_2), blood glucose, sleep disorder, etc. [59,60]. To improve and accelerate the quality of disease diagnosis, a few attempts have also been made to acquire, transmit, monitor, and process MBioSigs such as ECG, PPG, ballistocardiogram (BCG), electromyogram (EMG) signals, respiration, and body temperature [61]. Approximately 1.45 MB of computer memory is required to store only 12-lead ECG data of 1 min duration sampled at 1 kHz rate and with a 16 bit resolution, which is large enough, and the size keeps increasing with time and with addition of other biosignals. However, the area of compression of MBioSigs still remains underexplored [62]. In this section, an MBioSigs compression algorithm along with its MATLAB code is dicussed. ECG is considered as the most important medical signal, and PPG is considered as the second most important one. Hence, only MECG and PPG signals are considered here. Four standard ECG leads (lead III and the augmented limb leads aVR, aVL, aVF) out of 12 are linearly related to lead I and II, and hence these four leads are excluded from processing. Henceforth, MECG refers to eight ECG leads (I, II, V1–V6) in this section.

First, both the MECG and PPG signals are denoized. The clinical bandwidths of ECG and PPG signals are considered to be within 0.5–100 Hz and 0.5–15 Hz, respectively [62,63]. Hence, a bidirectional Butterworth band-pass filters are used for denoizing these signals. A notch filter is also used to remove the 50/60 Hz power-line noise from MECG.

Next, each lead of MECG and PPG is placed alongside to form a 2-D array or matrix, with the number of samples in each ECG lead and PPG are equal. The structure of the matrix is shown below.

A				**ECG Leads**					**PPG Signal**
Number of samples	I_1	II_1	$V1_1$	$V2_1$	$V3_1$	$V4_1$	$V5_1$	$V6_1$	PPG_1
	I_2	II_2	$V1_2$	$V2_2$	$V3_2$	$V4_2$	$V5_2$	$V6_2$	PPG_2
	I_3	II_3	$V1_3$	$V2_3$	$V3_3$	$V4_3$	$V5_3$	$V6_3$	PPG_3
	…	…	…	…	…	…	…	…	…
	I_r	II_r	$V1_r$	$V2_r$	$V3_r$	$V4_r$	$V5_r$	$V6_r$	PPG_r

Now, the matrix is decomposed using singular value decomposition (SVD) technique. SVD is a widely used technique of matrix factorization [64]. It factorizes a matrix into the product of another three matrices: a diagonal matrix (*S*), an orthogonal matrix (*U*), and the transpose of an orthogonal matrix (*V*). Therefore, we can write

$$A_{r \times c} = U_{r \times r}\, S_{r \times c}\, V_{c \times c}^{\mathrm{T}}, \tag{3.33}$$

where A is the matrix containing the MBioSigs, $U^{\mathrm{T}}U = I$, $V^{\mathrm{T}}V = I$, and the columns of $U_{r\times r} = [u_1, u_2, \ldots u_r]$ and $V_{c\times c} = [v_1, v_2, \ldots v_r]$ matrices are the left and right singular vectors, respectively. The matrix S is diagonal with real positive entries, which are called the singular values (σ). Singular values are the square roots of eigenvalues from either U or V matrix. The total information-energy of the A matrix can be mathematically expressed as

$$E_{\mathrm{T}} = \sigma_1^2 + \sigma_2^2 + \sigma_3^2 + \cdots + \sigma_c^2, \tag{3.34}$$

where $\sigma_1^2 \geq \sigma_2^2 \geq \sigma_3^2 \geq \cdots \geq \sigma_c^2 \geq 0$.

A very striking feature of SVD is that it is able to extract the fundamental structural modes of a composite signal, which can be utilized to analyze a signal exhibiting quasi-periodic nature such as ECG. If the correlation of the data present in a matrix is high, then most of the energy of the data is expected to be concentrated over the first few singular values. Hence, the original data matrix can be approximated with a low and acceptable level of distortion by discarding the rest of the singular values. From Equation (3.34), it can be noted that the singular values (σ) are arranged in the decreasing order of magnitude. Therefore, truncating the singular values to a lower number helps in achieving data compression. The compression performance is inversely proportional to the singular value truncation factor (β). If we consider only eight ECG leads and the PPG signal, then β could be varied from 1 to 9. Compression performance is high at a low value of β and vice-versa. Also, the error between the original and reconstructed signals is high at a low value of β and vice-versa. The truncation factor of the singular values solely depends on the tolerable level of data loss upon reconstruction. After truncation, the truncated *U*, *S*, and *V* matrices are to be compressed using either lossless or near-lossless compression techniques. The technique described in Section 3.4.1 could also be used.

During reconstruction, first, *U*, *S*, and *V* matrices are to be decompressed using an approach, which is the reverse of the compression technique, and finally, the decompressed *U*, *S*, and *V* matrices are multiplied using Equation (3.14) to reform the MBioSigs matrix. It is expected that the module would perform better when MBioSigs are collected simultaneously from the same subject and at the same sampling rate.

```
MATLAB® code

[U, S, V] = svd(A); % SVD computation of 'A' matrix. 'A' matrix contains
the %denoised MBioSigs
B=5; % Singular value truncation factor, which varies from 1 to 9
Ssmaller = zeros(B,B); % Truncated S matrix
for i=1:B
  Ssmaller(i,i)=S(i,i);
end
Usmaller = U(:,1:B); % Truncated U matrix
Vsmaller = V(:,1:B); % Truncated V matrix
% Now, compressed the Ssmaller, Usmaller and Vsmaller matrices'
coefficients
% in lossless or near-lossless fashion

% During reconstruction, first, decompress the Ssmaller, Usmaller and
% Vsmaller matrices' coefficients, and then..
A_rec = Usmaller*Ssmaller*Vsmaller'; % Signal reconstruction


% Plot the columns of A and A_rec matrices to verify the original and
% reconstructed signals.
```

If the coefficients of the 'Ssmaller', 'Usmaller', and 'Vsmaller' matrices of the above-written MATLAB code could be compressed and reconstructed in a lossless or near-lossless manner, then the reconstructed MBioSigs would exactly look like as shown in Figures 3.21 and 3.22 at $\beta = 4$ and 9, respectively.

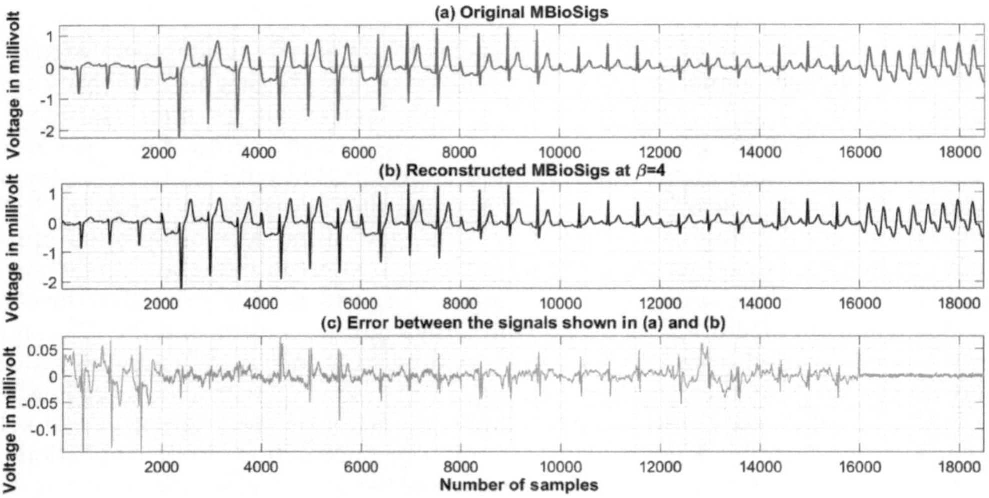

FIGURE 3.21
(a) Original MBioSigs, (b) reconstructed MBioSigs at $\beta = 4$, and (c) error between the signals shown in (a) and (b).

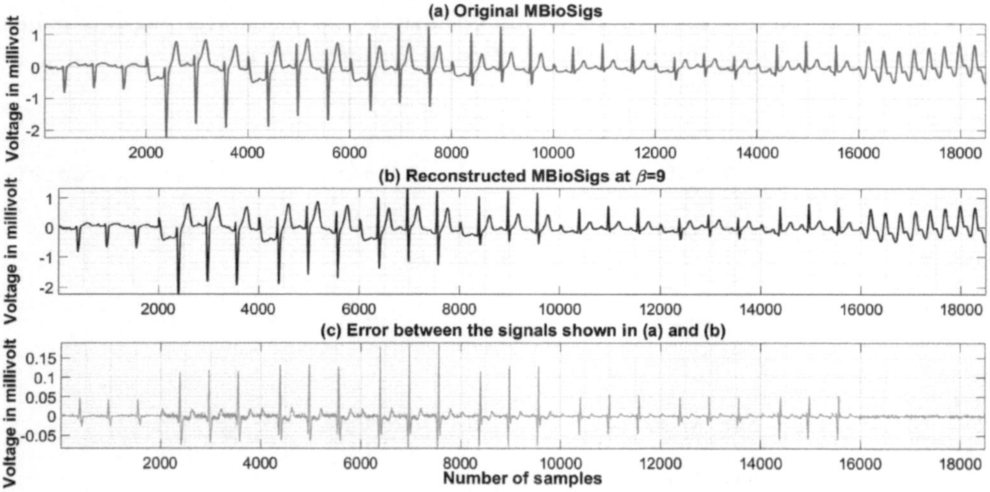

FIGURE 3.22
(a) Original MBioSigs, (b) reconstructed MBioSigs at $\beta = 9$, and (c) error between the signals shown in (a) and (b).

3.5 Guaranteeing the Reconstruction Quality in Compression

Maintaining the quality of the reconstructed signal at a predetermined level is a very important criterion of any biosignal compression algorithm. A strict lossless compression algorithm does not lose any clinical information. Hence, there is no point in worrying about the quality [10]. On the contrary, the theme of any transformation-based biosignal compression algorithm is to convert the original biosignal into some other domain, then discard the comparatively less significant coefficients using a threshold-based technique, and finally, encode the rest of the coefficients in a lossless or near-lossless fashion such as Huffman coding, RLE. Therefore, the distortion or loss, which enters into the signal upon reconstruction, is mostly due to the transformation and thresholding operations. In the cases where the remaining coefficients after thresholding out the less significant ones are compressed using lossless compression methods, it is possible to estimate the amount of distortion which would be present in the reconstructed signal accurately, and near-perfectly if the coefficients are compressed using near-lossless techniques. PRD, PRDN, or any other metrics, which are discussed in Section 3.3.1, could be used as the measures of distortion. WEDD could also be used but only for ECG.

MATLAB code of such a quality-guaranteed ECG compression algorithm using DCT and taking PRD as the quality controlling metric is given below, and the flow chart of the algorithm is shown in Figure 3.23. Please note that the signal MUST be filtered for the elimination of high- and low-frequency noises prior to compression. Here, in the MATLAB code, it has been assumed that the ECG signal has been refined using a fourth-order bidirectional Butterworth band-pass filter having lower and upper cutoff frequencies 0.5 Hz and 100 Hz, respectively. The frequency response of such a band-pass filter is shown in Figure 3.24.

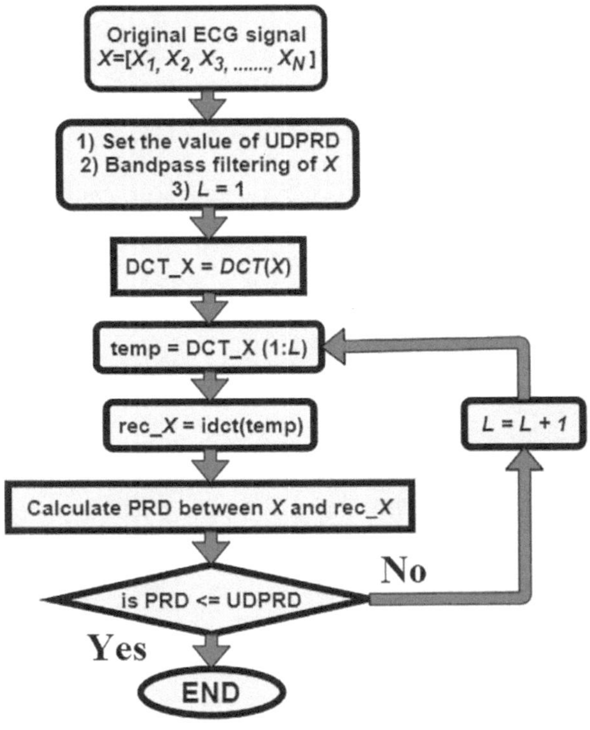

FIGURE 3.23
Flow chart of the DCT-based quality-guaranteed ECG compression algorithm.

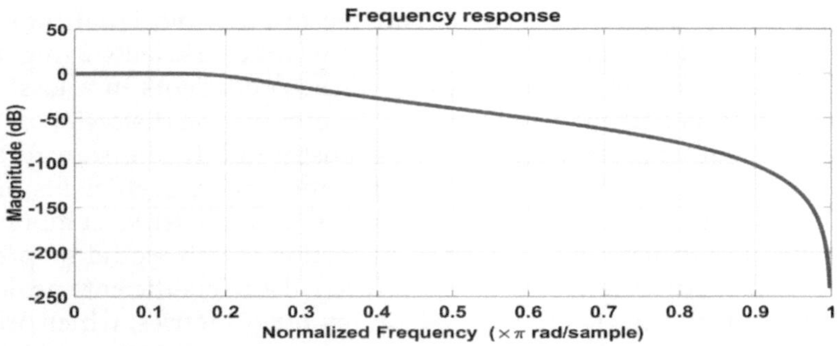

FIGURE 3.24
Frequency response of a 4th-order Butterworth band-pass filter (lower and upper cutoff frequencies 0.5 and 100 Hz, respectively).

MATLAB® code

```
X=original biosignal; % X array contains the original signal
UDPRD=input('\nEnter the value of maximum tolerable PRD = ');
samf=input('Enter the sampling frequency of the signal in Hz = ');
[b,a]=butter(4,[0.5*2/samf 100*2/samf],'bandpass'); % Change these
    filtering parameters as per the characteristics of the chosen biosignal
X=filtfilt(b,a,X);
dct_X=dct(X); % Discrete cosine transform of the original signal
temp(1:length(dct_X))=0;
L=1; %Minimum number of DCT coefficients, which needed to maintain the
given PRD value, initialized with 1
while (1)
  temp(1:L)=dct_X(1:L);
  rec_X=idct(temp);
  % PRD calculation between the original and the reconstructed signals
  n=sum((X-rec_X).*(X-rec_X));
  d=sum(X.*X);
  PRD=sqrt(n/d)*100;
  if PRD<=UDPRD
    break;
  else
    L=L+1;
  end
  if L>length(dct_X)
    fprintf('\nWARNING: It is not possible to maintain the given PRD
value, try with other PRD values');
    break;
  end
end
fprintf('\nExpected PRD = %0.2f',PRD);
fprintf('\nTotal number of DCT coefficients = %d',length(dct_X));
fprintf('\nNumber of coefficients required = %d',L);
% Now compress the coefficients of the 'temp' array in a lossless fashion.
% During signal reconstruction, decompress the 'temp' array in a lossless
  fashion and then apply inverse DCT on %'temp'
plot(X);hold on;plot(rec_X);hold on;
```

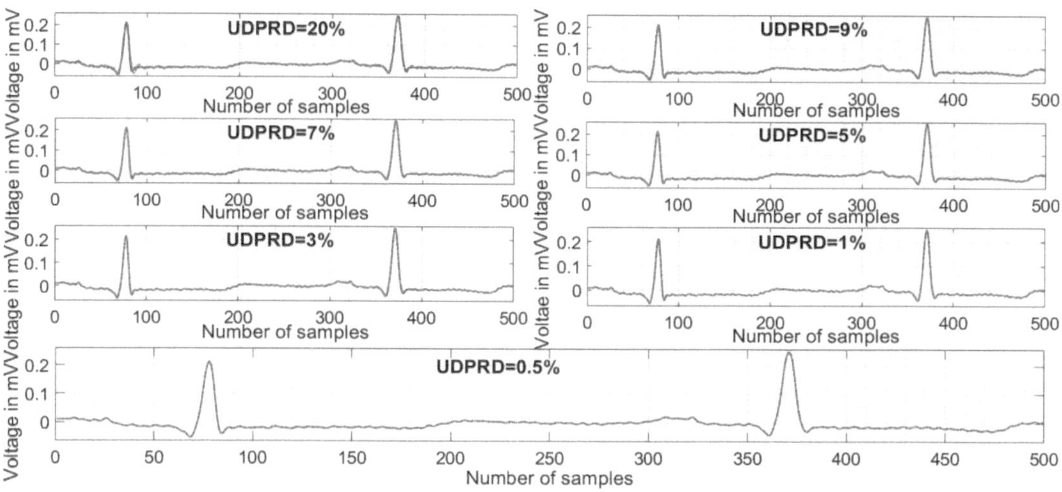

FIGURE 3.25
Original (MIT-BIH arrhythmia database, record no. 100) and reconstructed ECGs plotted on top of each other for different values of UDPRD.

If the coefficients of the 'temp' array of the above written MATLAB code could be compressed and reconstructed in a lossless manner, then the original and reconstructed ECGs would exactly look like as shown in Figure 3.25 for different values of maximum-tolerable-PRD or user-defined-PRD (UDPRD).

The time complexity of the algorithm depends on the number of samples present in the signal as well as the chosen value of PRD, and in the MATLAB code, the number of DCT coefficients are reduced based on their length; i.e., initially only one DCT coefficient is taken, and then the number is increased one by one until the PRD value comes down below the predefined measure; other thresholding methods could also be used.

3.6 Open Access ECG Database for Compression Algorithm Testing

Currently, there are few open access databases freely available which have been utilized in various medical signal processing research studies. The most popular among them is Physionet. Some of the databases under this mother repository have become established benchmarks for comparison of ECG compression algorithms. First, the most widely used single-lead ECG data for compression performance testing is MIT-BIH arrhythmia data, popularly abbreviated as mitdb or sometimes mita. This database contains annotated 48 records, each of nearly 30 min duration from two-channel (modified lead II and one chest lead) ambulatory Holter ECG collected from arrhythmia patients at the Beth Israel Hospital Arrhythmia Laboratory. The data is collected at 360 Hz at 11 bit resolution. This database is also considered as benchmark for ECG rhythm and research related to abnormality detection.

Another database, MIT BIH Compression Test, popularly known as cdb, was specifically developed for ECG compressors. This data contains unannotated 168 short segments of

ECG records, collected at 250 Hz sampling. However, till date, the use of this database is comparatively less, probably due to the unviability of annotations.

For multilead ECG compression performance, the researchers have used PTB Diagnostic ECG database, popularly known as ptddb, which contains 15 channel (12 standard leads and 3 Frank vectorcardiographic leads) ECG records containing various cardiac abnormalities.

For research on PPG compression, until now there is no established benchmark database available. Some researchers have used Multiparameter Intelligent Monitoring in Intensive Care (MIMIC) database which is a collection of 5–6 critical parameters like ECG, respiration, PPG, etc.

3.7 Conclusion

In this chapter, some of the major approaches used in ECG and PPG compression in the last decade are described. Biomedical signal compression has been an active area of research for the last 40 years. Until now, numerous methods have been successfully implemented for ECG data compression, while PPG compression is a relatively new area of research, mainly due to its wider application in cardiovascular health monitoring being exercised in last two decades. Initial application of ECG compression was mainly confined to continuous data recorders. However, with expansion of ambulatory monitoring and wireless sensor networks in healthcare, the need for biosignal compression has gradually been expanded in continuous health monitoring using wearable biomedical systems. Advent of low-power microcontrollers with limited on-chip buffer has made it possible to implement lightweight compression algorithms for real-time systems [65]. Compressed sensing is one such technique that has been utilized [66,67] for energy efficient wearable biomedical systems.

There is a trend of imaging-based diagnosis (e.g., echocardiography) replacing the conventional paper-based records for more in-depth information. However, ECG and PPG still continue to be the low-cost and cost-effective diagnostic tools in the coming years. Thus, biomedical signal compression will continue as one of the potential of areas of research.

Acknowledgments

Rajarshi Gupta acknowledges Dipankar Das and Subhajit Das, M. Tech students of Instrumentation & Control Engineering, 2016, from University of Calcutta for the test results of Figure 3.17 of this chapter.

References

1. Y. Zigel, A. Cohen, and A. Katz, The weighted diagnostic distortion (WDD) measure for ECG signal compression, *IEEE Trans. Biomed. Eng.*, vol. 47, no. 11, 1422–1430, 2000.
2. M. S. Manikandan and S. Dandapat, Wavelet energy based diagnostic distortion measure for ECG, *Biomed. Signal Process. Control*, vol. 2, no. 2, 80–96, 2007.

3. S. Padhy, L. N. Sharma, and S. Dandapat, Multilead ECG data compression using SVD in multiresolution domain, *Biomed. Signal Process. Control*, vol. 23, 10–18, 2016.
4. J. P. Abenstein and W. J. Tompkins, A new data-reduction algorithm for real-time ECG analysis, *IEEE Trans. Biomed. Eng.*, vol. BME-29, no. 1, 43–48, 1982.
5. S. Banerjee, R. Gupta, and M. Mitra, Delineation of ECG characteristic features using multiresolution wavelet analysis method, *Measurement*, vol. 45, no. 3, 474–487, 2012.
6. L. J. Hadjileontiadis, Biosignals and compression standards, in *M-Health Emerging Mobile Health Systems*, R.S.H. Istepanian, S. Laxminarayan, and C.S. Pattichis (Eds.), Springer, New York, 2006, pp. 277–292.
7. J. R. Cox, F. M. Nolle, H. A. Fozzard, and G. C. Oliver, AZTEC, a preprocessing program for real time ECG rhythm analysis, *IEEE Trans. Biomed. Eng.*, vol. BME-15, no. 2, 128–129, 1968.
8. W. C. Mueller, Arrhythmia detection software for an ambulatory ECG monitor, *Biomed. Sci. Instrum.*, vol. 14, 81–85, 1978.
9. U. E. Ruttimann and H. V. Pipberger, Compression of the ECG by prediction or interpolation and entropy coding, *IEEE Trans. Biomed. Eng.*, vol. 26, no. 11, 613–623, 1979.
10. S. K. Mukhopadhyay, S. Mitra, and M. Mitra, A lossless ECG data compression technique using ASCII character encoding, *Comput. Electr. Eng.*, vol. 37, no. 4, 486–497, 2011.
11. S. K. Mukhopadhyay, M. Mitra, and S. Mitra, ECG signal processing: Lossless compression, transmission via GSM network and feature extraction using Hilbert transform, in *IEEE EMBS Special Topic Conference on Point-of-Care (POC) Healthcare Technologies: Synergy Towards Better Global Healthcare, PHT 2013*, India, pp. 85–88.
12. S. K. Mukhopadhyay, S. Mitra, and M. Mitra, An ECG signal compression technique using ASCII character encoding, *Measurement*, vol. 45, no. 6, 1651–1660, 2012.
13. S. K. Mukhopadhyay, S. Mitra, and M. Mitra, ECG signal compression using ASCII character encoding and transmission via SMS, *Biomed. Signal Process. Control*, vol. 8, no. 4, 354–363, 2013.
14. S. K. Mukhopadhyay, M. Mitra, and S. Mitra, An ECG data compression method via R-Peak detection and ASCII character encoding, in *International Conference on Computer, Communication and Electrical Technology, ICCCET 2011*, India, pp. 136–141.
15. S. K. Mukhopadhyay, S. Mitra, and M. Mitra, A combined application of lossless and lossy compression in ECG processing and transmission via GSM-based SMS, *J. Med. Eng. Technol.*, vol. 39, no. 2, 105–122, 2015.
16. S. K. Mukhopadhyay, M. Mitra, and S. Mitra, An ECG data compression method via standard deviation and ASCII character encoding, in *Proceedings - 2nd International Conference on Emerging Applications of Information Technology, EAIT 2011*, India, pp. 67–70.
17. S. K. Mukhopadhyay, M. O. Ahmad, and M. N. S. Swamy, An ECG compression algorithm with guaranteed reconstruction quality based on optimum truncation of singular values and ASCII character encoding, *Biomed. Signal Process. Control*, vol. 44, 288–306, 2018.
18. R. Gupta and M. Mitra, An ECG compression technique for telecardiology application, in *Proceedings - 2011 Annual IEEE India Conference: Engineering Sustainable Solutions, INDICON-2011*, India.
19. P. Bera and R. Gupta, Hybrid encoding algorithm for real time compressed electrocardiogram acquisition, *Measurement*, vol. 91, 651–660, 2016.
20. S. Roy and R. Gupta, Short range centralized cardiac health monitoring system based on ZigBee communication, in *IEEE Global Humanitarian Technology Conference - South Asia Satellite, GHTC-SAS 2014*, India, pp. 177–182.
21. R. Gupta and M. Mitra, Wireless electrocardiogram transmission in ISM band: An approach towards telecardiology, *J. Med. Syst.*, vol. 38, no. 10, 2014.
22. P. Bera and R. Gupta, Real-time compression of electrocardiogram using dynamic bit allocation strategy, in *IEEE 1st International Conference on Control, Measurement and Instrumentation, CMI 2016*, India, pp. 21–25.
23. S. M. Jalaleddine, C. G. Hutchens, R. D. Strattan, and W. A. Coberly, ECG data compression techniques—A unified approach, *IEEE Trans. Biomed. Eng.*, vol. 37, no. 4, 329–342, 1990.

24. H. Lee and K. M. Buckley, ECG data compression using cut and align beats approach and 2-D transforms, *IEEE Trans. Biomed. Eng.*, vol. 46, no. 5, 556–564, 1999.
25. J. Pan and W. J. Tompkins, A real-time QRS detection algorithm, *IEEE Trans. Biomed. Eng.*, vol. BME-32, no. 3, 230–236, 1985.
26. B. U. Kohler, C. Hennig, and R. Orglmeister, The principles of software QRS detection, *IEEE Eng. Med. Biol.*, vol. 6, no. 1, 42–57, 2002.
27. M. S. Manikandan and K. P. Soman, A novel method for detecting R-peaks in electrocardiogram (ECG) signal, *Biomed. Signal Process. Control*, vol. 7, no. 2, 118–128, 2012.
28. E. B. Mazomenos, et al., A low-complexity ECG feature extraction algorithm for mobile healthcare applications, *IEEE J. Biomed. Health Inform.*, vol. 17, no. 2, 459–469, 2013.
29. S. Mallat, *A Wavelet Tour of Signal Processing*. Amsterdam: Elsevier Inc., 2009.
30. P. S. Addison, Wavelet transforms and the ECG: A review, *Physiol. Meas.*, vol. 26, no. 5, 2005.
31. M. L. Hilton, Wavelet and wavelet packet compression of electrocardiograms, *IEEE Trans. Biomed. Eng.*, vol. 44, no. 5, 394–402, 1997.
32. Z. Lu, D. Y. Kim, and W. A. Pearlman, Wavelet compression of ECG signals by the set partitioning in hierarchical trees algorithm, *IEEE Trans. Biomed. Eng.*, vol. 47, no. 7, 849–856, 2000.
33. B. A. Rajoub, A. Alshamali, and A. S. Al-Fahoum, An efficient coding algorithm for the compression of ECG signals using the wavelet transform, *IEEE Trans. Biomed. Eng.*, vol. 49, no. 4, 355–362, 2002.
34. D. Biswas, E. B. Mazomenos, and K. Maharatna, ECG compression for remote healthcare systems using selective thresholding based on energy compaction, in *International Symposium on Signals, Systems, and Electronics (ISSSE)*, Germany, 2012.
35. R. Benzid, F. Marir, A. Boussaad, M. Benyoucef, and D. Arar, Fixed percentage of wavelet coefficients to be zeroed for ECG compression, *Electron. Lett.*, vol. 39, no. 11, 830, 2003.
36. J. Chen and S. Itoh, A wavelet transform-based ECG compression method guaranteeing desired signal quality, *IEEE Trans. Biomed. Eng.*, vol. 45, no. 12, 1414–1419, 1998.
37. Á. Alesanco and J. García, Automatic real-time ECG coding methodology guaranteeing signal interpretation quality, *IEEE Trans. Biomed. Eng.*, vol. 55, no. 11, 2519–2527, 2008.
38. M. S. Manikandan and S. Dandapat, Wavelet threshold based ECG compression using USZZQ and Huffman coding of DSM, *Biomed. Signal Process. Control*, vol. 1, no. 4, 261–270, 2006.
39. C.-T. Ku, K.-C. Hung, T.-C. Wu, and H.-S. Wang, Wavelet-based ECG data compression system with linear quality control scheme, *IEEE Trans. Biomed. Eng.*, vol. 57, no. 6, 1399–1409, 2010.
40. B. S. Kim, S. K. Yoo, and M. H. Lee, Wavelet-based low-delay ECG compression algorithm for continuous ECG transmission, *IEEE Trans. Inf. Technol. Biomed.*, vol. 10, no. 1, 77–83, 2006.
41. H. L. Chan, Y. C. Siao, S. W. Chen, and S. F. Yu, Wavelet-based ECG compression by bit-field preserving and running length encoding, *Comput. Methods Programs Biomed.*, vol. 90, no. 1, 1–8, 2008.
42. H. Zou and A. H. Tewfik, Parametrization of compactly supported orthonormal wavelets, *IEEE Trans. Signal Process.*, vol. 41, no. 3, 1428–1431, 1993.
43. M. Blanco-Velasco, F. Cruz-Roldán, J. I. Godino-Llorente, and K. E. Barner, Wavelet packets feasibility study for the design of an ECG compressor, *IEEE Trans. Biomed. Eng.*, vol. 54, no. 4, 766–769, 2007.
44. S. C. Tai, C. C. Sun, and W. C. Yan, A 2-D ECG compression method based on wavelet transform and modified SPIHT, *IEEE Trans. Biomed. Eng.*, vol. 52, no. 6, 999–1008, 2005.
45. X. Wang and J. Meng, A 2-D ECG compression algorithm based on wavelet transform and vector quantization, *Digital Signal Process.*, vol. 18, no. 2, 179–188, 2008.
46. R. Das, P. Bera, and R. Gupta, Electrocardiogram compression technique using DWT-based residue encoder with desired reconstruction quality, in *Fifth International Conference on Emerging Applications of Information Technology (EAIT)*, India, 2018, pp. 1–5.
47. F. Castells, P. Laguna, L. Sornmo, A. Bollmann, and M. R. Jos, Principal component analysis in ECG signal processing, *EURASIP J. Adv. Signal Process.*, 074580, 2007.
48. R. Gupta, Quality aware compression of electrocardiogram using principal component analysis, *J. Med. Syst.*, vol. 40, no. 5, 1–11, 2016.

49. Physionet.org. [Online]. Available: www.physionet.org.
50. S. Alam and R. Gupta, Zonal complexity based measure for lossy compression of photoplethysmogram using delta encoding, in *IEEE International Conference on Signal Processing, Informatics, Communication and Energy Systems, SPICES 2015*, India, pp. 1–5.
51. R. Gupta, Lossless compression technique for real-time photoplethysmographic measurements, *IEEE Trans. Instrum. Meas.*, vol. 64, no. 4, 975–983, 2015.
52. S. K. Mukhopadhyay, M. O. Ahmad, and M. N. S. Swamy, ASCII-character-encoding based PPG compression for tele-monitoring system, *Biomed. Signal Process. Control*, vol. 31, 470–482, 2017.
53. S. Dhar, S. K. Mukhopadhyay, S. Pal, and M. Mitra, An efficient data compression and encryption technique for PPG signal, *Measuremet*, vol. 116, 533–542, 2018.
54. K. S. Chong, K. B. Gan, and E. Zahedi, Development of a real-time adaptive delta compression algorithm for photoplethysmography system, *IEEJ Trans. Electr. Electron. Eng.*, vol. 13, no. 10, 1454–1460, 2018.
55. K. A. Reddy, B. George, and V. J. Kumar, Use of Fourier series analysis for motion artifact reduction and data compression of photoplethysmographic signals, *IEEE Trans. Instrum. Meas.*, vol. 58, no. 5, 1706–1711, 2009.
56. S. Alam, R. Gupta, and J. Bera, Quality controlled compression technique for Photoplethysmogram monitoring applications, *Meas. J. Int. Meas. Confed.*, vol. 130, 236–245, 2018.
57. D. Sadhukhan, S. Pal, and M. Mitra, Adaptive band limit estimation based PPG data compression for portable home monitors, *Meas. J. Int. Meas. Confed.*, vol. 134, 153–165, 2019.
58. L. Yan, et al., A 13 μa analog signal processing IC for accurate recognition of multiple intra-cardiac signals, *IEEE Trans. Biomed. Circuits Syst.*, vol. 7, no. 6, 785–795, 2013.
59. H. J. Baek, G. S. Chung, K. K. Kim, and K. S. Park, A smart health monitoring chair for non-intrusive measurement of biological signals, *IEEE Trans. Inf. Technol. Biomed.*, vol. 16, no. 1, 150–158, 2012.
60. P. Fonseca, et al., Validation of photoplethysmography-based sleep staging compared with polysomnography in healthy middle-aged adults, *Sleep*, vol. 40, no. 7, 2017. doi: 10.1093/sleep/zsx097.
61. A. M. R. Dixon, E. G. Allstot, D. Gangopadhyay, and D. J. Allstot, Compressed sensing system considerations for ECG and EMG wireless biosensors, *IEEE Trans. Biomed. Circuits Syst.*, vol. 6, no. 2, 156–166, 2012.
62. S. K. Mukhopadhyay, M. O. Ahmad, and M. N. S. Swamy, SVD and ASCII character encoding-based compression of multiple biosignals for remote healthcare systems, *IEEE Trans. Biomed. Circuits Syst.*, vol. 12, no. 1, 137–150, 2018.
63. W. H. Lin, H. Wang, O. W. Samuel, G. Liu, Z. Huang, and G. Li, New photoplethysmogram indicators for improving cuffless and continuous blood pressure estimation accuracy, *Physiol. Meas.*, vol. 39, no. 2, 1–13, 2018.
64. S. Banerjee and A. Roy, *Linear Algebra and Matrix Analysis for Statistics*. Boca Raton, FL: CRC Press, 2014.
65. C. J. Deepu and Y. Lian, A Joint QRS detection and data compression scheme for wearable sensors, *IEEE Trans. Biomed. Eng.*, vol. 62, no. 1, 165–175, 2015.
66. Y. Wang, S. Doleschel, R. Wunderlich, and S. Heinen, Evaluation of digital compressed sensing for real-time wireless ECG system with bluetooth low energy, *J. Med. Syst.*, vol. 40, no. 7, 2016. doi: 10.1007/s10916-016-0526-1.
67. B. Liu, Z. Zhang, G. Xu, H. Fan, and Q. Fu, Energy efficient telemonitoring of physiological signals via compressed sensing: A fast algorithm and power consumption evaluation, *Biomed. Signal Process. Control*, vol. 11, no. 1, 80–88, 2014.

4

Health Monitoring Based on Internet of Things (IoT)

Domenico Balsamo
Newcastle University

Saptarshi Das
University of Exeter

CONTENTS

4.1 Energy-Constrained IoT Systems

4.1.1 Introduction

The Internet of Things (IoT) is a new networking paradigm, enabling wide-ranging opportunities by allowing the internet to encompass a large number of smart devices through standard communication protocols, such as internet protocol (IP). This provides services to end users and helps to engineer new solutions to societal-scale problems [1].

Fundamental to realizing this are networked devices, collections of tens to thousands of nodes, each of which are organized into a part of a larger cooperative network. Such devices encompass ultralow-power embedded devices used as IoT sensing devices, through to the low-power mobile platforms which form IoT edge devices and data aggregators. Each IoT device is typically equipped with sensors to detect physical phenomena and gather information about their environment; actuators to take actions on the monitored environment; diverse computing units, such as microcontrollers (MCUs), central processing units (CPUs), graphics processing unit (GPU), digital signal processors (DSP) etc., to process data; memory for data storage; and finally, radio frequency (RF) transceivers for communication.

For example, smart healthcare is fuelling interest in IoT applications which require embedded sensors and actuators in patients to monitor physiological status. These are typically utilized to collect and process data and gather useful information to make suitable actions and decisions regarding the patients [2].

Technological advances and changes in the social perception of technology are also fuelling interest in flexible, wearable, and implantable devices. These target a range of application domains, including fitness, assisted living, and healthcare. For reasons of installation ease, location, or aesthetics, many of the networked sensing devices which underpin these applications do not have access to a wired electricity source and instead rely on batteries as their power source. Battery-powered devices inevitably experience market demands requiring long lifetime and small physical dimensions and weight.

While different types of wireless communication technologies and protocols have already been put in place to support IoT systems, providing reliable power supply for autonomous devices deployed in remote locations still remains a major challenge. Scaling CMOS device geometry has far outpaced the scaling of energy densities in batteries, meaning that power supply is often the largest and most expensive part of IoT devices. In fact, the high cost and disruption associated with replacing the batteries is limiting the large-scale deployment of IoT systems [3].

As a result, research effort has been initiated toward replenishing batteries with charges using the principle of energy harvesting (EH), defined as the process of harnessing electrical energy from alternative sources such as light, wind, heat vibration, and movement, and using it to power IoT sensing devices [4]. This chapter provides a comprehensive review on IoT systems with focus on energy-constrained IoT devices. In Section 4.1.1, EH principles are presented followed by a discussion on EH IoT system design (Section 4.1.3). This chapter is concluded with a discussion on standard IoT protocols and their properties in Section 4.1.4.

4.1.2 EH Principles

In this section, EH principles are discussed with emphasis on the technologies that are typically used in EH IoT system design.

EH simply refers to exploitation of any source to create electricity. It is highly beneficial for widely distributed and hard-to-reach networks, as ambient resources in a device's environment are utilized to extract energy. There are a variety of EH techniques currently in use, whereby an energy budget is autonomously (re)constituted to replenish the storage without requiring maintenance or human supervision. In general, EH mechanisms are commonly categorized into five groups, namely mechanical, thermal, fluid flow, radiant, and wireless EH [5].

Solar power, harnessed using the principle of the photovoltaic (PV) effect, is widely used in rural sites to generate electricity. For indoor applications, specialized PV cells, better suited to diffused lights, convert artificial propagation into usable electrical energy. However, this energy is restricted by size of the PV cell, which limits PV utilization, especially in indoors due to the small form factors of sensor nodes and the limited operation area. Similar to solar EH, wind and hydro-power, i.e., flow energy, offers an alternative to operate wireless IoT devices and larger systems, in a self-sufficient way. Any mechanical movement (e.g., vibration, pressure variation) can be converted into a reliable energy source, thanks to piezoelectric materials. Similarly, thermoelectric generators (TEG) take advantage of high temperature gradients, assuring battery-less operation of emerging wireless devices. TEGs have been extensively utilized in diverse domains such as wearable electronics. In urban areas on the other hand, a recent technology, in which the RF signals in abundance are exploited to produce energy, has started replacing the conventional methods of power provisioning.

A specialized version of EH, namely wireless EH or wireless power transfer (WPT), which refers to exploitation of RF signals, enables remote energization of wireless devices at a distance. The recently emerged WPT technology therefore stands highly promising to mitigate battery constraints and eliminate the need for maintenance. It has been extensively studied to be put into practice for sustaining IoT services in diverse domains [6].

As shown in Table 4.1, the power density of the exploitable resources is typically low and varies depending on the harvester efficiency, deployment location, and many other unforeseeable and uncontrollable factors. At the same time, the workload profile of IoT

TABLE 4.1

Existing EH Techniques [4,7,8]

Category	Source	Polarity	Efficiency	Harvested Power	Characteristics
Radiant	Light	DC	~10%–30%	100 mW/cm² (outdoor, solar)	Not always available, hard to deploy outdoor
				100 μW/cm² (office, diffused light)	Requires maximum power point tracking
Fluid flow	Wind	AC	~39%	35 μW/cm² (@<1 m/s, wind)	Available day and night but has fluctuating density
			~41%	3.5 mW/cm² (@8:4 m/s, generator)	Requires impedance matching
Mechanical	Vibration, human activity	AC	~25%–50%	~800 mW (machines - kHz)	Compact configuration and lightweight
				~4 μW/cm³ (motion - Hz)	High fluctuations at the output
					Requires rectifier and step-up/step-down circuits
Thermal	Thermal gradient	DC	~5%	~1–10 mW/cm² (industrial)	Low maintenance cost, scalable
					Requires efficient heat sinking
Wireless	RF	AC	~50%	0.1 μW/cm² (GSM, 900 MHz)	Allows mobility, abundant in urban areas
				1 μW/cm² (Wi-Fi, 2.45 GHz)	Requires impedance matching

systems is typically 'bursty' in nature, meaning that they remain in a low-power mode or 'sleep' mode for most of the time, waking up for taking measurements (periodically or when something happens), performing computation, and communication. This mismatch between energy availability and power utilization has to be taken into account at the time of design. In the next section, different approaches and techniques are presented to address this research challenge.

4.1.3 EH IoT System Design

A primary challenge in developing EH-powered systems is uncontrollable nature of the source, which usually exhibits temporal and spatial variability. This is shown in Figure 4.1 where the instantaneous powers harvested from three photovoltaic cells (or PV modules) exhibit significant time and spatial variability depending on the location and weather conditions. The output voltages from two micro-wind generators are also shown, with high power-cycle frequency when compared to the PV modules and also spatial variability.

To accommodate this, EH systems typically incorporate large energy storage, such as supercapacitors or rechargeable batteries, to sustain computation and cope with the mismatch between the source and load. Supercapacitors are often preferred as energy storage due to the better cycle life and electrical characteristics when compared to traditional batteries, despite their lower energy densities.

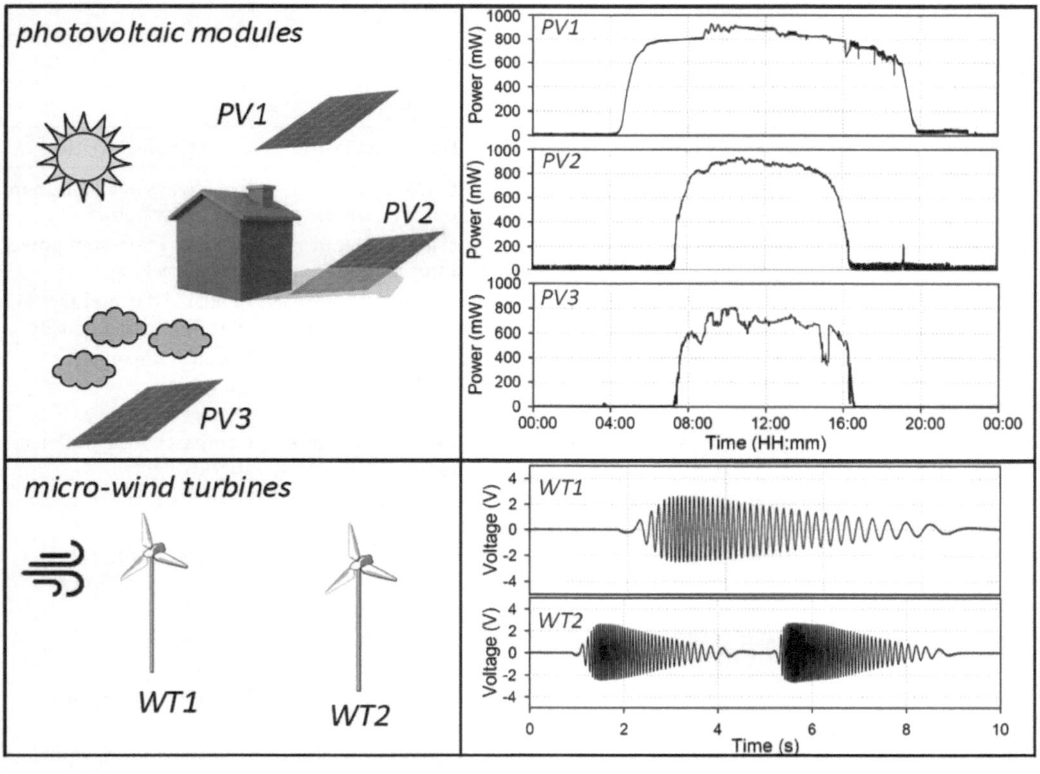

FIGURE 4.1
Example outputs from EH systems: power from three PV modules located in three different sites over a day, voltage from two wind harvesters over one 'gust'.

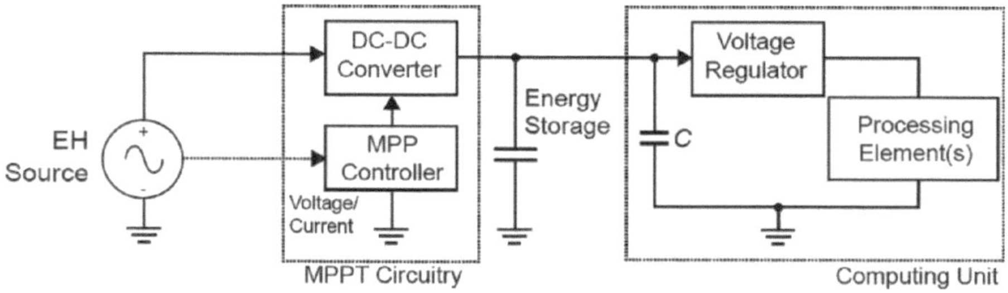

FIGURE 4.2
MPPT system architecture for energy-neutral systems.

Additionally, these systems also require a maximum power point tracking (MPPT) unit to maximize power extraction under variable environmental conditions. For example, MPPT techniques are used with systems powered by PV modules to continually maximize the output power which depends on solar radiation and temperature. As shown in Figure 4.2, the MPPT unit typically relies on a switching-mode DC–DC converter, a maximum power point (MPP) controller that monitors the voltage and current, and controls the charging current to the energy storage, which in turn provides energy to the computing unit. This additional energy storage also decouples the load from the EH source dynamics, managing the mismatch between the maximum power point current and the load current.

This system architecture enables energy-neutral operation, which attempts to balance the long-term energy consumption against the harvested energy [9]. A system is energy-neutral over a period of time T, if the energy stored in the storage, after this time, is greater than or equal to the initial energy stored. This can be expressed as

$$\int_0^T P_h(t)\,dt \simeq \int_0^T P_c(t)\,dt, \tag{4.1}$$

where $P_h(t)$ is the instantaneous power harvested at time t, and $P_c(t)$ is the power consumed by the computing unit (e.g., an MCU) at that time. Equation (4.1) is met with if

$$E \simeq E_0, \tag{4.2}$$

where E is the energy stored in the buffer after time T and E_0 is initial energy stored at time 0. Energy-neutral operation in IoT devices is typically achieved if a desired functionality can be supported perpetually (i.e., infinite lifetime), by balancing the consumed energy against that harvested over T.

Energy-neutral systems have been largely used in domains such as wireless sensor networks (WSNs), where Equation (4.2) is met with by adaptively adjusting the workload and hence the power consumption [10,11]. Methods for achieving workload adaptation include adjusting device activity (e.g., changing sample/transmit duty cycles [12]), adaptive sleep [13], or participation in network activity (e.g., packet routing) [14].

For IoT devices that have constrained dimensions, it is desirable and convenient to power them directly from the harvester without any additional storage other than decoupling capacitance C. In this application domain, attempting to use a storage inevitably leads to high losses due to power harvesting costs, self-discharge, and converter inefficiencies. However, this makes systems susceptible to frequent power interruptions and resets caused by the intermittent source.

This issue inherently promotes multiple sources, i.e., hybrid, EH procedures, where numerous systems have already been equipped with distinct, yet collaborative EH mechanisms to overcome the ongoing constraints [8]. Extracting energy from multiple resources reduces the variance on a single EH's output enabling power neutrality and thus contributes to non-intermittent behavior [15]. Power neutrality refers to the possibility of adjusting the system's performance dynamically such that instantaneous harvested power matches the instantaneous power requirement, negating the need for additional storage. This can be achieved by tracking the available harvested power [16].

However, power-neutral operation is applicable only when the average harvested power is comparable (i.e., equal or bigger) to the required power for sustained system operation (gray area in Figure 4.3). This can be formalized as

$$P_h(t) \geq \min\left(P_c(t)\right), \forall t. \tag{4.3}$$

If this is the case, the computing unit can be directly coupled with the variable EH source as shown in Figure 4.4. Here, the processing element tracks the instantaneous harvested power by monitoring the voltage across the decoupling capacitance C (shown by dotted arrow).

In order to achieve power neutrality, different control schemes are used to adjust the power consumption dynamically depending on the type of processing unit. In smaller-sized single-core MCUs, dynamic frequency scaling (DFS) can be used in which the clock frequency of the processing element is adjusted to the specific power constraints, whilst with multicore systems more sophisticated techniques exist, such as dynamic voltage and frequency scaling (DVFS), where the supply voltage of the processing unit is also adjusted, and core hot-plugging technique to dynamically switch cores on or off. Different runtime approaches to manage and adapt these controls have been proposed for power-neutral management on both single and multicore systems [17,18].

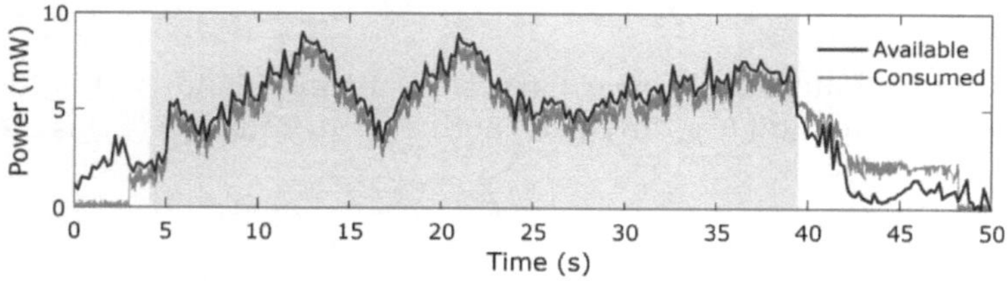

FIGURE 4.3
An EH system under power neutrality, showing the available harvested power and the consumed power.

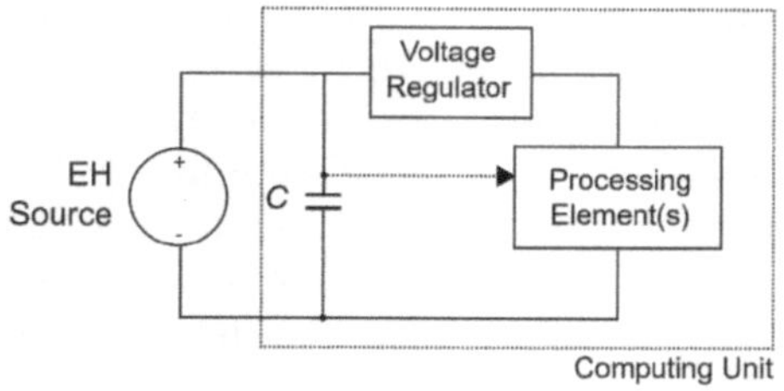

FIGURE 4.4
Storage-less architecture for power-neutral systems.

As already mentioned, when considering systems powered by energy harvesters, it is important to accommodate the uncontrollable, and often intermittent, nature of the source. Various approaches for intermittent computing exist in literature to enable system state retention [19,20]. Broadly, they can be classified into two categories: software-based approaches, in which the system state is saved into a non-volatile memory (NVM) before a power failure occurs and restored once the power supply recovers, and architectural approaches, in which the entire system is designed is to be non-volatile (e.g., non-volatile processors [21]). In the next section, wireless communication is discussed with focus on different protocols and techniques currently available for IoT systems.

4.1.4 IoT Wireless Protocols

As discussed in Section 4.1.1, each IoT device typically includes different sub-systems. Among these, the communication unit (i.e., the radio transceiver) typically dominates the overall energy consumption. In this section, the IoT device consumption is described in respect of the communication technologies and protocols commonly utilized [22].

In the first place, it is useful to remark that IoT systems typically transmit small data packets from the IoT sensing device to the data aggregator, such as a gateway or coordinator, i.e., from a few bytes up to hundreds of bytes. These packets are typically transmitted using a low data rate in order to increase the system lifetime when battery-powered.

Figure 4.5 shows how the communication is typically organized between the IoT sensing device and the dedicated server in which data are then processed to gather useful information. Here, the internet protocol is only available between the data aggregator and the server, due to the limited resources of each IoT device, which cannot support the standard internet protocol. As a result, IoT sensing devices use internal communication protocols to transmit packets to the data aggregator. The data aggregator then forwards these to the server using the internet protocol. In the following, a classification of different types of wireless networks and relative protocols are presented. Based on their coverage range, wireless networks can be defined as wireless personal area networks (WPANs) for short-range coverage (within a few meters) (i.e., protocols based on Bluetooth low energy (BLE) or IEEE 802.15.1); wireless local area networks (WLANs) for coverage range up to hundreds

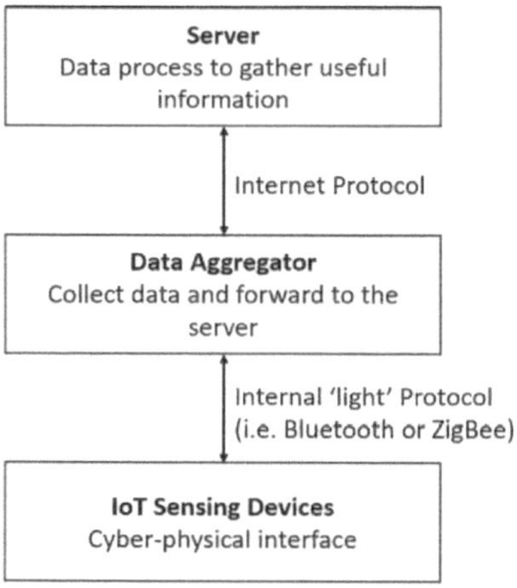

FIGURE 4.5
Typical communication between IoT devices.

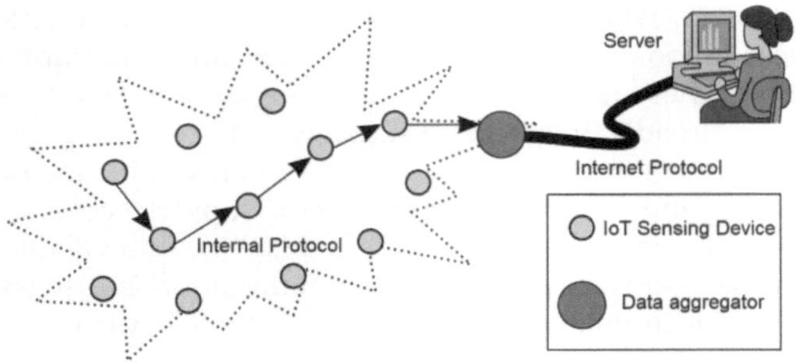

FIGURE 4.6
An example of mesh network typically used for WSNs.

TABLE 4.2
Examples of Available Wireless Transceivers for IoT Devices

		Current Consumption (mA)	
Technology	**Transceiver**	**Tx**	**Rx**
Bluetooth	Texas Instruments CC2640b	6 (0 dBm)	6
Zigbee	Texas Instruments CC2630b	6 (0 dBm)	6
Wi-Fi	Texas Instruments CC3200b	229	59
LoRa	Sentech SX1272	28 (+13 dBm)	10

of meters (i.e., Wi-Fi technology, IEEE 802.11); and wireless wide area networks (WWANs) which offer citywide coverage range, including cellular networks [23].

Another important aspect with wireless networks is their topologies, i.e., star or mesh type. In star networks, each IoT sensing device can communicate directly with the data aggregator using a point-to-point communication, whilst a mesh network refers to a rich interconnection among multiple sensing devices and one or more data aggregators. Mesh topologies are necessary when the protocol adopted is, for example, a short-range protocol and the total coverage area is bigger than this range. As a result, the network typically relies on a multi-hop structure (shown in Figure 4.6), where IoT sensing devices can communicate with their neighbors, increasing the overhead and thus having an impact on the energy efficiency [24]. In the following, some existing and forthcoming protocols are discussed, considering their coverage range, network topology, and data rate (shown in Table 4.2).

4.1.4.1 Bluetooth Low Energy

Bluetooth Low Energy (BLE) has gained significant momentum over the past few years becoming the de-facto standard for applications such as body area networks (BANs) for smart healthcare applications [25,26]. BLE is a standard for WPAN in the 2.4GHz unlicensed industrial, scientific, and medical (ISM) frequency bands. This protocol uses a short-range radio that guarantees a coverage range up to tens of meters and relatively low data rate (1,000 kbps).

In contrast with its previous versions, BLE is specifically designed as a low-power solution for applications requiring short-range IoT devices such as smart watches and other types of wearable devices. BLE can be operated by a transmission power between 0.01 and 10mW. Its design only enables star topology of the master–slave type.

BLE employs frequency hopping over 37 channels for (bidirectional) communication and 3 for (unidirectional) advertising, with a bit rate of 1 Mbps. For the discovery mechanism, slaves send packets to the master using these advertising channels, which are scanned by the master. However, its star topology limits network coverage range and precludes end-to-end path diversity. In contrast to this, other competing technologies, such as IEEE 802.15.4-based standards (like Zigbee), overcome such constraints by supporting mesh network topology.

4.1.4.2 IEEE 802.15.4

The IEEE 802.15.4 is a standard that specifies the physical layer (PHY) and the sub-layer for Medium Access Control (MAC) for low-data-rate WPAN (LR-WPAN) [27]. This standard has been intensively used for IoT applications for healthcare, wellness, and fitness. IEEE 802.15.4-based technologies (e.g., Zigbee Health Care) are typically employed for these IoT applications due to their low power consumption and cost, their ability to manage large number of nodes, and their operability among multiple IoT platforms. They also provide high-level security, encryption, and authentication services.

IEEE 802.15.4 supports three frequency channel bands and utilizes a direct sequence spread spectrum (DSSS) method. Based on the used frequency channels, the physical layer transmits and receives data over three data rates: 250 kbps at 2.4 GHz, 40 kbps at 915 MHz, and 20 kbps at 868 MHz. IEEE 802.15.4 MAC utilizes the carrier sense multiple access with collision avoidance (CSMA/CA) protocol to reduce potential collisions.

IEEE 802.15.4 defines two types of devices:

- Full Function Device (FFD): high-cost devices with high power consumption (typically main powered) and with extended functions and services
- Reduced Function Device (RFD): low-cost devices with low power consumption (typically battery-powered or supplied by EH sources) and limited functionalities.

There are many IEEE 802.15.4-based standards, such as Zigbee, 6LoWPAN, IPv6, Thread, etc. Some of them are still under development and standardization. In the following, some insights are provided on Zigbee standard which is the most adopted one for healthcare applications.

4.1.4.3 Zigbee Standard

Zigbee defines a complete open-global standard for reliable, cost-effective, low-power, wirelessly networked products addressing monitoring and control. Zigbee is built on the top of the IEEE 802.15.4 standard by adding the network, security, and application framework layers [28,29]. Zigbee's typical coverage ranges up to hundreds of meters and defines the specification for a low-rate mesh network topology.

Specifically, the network layer includes some important networking functions such as the following:

- **Starting a network:** the ability to successfully establish a new network
- **Joining and leaving a network:** the ability to gain membership (join a network) or relinquish membership (leave a network)
- **Configuring a new device:** the ability to sufficiently configure the stack for operation as required

- **Addressing:** the ability of a Zigbee data aggregator (i.e., coordinator) to assign addresses to devices joining the network
- **Synchronization within a network:** the ability for a device to achieve synchronization with another device either through tracking beacons or by polling
- **Routing:** routing frames to their intended destinations.

Additionally, Zigbee standard defines three types of logical devices (coordinator, router, and end device):

- Coordinator is the main governor of the network, with the capability of starting the network, setting the security level, and enabling the relevant actions associated to full functioning of the network.
- Router (FFD-associated device) establishes the connection from the coordinator to other router or from router to end devices, enabling a wide range of communication.
- End device (RFD-associated device) is a low-powered device with reduced functionalities; i.e., they never collect information from other networks.

Zigbee standard has been extensively used for IoT applications related to smart health, wellness, and fitness for improving personal healthcare. On this purpose, the Zigbee Alliance has joined forces with the Continua Health Alliance, a non-profit, open industry coalition of the finest healthcare and technology companies collaborating to improve the quality of personal healthcare. Zigbee Health Care defines a global standard to enable secure and reliable monitoring and management of non-critical, low-acuity healthcare services targeted at chronic disease, aging independence and general health, wellness, and fitness. Leading healthcare and technology companies are supporting the development of Zigbee Health Care, including Motorola, Phillips, NXP Semiconductors, and RF Technologies [30].

4.1.4.4 Wi-Fi

Wi-Fi is the de-facto standard for WLAN and is based on the IEEE 802.11 standard, which operates in the 2.4 and 5 GHz ISM bands. Wi-Fi is designed to provide high-speed (i.e., up to 10 Mbps) wireless links for devices that can access the Internet directly in the range up to hundreds of meters, via access points (APs) using a star network topology.

Due to the higher power requirements of the transceivers compared to BLE and Zigbee, Wi-Fi is not yet a reference standard for IoT applications. However, deploying IoT devices with Wi-Fi standard has a distinct advantage: these devices can utilize the well-established Wi-Fi APs in buildings and cities, reducing additional costs on new gateway infrastructure. This is an interesting research area that still requires further investigation [31].

4.1.4.5 LoRa and SigFox

LoRa is an emerging proprietary low-power long-range wide area network (WAN) technology using the unlicensed spectrum in the sub-GHz band [32–34]. Specifically, LoRa operates in the 868–915 MHz ISM band allowing bit rates between 0.37 and 46.9 kbps and promising ranges up to 20–25 km. This reduces the need of a large number of gateways to drive down the infrastructure costs.

LoRaWAN, the MAC protocol for WANs, is based on the ALOHA protocol and is ideal for applications with low-traffic and sporadic communication requirements. One of the features making LoRa attractive is its energy efficiency for uplink communication while achieving a long range. In LoRaWAN, the distributed battery-operated sensor nodes send data directly to an always-on gateway. The energy efficiency comes by low duty cycling of the main radio transceiver when idle.

Another emerging proprietary low-power WAN technology is SigFox, which uses as well the unlicensed spectrum in the sub-GHz band. It offers low-power and low-cost transceivers, enabling IoT devices to achieve long-range communication with gateways kilometers apart with a star topology.

4.2 Biomedical Signal Processing and Machine Learning Challenges in IoT Research

It is now an interesting question to investigate the potential of emerging IoT technologies to revolutionize the future healthcare and biomedical monitoring systems. Various IoT healthcare technologies have emerged recently, e.g., cloud computing, grid computing, big data, wireless networks, ambient intelligence, augmented reality and wearable sensors, etc. Various countries like Australia, Denmark, the European Commission, Finland, Kenya, Mauritius, Saudi Arabia, Scotland, Sweden, Switzerland, the US have already published their *e*-Health strategy and evidence-based policy and regulations [35]. Automated machine interpretation of big IoT data is becoming a huge challenge in many industrial and service sectors beside healthcare technologies, e.g., smart metering, smart transportation, smart supply chain, smart agriculture, smart grid, smart traffic light system, e-commerce, smart cities, retail and logistics, etc., posing various challenging research areas like privacy, data mining and pattern recognition, visualization, data integration, etc. Existing data analytics systems are real-time analytics, off-line analytics, memory-level analytics, business intelligence (BI) analytics, massive analytics which employs various classification, clustering, rule association, and time series methods like Bayesian networks, support vector machine (SVM), *k*-nearest neighbor (*k*-NN), fuzzy logic, and agglomerative and hierarchical clustering [36].

4.2.1 IoT Network for Healthcare

The concepts of distributed sensing and IoT have been extended in healthcare technologies to give rise to the IoT network for healthcare or the IoThNet [35]. The IoThNet collects various vital signals like blood pressure, body temperature, electrocardiogram (ECG), oxygen saturation etc. from various sensors like ECG, electroencephalogram (EEG), electromyogram (EMG), blood pressure sensors, etc.

4.2.2 IoT-Based Healthcare Services

Various concepts of IoT-based healthcare services have been introduced in [35], e.g., ambient assisted living, the internet of m-health things (m-IoT), adverse drug reaction, community healthcare, children health information, wearable device access, semantic medical access, indirect emergency healthcare, embedded gateway configuration, embedded context prediction, as discussed in [35].

4.2.3 IoT-Based Healthcare Applications

There are many applications in biomedical health monitoring and intervention where IoT can have potential applications, e.g., glucose level sensing, ECG monitoring, blood pressure monitoring, body temperature monitoring, oxygen saturation monitoring, rehabilitation system, medication management, wheelchair management, imminent healthcare solutions, healthcare solutions using smartphones, as discussed in [35]. Using various types of sensor data, one can monitor various medical conditions, e.g., diabetes, wound analysis, heart rate monitoring, blood pressure monitoring, body temperature monitoring, designing rehabilitation system, medication management, wheelchair management, oxygen saturation monitoring, eye disorder, skin infection, asthma, chronic pulmonary disease, cough detection, allergic symptoms, remote surgery, etc. Various smartphone apps have been designed for such monitoring purpose, e.g., Health Assistant, Healthy Children, Google Fit, Calorie Counter, Water Your Body, Noom Walk Pedometer, Period Calendar, Period Tracker, Instant Heart Rate, Cardiax Mobile ECG, ECG Self-monitoring, ElektorCardioscope, Runtastic Heart Rate, Heart Rate Monitor, Cardiomobile, Blood Pressure Watch, Finger Blood Pressure Prank, OnTrack Diabetes, Finger Print Thermometer, Body Temperature, Medisafe Meds & Pill Reminder, Dosecast Medication Reminder, Rehabilitation Game, iOximeter, Eye Care Plus, SkinVision, Asthma Tracker and Log, PulmonaryRehabilitationl Test Your Hearing, uHear, Real Noise 3, Relax Melodies, Sleep Aid, Fall Detector, Fade, iFall, Calm, Mayo Clinic Meditation, and Daily Yoga amongst many others.

4.2.4 Security in IoT-Based Healthcare

While designing an IoT-based health monitoring system, one need to be careful about security requirements like confidentiality, integrity, authentication, availability, data freshness, nonrepudiation, authorization, resiliency, fault tolerance, self-healing, etc. There are various security challenges in designing such systems, e.g., computational limitations, memory limitations, energy limitations, mobility, scalability, communications media, multiplicity of devices, dynamic network topology, multiprotocol network, dynamic security updates, tamper-resistant packages, etc. There have been various studies on threat modeling, and different attack taxonomies have been proposed, e.g., interruption, interception, modification, fabrication, replay, user compromise, hardware compromise, software compromise, standard protocol compromise, network protocol stack attack, etc. as discussed in [35].

4.2.5 IoT-Based ECG Signal Processing

Satija et al. [37] proposed a method for an IoT system aware of ECG signal quality. It has been tuned to filter out the baseline wander which goes up to 1 Hz. This could degrade the P-QRS-T waves of ECG signals, and a filtering technique based on discrete Fourier transform (DFT) has been proposed for abrupt change detection. The baseline wander and abrupt change detection have been designed using a threshold-based method. Next, ECG signal absence detection system has been developed using a decision rule in case of loss of contact between the IoT device and the patient's body. Similarly, high-frequency noise detection is also an important aspect of such monitoring due to muscle artifact, motion artifacts, instrument noise, etc. A signal quality index (SQI) has been proposed as a linear combination of individual SQIs for abrupt change and abrupt baseline wander (ABW), loss of connection or flat line (FL), and high-frequency (HF) noise as

$$\text{SQI} = \text{SQI}_{\text{ABW}} + \text{SQI}_{\text{FL}} + \text{SQI}_{\text{HF}}. \tag{4.4}$$

Such an index captures various aspects of the signal quality and may corrupt the signal to different extents on a case-by-case basis for a specific wearable IoT device. Based on the SQI, a signal quality grade (SQG) is proposed as

$$\text{SQG} = \begin{cases} \text{Good} & \text{SQI} = 0 \\ \text{Intermediate} & \text{SQI} = 0.5 \\ \text{Bad} & \text{SQI} > 0 \end{cases}. \tag{4.5}$$

Finally, a sensitivity score has been proposed to compare the quality of the ECG signals given by

$$\text{Sensitivity} = \frac{\text{TP}}{\text{FN+TP}} \times 100\%, \tag{4.6}$$

where TP denotes the true positive correctly detected segments and FN denotes false negatives missed segments. The method has been benchmarked on the PhysioNet and MIT-BIH arrhythmia databases using standard QRS detector with template matching and R-peak detectors and sub-band power using higher order moments.

4.2.6 IoT-Based Machine Learning Method for Heart Disease Detection

Machine learning challenges for automated analysis of IoT data were discussed in [38], where it is highlighted that we need to first understand the product and IT maturity. Also, the key factor is the understanding of the capabilities of current IoT technologies like monitoring, control, optimization, autonomy, etc. As an example, in the case of ECG signal monitoring using IoT, commonly used clinical features have been used like QT interval, QRS complex, J point to the T peak, T peak to T end, etc. [39]. These features were then used in various machine learning algorithms such as *k*-NN, SVM, random forest, adaboost, etc. using various performance measures like accuracy on test data, training time, and classification time as a function of increased number of training samples.

Kumar and Gandhi [40] have used logistic regression to classify heart disease using seven variables, viz. respiratory rate (X_1), heart rate (X_2), blood pressure – systolic range (X_3), blood pressure – diastolic range (X_4), body temperature (X_5), blood sugar – fasting (X_6), blood sugar – post meal (X_6). The multiple logistic regression is given by

$$\text{logit}(p) = b_0 + b_1X_1 + b_2X_2 + \cdots + b_7X_7 = \ln\left(\frac{p}{1-p}\right). \tag{4.7}$$

Here, p denotes the probability of presence of the dependent variable, i.e., heart disease (in 0 or 1), and b denotes the respective coefficients for each variable. The odds ratio is expressed as

$$\text{odds} = \frac{p}{1-p} = \frac{\text{probability of presence of the variable}}{\text{probability of absence of the variable}}. \tag{4.8}$$

The probability of calculating the presence of the particular event is expressed as

$$p = \frac{\exp(b_0 + b_1X_1 + b_2X_2 + \cdots + b_7X_7)}{1 + \exp(b_0 + b_1X_1 + b_2X_2 + \cdots + b_7X_7)}. \tag{4.9}$$

The IoT-based health monitoring model has been developed in the Apache Mahout and elastic MapReduce framework using cloud computing. For each variable, initially, coefficients, standard error, *Z*-score, *p*-value, and odds ratio were calculated. Next, the confusion matrix was derived, and a more detailed analysis of the classification performance was made using the following metrics apart from the sensitivity:

$$\text{Specificity} = \frac{\text{TN}}{\text{FP+TN}}, \tag{4.10}$$

$$\text{Positive Likelihood Ratio (PLR)} = \frac{\text{Sensitivity}}{100 - \text{Specificity}}, \tag{4.11}$$

$$\text{Negative Likelihood Ratio (NLR)} = \frac{100 - \text{Sensitivity}}{\text{Specificity}}, \tag{4.12}$$

$$\text{Positive Predictive Value (PPV)} = \frac{\text{TP}}{\text{TP+FP}}, \tag{4.13}$$

$$\text{Negative Predictive Value (NPV)} = \frac{\text{TN}}{\text{TN+FN}}, \tag{4.14}$$

$$\text{Disease Prevalence (DP)} = \frac{\text{TP+FN}}{\text{TP+FP+TN+FN}}. \tag{4.15}$$

Here, TN and FP denote the true negative and false positive rates, and the classification metrics are all measured in percentage. These measures indicate different aspects of the receiver operator characteristic (ROC) curve. For continuous monitoring through IoT devices, using the measures and the important variables, approximate age-based inference mechanism has also been developed for newborns, children with an age of few months to years and adults.

4.2.7 Community-Based IoT Personalized Healthcare System

Instead of having IoT-based single-station monitoring, it is possible to have distributed sensors at different sites for simultaneous monitoring over large geographical area. As an example, Catherwood et al. [41] reported a trial for urinary tract information where the diagnostic system test results were sent across communication channels. As a result, various communication performance metrics have been developed for the community-based healthcare system deployed at rural, city, and sub-urban areas. The commonly used communication metrics in such applications are signal to noise ratio (SNR), log-distance path loss $\left(\overline{\text{PL}}\right)$, path loss exponent ($n$) which depend on the transmitter–receiver separation distance (d), carrier frequency, environment, obstruction, etc. and are given by

$$\text{SNR} = P_{\text{signal}}/P_{\text{noise}}, \quad \text{SNR(dB)} = 10\log_{10}\left(P_{\text{signal}}\right) - 10\log_{10}\left(P_{\text{noise}}\right), \tag{4.16}$$

$$\overline{\text{PL}}(d) = \text{PL}(d_0) + 10n\log_{10}(d/d_0), \tag{4.17}$$

$$n = \overline{\text{PL}}(d) - \text{PL}(d_0)/10\log_{10}(d/d_0). \tag{4.18}$$

Here, d_0 is the closest in reference distance, determined from measurements close to the transmitter.

4.2.8 IoT-Based Biosignal Compression

In order to boost the battery life, often it is wise to get rid of the redundant information in the biosignals using compression methods. These compressed signals like heart rate, respiratory rate, ECG, and photoplethysmographic signals can then be transmitted over long distance and reconstructed again for visualization and interpretation. Various signal compression algorithms for IoT-based health monitoring systems have been discussed in Hooshmand et al. [42]. Amongst many available methods based on online dictionary (passband filtering, peak detection, segment extractor, pattern matching, codebook manager, etc.), gain-shape vector quantization (GSVQ), principal component analysis (PCA), autoencoders, compressive sensing (simultaneous orthogonal matching pursuit, block sparse Bayesian learning, etc.), discrete cosine transform (DCT), discrete wavelet transform (DWT), and lightweight temporal compression (LTC) methods are notable. A benchmarking study has been reported using root mean squared error (RMSE) and compression efficiency.

4.3 Recent Research Trends in IoT-Based Healthcare Monitoring

In the biomedical engineering and healthcare technology field, IoT is emerging as a prominent research discipline. In the Scopus search, some patterns were retrieved for the first 2,000 highly cited papers on this topic. The word clouds in Figure 4.7 show that in the papers' titles, few frequently used words are IoT, healthcare, system, monitoring along with some relatively less used words like smart, networks, security, devices, sensor, wearable, data. These word frequencies show the relative research efforts and different sub-areas in

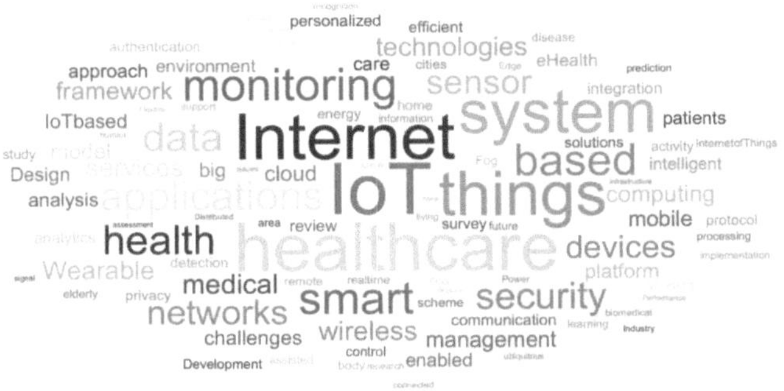

FIGURE 4.7
Word clouds of the titles of recent IoT-based research papers. (Source: Scopus.)

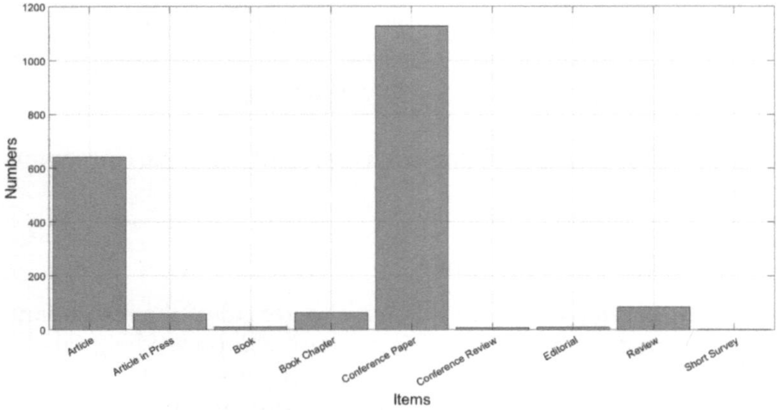

FIGURE 4.8
Types of recent papers on IoT applications in healthcare. (Source: Scopus.)

this field which are actively growing. Figure 4.8 also shows a bar chart of recently published works which shows there were twice more conference papers published over journal articles, followed by review articles and book chapters.

Similar to exploring the article titles, Figure 4.9 shows the indexed and author's keywords in the top 2,000 papers in this active research area. As opposed to the title-based text analysis, often the keyword-based analysis is capable of capturing finer technical details or an emphasis of fewer most popular technologies. In both the word clouds, apart from IoT, the most frequent words were computing, systems, networks, sensor, cloud, etc.

FIGURE 4.9
Indexed and author's keywords for recent papers on IoT applications in healthcare. (Source: Scopus.)

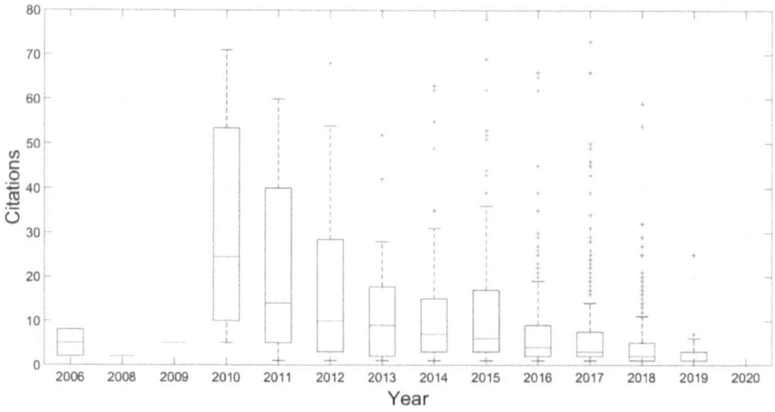

FIGURE 4.10
Year-wise citations of recent papers on IoT applications in healthcare. (Source: Scopus.)

which shows the finer details of the computing and communication technologies gradually increasing over the years. While exploring the citation patterns in this research field, it is clear that the citations of the papers on IoT application in healthcare jumped down suddenly from 2010 (Figure 4.10). The gradual decrease in the citation almost follows an exponential trend, because the citation dynamics is usually a cumulative phenomenon and as such recent technological advancements take years to get assimilated in industrial and academic practice. Also, there are more outliers in the papers from 2015 onward which shows there are few highly impactful papers compared to the citation of average papers published in the recent years.

Although many significant steps have been taken in the field of IoT-based health monitoring, still there are many open research areas that need to be addressed in future. This includes standardization, IoT healthcare platforms, cost analysis, app development process, technology transition, low-power protocol, network type, scalability, continuous monitoring, new diseases and disorders, identification, business model, quality of service (QoS), data protection, mobility, edge analytics, and ecological impact. Amongst these challenges, data protection through resource efficient security, physical security, secure routing, data transparency, secure handling IoT big data, etc. is very crucial to restrict illicit access.

4.4 Industrial IoT Data Analytics: A Case Study

An openly available dataset has been considered for a case study for industrial IoT [43]. The dataset had 16,382 datapoints with nine variables like demand response, area, season, energy, cost, pair number, and distance. This is an example of industrial data collected using IoT devices. Here we explore the association of the variables in such a dataset. The dataset was corrupted with few outliers. An automated outlier detection algorithm based on interquartile range (IQR) was employed to filter out the extreme values, and finally 14,203 datapoints were included for the data analysis. The correlation analysis amongst the nine variables are shown in Figure 4.11 along with the estimates of Kendall's correlation

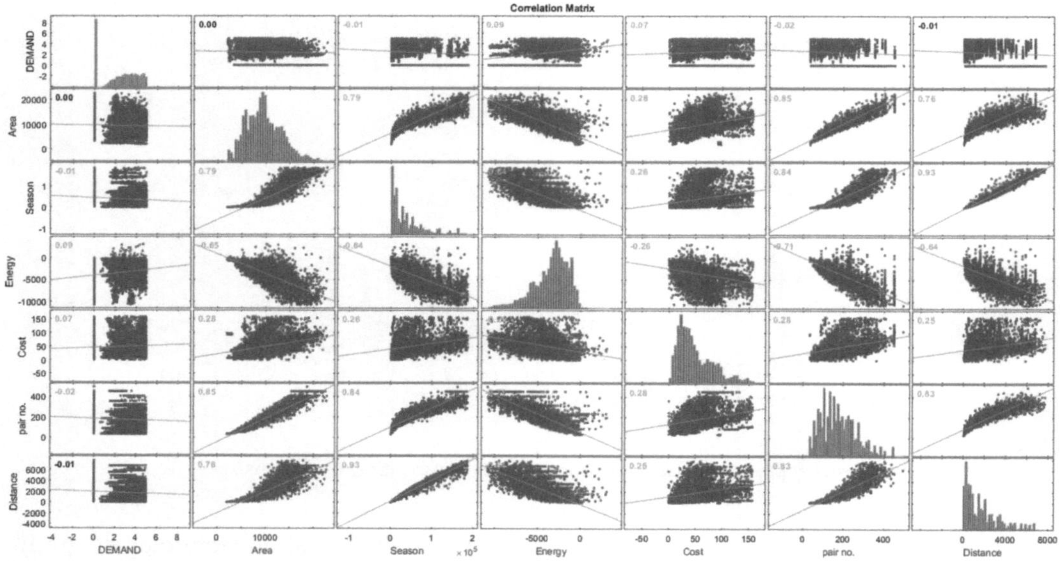

FIGURE 4.11
Correlation plot amongst the variables for the industrial IoT dataset.

coefficient for each pair of variables. It is evident that the season and distance variables are highly interdependent on each other ($r=0.93$), followed by pair number vs. area ($r=0.85$) and pair number vs. season ($r=0.84$). It is apparent that most of the univariate histograms are highly skewed and contain multiple modes; therefore the Pearson's correlation has not been investigated; rather rank-based correlation measures are better suitable for identifying variable interdependencies. Rest of the variables do not show any strong correlation pattern as such.

In industrial IoT setting, such statistical analysis is highly relevant. Because of the abundant and cheap IoT devices, it is likely to gather huge amount of information from various sensors, which can then be represented in compact manner using low-dimensional projections. Next, we explore an unsupervised dimensionality reduction method applied on this data. We here employ the *t*-distributed stochastic neighbor embedding (*t*-SNE) method for a low-dimensional representation of this industrial IoT dataset [43]. The commonly used *t*-SNE hyperparameters are the distance measures and perplexity parameter. Figure 4.12 shows the 2D projection of the *t*-SNE components using two principal components (PCs) with four different distance measures – Euclidean, Mahalanobis, Chebyshev, and cosine while keeping the perplexity parameter fixed at 30. Next, we explore the effect of perplexity variation (5, 30, 50, 100) on the big industrial IoT data in Figure 4.13 with respect to the fixed Euclidean distance measure. It is interesting to see that there are very clearly identifiable groups or non-overlapping clusters in the original 7D industrial IoT data in terms of the original features – demand, area, season, energy, cost, pair number, and distance in Figure 4.11. However, in the corresponding 2D projections in the *t*-SNE space, there are many groups, which share similarities with respect to certain distance measures as shown in Figures 4.12 and 4.13. Therefore, such methods could be applied for visualization and automated clustering or grouping of large volume of industrial IoT datasets recorded by different types of biomedical sensors. However, the only concern for such analysis is for large number of datapoints, *t*-SNE becomes computationally demanding to calculate the

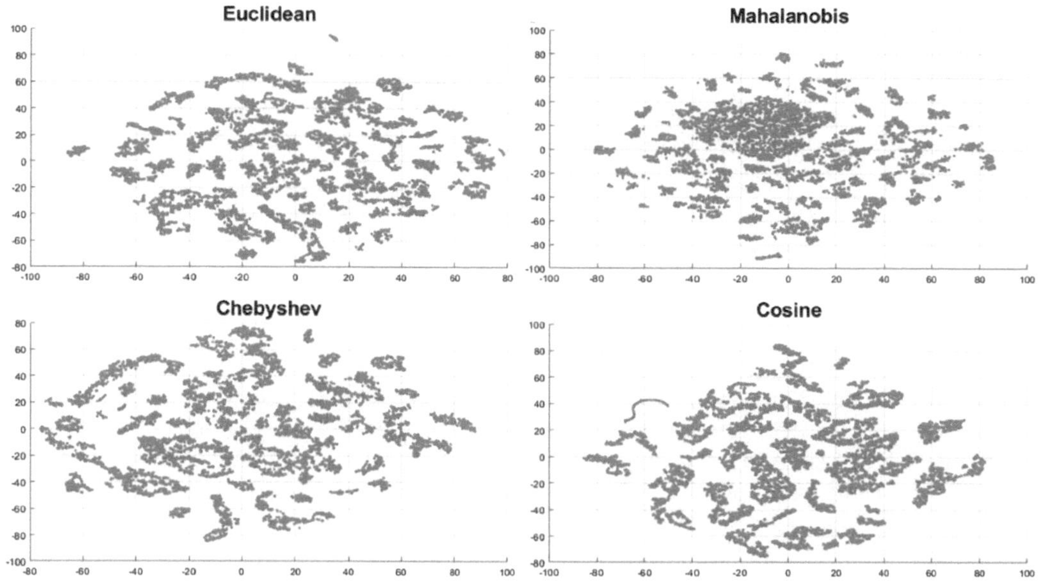

FIGURE 4.12
t-SNE plots for the industrial IoT dataset with four different distance measures.

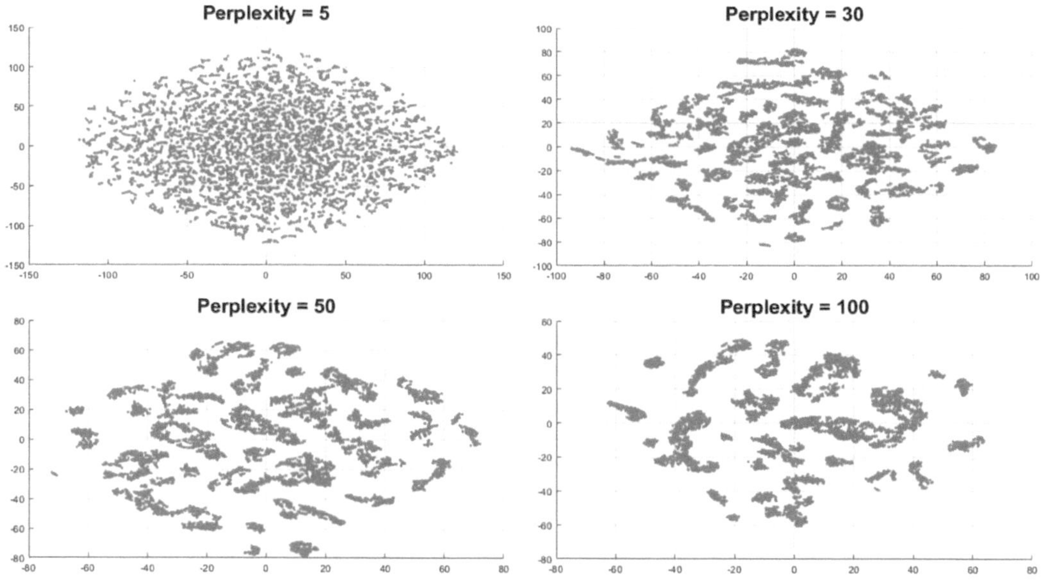

FIGURE 4.13
t-SNE plots for the industrial IoT dataset with four different perplexities.

distance-based similarity index between data points. As an example, for dimensionality reduction of the present industrial IoT data, it takes approximately 125 s, 127 s, 127 s, 122 s, respectively, for the four distance measures in Figure 4.12, whereas for a different perplexity parameter, it takes 127 s, 126 s, 135 s, 165 s, respectively, of computation time on a

64-bit Windows PC with 64 GB memory and an AMD Ryzen 7, 3.6 GHz processor. It is also evident from Figures 4.12 and 4.13 that although the variation of the distance measures does not have a very clear impact on the cluster formation, an increase in the perplexity parameter clearly gives rise to fewer number of compact clusters in the projected 2D *t*-SNE component space which share some degree of similarity with the original 7D industrial IoT dataset.

4.5 Conclusion

In this chapter, a comprehensive review on IoT systems with focus on energy-constrained IoT devices is presented. Specifically, EH principles are first discussed, and then a taxonomy describing and classifying the landscape of EH computing systems is presented. This chapter is then concluded with a discussion on standard IoT protocols and their properties, presenting the most used standards such as Zigbee, Bluetooth, and LoRa.

Following this, a review of various existing technologies and open challenges of using IoT technologies in future healthcare services is presented. Amongst these, energy-constrained IoT technologies and IoT data processing using signal processing and machine learning techniques are especially highlighted. It is apprehended that in future many more healthcare service industries will pose their IoT big data analytics challenges in open competitions like Kaggle which will attract invention of new solution methods and digital technologies from future computer scientists, mathematicians, and engineers by collaborative research efforts.

References

1. Jayakumar, H., Raha, A., Kim, Y., Sutar, S., Lee, W.S., and Raghunathan, V., 2016, January. Energy-efficient system design for IoT devices. In *2016 21st Asia and South Pacific Design Automation Conference (ASP-DAC)* (pp. 298–301). IEEE, Macau, China.
2. Alansari Z., Soomro S., Belgaum M.R., and Shamshirband S., 2018. The rise of Internet of Things (IoT) in big healthcare data: Review and open research issues. In *Progress in Advanced Computing and Intelligent Engineering. Advances in Intelligent Systems and Computing*, K. Saeed, N. Chaki, B. Pati, S. Bakshi, D. Mohapatra (Eds.) vol 564. Singapore: Springer. https://doi.org/10.1007/978-981-10-6875-1_66.
3. Jayakumar, H., Lee, K., Lee, W.S., Raha, A., Kim, Y., and Raghunathan, V., 2014, August. Powering the Internet of Things. In *Proceedings of the 2014 International Symposium on Low Power Electronics and Design* (pp. 375–380). ACM, La Jolla, CA.
4. Yau C.W., Kwok T.TO., Lei C.U., Kwok Y.K., 2018. Energy harvesting in Internet of Things. In *Internet of Everything. Internet of Things (Technology, Communications and Computing)*. B. Di Martino, K.C., Li, L. Yang, A. Esposito (Eds.) Singapore: Springer, Singapore. https://doi.org/10.1007/978-981-10-5861-5_3.
5. Priya, S. and Inman, D.J. eds., 2009. *Energy Harvesting Technologies* (Vol. 21, p. 2). New York: Springer.
6. Cetinkaya, O. and Akan, O.B., 2017. Electric-field energy harvesting from lighting elements for battery-less Internet of Things. *IEEE Access*, 5, 7423–7434.

7. Ku, M.L., Li, W., Chen, Y., and Liu, K.R., 2016. Advances in energy harvesting communications: Past, present, and future challenges. *IEEE Communications Surveys & Tutorials*, 18(2), 1384–1412.
8. Akan, O.B., Cetinkaya, O., Koca, C., and Ozger, M., 2018. Internet of hybrid energy harvesting things. *IEEE Internet of Things Journal*, 5(2), 736–746.
9. Kansal, A., Hsu, J., Zahedi, S., and Srivastava, M.B., 2007. Power management in energy harvesting sensor networks. *ACM Transactions on Embedded Computing Systems* (*TECS*), 6(4), 32.
10. Mitcheson, P.D., Yeatman, E.M., Rao, G.K., Holmes, A.S., and Green, T.C., 2008. Energy harvesting from human and machine motion for wireless electronic devices. *Proceedings of the IEEE*, 96(9), 1457–1486.
11. Cetinkaya, O. and Akan, O.B., 2017. Electric-field energy harvesting in wireless networks. *IEEE Wireless Communications*, 24(2), 34–41.
12. Buchli, B., Sutton, F., Beutel, J., and Thiele, L., 2014, November. Dynamic power management for long-term energy neutral operation of solar energy harvesting systems. In *Proceedings of the 12th ACM Conference on Embedded Network Sensor Systems* (pp. 31–45). ACM, Memphis, TN.
13. Anastasi, G., Conti, M., and Di Francesco, M., 2009. Extending the lifetime of wireless sensor networks through adaptive sleep. *IEEE Transactions on Industrial Informatics*, 5(3), 351–365.
14. Heo, J., Hong, J., and Cho, Y., 2009. EARQ: Energy aware routing for real-time and reliable communication in wireless industrial sensor networks. *IEEE Transactions on Industrial Informatics*, 5(1), 3–11.
15. Balsamo, D., Weddell, A.S., Merrett, G.V., Al-Hashimi, B.M., Brunelli, D., and Benini, L., 2015. Hibernus: Sustaining computation during intermittent supply for energy-harvesting systems. *IEEE Embedded Systems Letters*, 7(1), 15–18.
16. Balsamo, D., Das, A., Weddell, A.S., Brunelli, D., Al-Hashimi, B.M., Merrett, G.V., and Benini, L., 2016. Graceful performance modulation for power-neutral transient computing systems. *IEEE Transactions on Computer-Aided Design of Integrated Circuits and Systems*, 35(5), 738–749.
17. Fletcher, B.J., Balsamo, D., and Merrett, G.V., 2017, March. Power neutral performance scaling for energy harvesting MP-SoCs. In *Proceedings of the Conference on Design, Automation & Test in Europe* (pp. 1520–1525). European Design and Automation Association, Lausanne, Switzerland.
18. Balsamo, D., Fletcher, B.J., Weddell, A.S., Karatziolas, G., Al-Hashimi, B.M., and Merrett, G.V., 2019. Momentum: Power-neutral performance scaling with intrinsic MPPT for energy harvesting computing systems. *ACM Transactions on Embedded Computing Systems* (*TECS*), 17(6), 93.
19. Rodriguez Arreola, A., Balsamo, D., Das, A.K., Weddell, A.S., Brunelli, D., Al-Hashimi, B.M. and Merrett, G.V., 2015, November. Approaches to transient computing for energy harvesting systems: A quantitative evaluation. In *Proceedings of the 3rd International Workshop on Energy Harvesting & Energy Neutral Sensing Systems* (pp. 3–8). ACM, Seoul, South Korea.
20. Sliper, S.T., Balsamo, D., Weddell, A.S., and Merrett, G.V., 2018, November. Enabling intermittent computing on high-performance out-of-order processors. In *Proceedings of the 6th International Workshop on Energy Harvesting & Energy-Neutral Sensing Systems* (pp. 19–25). ACM, Shenzhen, China.
21. Ma, K., Li, X., Swaminathan, K., Zheng, Y., Li, S., Liu, Y., Xie, Y., Sampson, J.J., and Narayanan, V., 2016. Nonvolatile processor architectures: Efficient, reliable progress with unstable power. *IEEE Micro*, 36(3), 72–83.
22. Al-Fuqaha, A., Guizani, M., Mohammadi, M., Aledhari, M., and Ayyash, M., 2015. Internet of things: A survey on enabling technologies, protocols, and applications. *IEEE Communications Surveys & Tutorials*, 17(4), 2347–2376.
23. Minoli, D., 2013. *Building the Internet of Things with IPv6 and MIPv6: the Evolving World of M2m Communications*. Hoboken, NJ: John Wiley & Sons.
24. Liang, N.C., Chen, P.C., Sun, T., Yang, G., Chen, L.J., and Gerla, M., 2006, October. Impact of node heterogeneity in ZigBee mesh network routing. In *2006 IEEE International Conference on Systems, Man and Cybernetics* (Vol. 1, pp. 187–191). IEEE, Taipei, Taiwan.

25. Gomez, C., Oller, J., and Paradells, J., 2012. Overview and evaluation of bluetooth low energy: An emerging low-power wireless technology. *Sensors*, 12(9), 11734–11753.
26. Yu, B., Xu, L., and Li, Y., 2012, June. Bluetooth Low Energy (BLE) based mobile electrocardiogram monitoring system. In *2012 IEEE International Conference on Information and Automation* (pp. 763–767). IEEE, Shenyang, China.
27. Dutta, D., 2019. IEEE 802.15. 4 as the MAC protocol for Internet of Things (IoT) applications for achieving QoS and energy efficiency. In *Advances in Communication, Cloud, and Big Data*, H.K. Deva Sarma, S. Borah, N. Dutta (Eds.) (pp. 127–132). Singapore: Springer.
28. Farahani, S., 2008. *ZigBee Wireless Networks and Transceivers: The Complete Guide for RF/Wireless Engineers*. Amsterdam: Newnes (an imprint of Butterworth-Heinemann Ltd).
29. Cetinkaya, O. and Akan, O.B., 2015, February. A ZigBee based reliable and efficient power metering system for energy management and controlling. In *2015 International Conference on Computing, Networking and Communications* (*ICNC*) (pp. 515–519). IEEE, Garden Grove, CA.
30. ZigBee Alliance health care. www.zigbee.org/zigbee-for-developers/applicationstandards/zigbee-health-care/. Accessed: 2019-04-08.
31. Tozlu, S., Senel, M., Mao, W., and Keshavarzian, A., 2012. Wi-Fi enabled sensors for Internet of Things: A practical approach. *IEEE Communications Magazine*, 50(6), 134–143.
32. Magno, M., Aoudia, F.A., Gautier, M., Berder, O., and Benini, L., 2017, March. WULoRa: An energy efficient IoT end-node for energy harvesting and heterogeneous communication. In *Proceedings of the Conference on Design, Automation & Test in Europe* (pp. 1532–1537). European Design and Automation Association, Lausanne, Switzerland.
33. Piyare, R., Murphy, A., Magno, M., and Benini, L., 2018. On-demand LoRa: Asynchronous TDMA for energy efficient and low latency communication in IoT. *Sensors*, 18(11), 3718.
34. Aoudia, F.A., Gautier, M., Magno, M., Le Gentil, M., Berder, O., and Benini, L., 2018. Long-short range communication network leveraging LoRa™ and wake-up receiver. *Microprocessors and Microsystems*, 56, 184–192.
35. Islam, S.R., Kwak, D., Kabir, M.H., Hossain, M., and Kwak, K.S., 2015. The Internet of Things for health care: A comprehensive survey. *IEEE Access*, 3, 678–708.
36. Marjani, M., Nasaruddin, F., Gani, A., Karim, A., Hashem, I.A.T., Siddiqa, A., and Yaqoob, I., 2017. Big IoT data analytics: Architecture, opportunities, and open research challenges. *IEEE Access*, 5, 5247–5261.
37. Satija, U., Ramkumar, B., and Manikandan, M.S., 2017. Real-time signal quality-aware ECG telemetry system for IoT-based health care monitoring. *IEEE Internet of Things Journal*, 4(3), 815–823.
38. Earley, S., 2015. Analytics, machine learning, and the Internet of Things. *IT Professional*, 17(1), 10–13.
39. Hijazi, S., Page, A., Kantarci, B., and Soyata, T., 2016. Machine learning in cardiac health monitoring and decision support. *Computer*, 49(11), 38–48.
40. Kumar, P.M. and Gandhi, U.D., 2018. A novel three-tier Internet of Things architecture with machine learning algorithm for early detection of heart diseases. *Computers & Electrical Engineering*, 65, 222–235.
41. Catherwood, P.A., Steele, D., Little, M., Mccomb, S., and McLaughlin, J., 2018. A community-based IoT personalized wireless healthcare solution trial. *IEEE Journal of Translational Engineering in Health and Medicine*, 6, 1–13.
42. Hooshmand, M., Zordan, D., Del Testa, D., Grisan, E., and Rossi, M., 2017. Boosting the battery life of wearables for health monitoring through the compression of biosignals. *IEEE Internet of Things Journal*, 4(5), 1647–1662.
43. www.kaggle.com/pitasr/industrialiot/discussion/35931.

5

Telemedicine Technology

Jayanta Mukhopadhyay
Indian Institute of Technology Kharagpur

CONTENTS

5.1 Introduction

Telemedicine is the remote delivery of healthcare services using telecommunication infrastructure, so that in-person visits of recipients of these services to the care providers are not required. The services could be in the form of providing clinical care, diagnosis, consultation of medical cases, etc. From the very beginning of wired and wireless communication technology, there were efforts for providing healthcare services from the distant end. In 1920s in USA, maritime sailors used to get clinical care from the shore through radio communication. Later in 1960s, NASA performed regular assessments of the health of astronauts in space from the earth and provided clinical advice whenever needed. In 1970s, paramedics in sparsely populated regions of Alaska and Canada used to get training and advice from city hospitals using satellite communication systems. Yet, for a long time, telemedicine services could not be established in large scale even in developed countries. There were several reasons for its very restricted use, such as high cost of technology, low reliability of functional components of data acquisition and telecommunication system, lack of skilled manpower for their daily operation and maintenance, etc. The scenario changed in the beginning of this century.

The digital revolution in late nineties of the last century made a quantum jump in removing the technological barriers mentioned above, and since then telemedicine services have been leaving more and more footprints around the world, not only in advanced countries [1–3], but also in developing countries like India [4], China [5], Thailand [6], etc. In 2015, the global telemedicine market was estimated at USD 17,878.7 million,[1] and since then it has been rapidly growing.

5.2 Core Technology

Digital revolution not only made the telecommunication technology reliable and affordable with increasing capacity of information exchange, but it also brought a new era of information technologies with the capability of growing in terms of data acquisition, storage, and processing, enabling them to be used as household commodities. Telemedicine is the result of convergence of all these different technologies, namely *medical instrumentation*, *information technology*, and *telecommunication technology*. In this section, we briefly discuss a few aspects and concepts related to these technologies.

5.2.1 Medical Instrumentation

Use of digital technology in medical instrumentation has increased the reliability and accuracy of diagnosis of several diseases. In many cases, it is not required to have the presence of a patient during the analysis of data acquired by these instruments. There are various medical imaging systems and modalities available for capturing radiological images of a patient such as *X-Ray*, *Computed Tomography* (*CT*), *Magnetic Resonance* (*MR*), *Positron Emission Tomography* (*PET*), etc. These images are available in the digital form through conversion from analog signal by *sampling* and *quantization*. Different types

[1] www.mordorintelligence.com/industry-reports/market-entry-telemedicine-industry-in-israel.

of biomedical signals, such as *electrocardiogram* (ECG), *electroencephalogram* (EEG), *pulse photoplethysmogram* (PPG), *electromyogram* (EMG), etc., are also converted to digital data. In addition to them, various other vital parameters, such as blood pressure, oxygen saturation, body temperature, body weight, etc., can be captured by digital gadgets and stored in the digital form. With these data, the associated patient's demographic and clinical information could also be stored and represented as an integrated unit, thus making the data processing and exchange of information much simpler. For exchanging such integrated medical information, we need to follow different international standards such as *HL7* (Health Level Seven International), *DICOM* (Digital Imaging and Communications in Medicine), etc.

In the context of telemedicine, sometimes it is required to transmit data from laboratories to distant places in real time, so that it is possible for an expert, such as a pathologist or a radiologist, to provide necessary advice and feedback readily by analyzing the data. In this case, the rate at which the data is generated during the acquisition becomes the determining factor in designing such a system. The rate of data generation is determined by two parameters of digitization, namely the *sampling rate* and the *level of quantization of samples*. The sampling rate is the number of samples to be sensed by the transducer per second. The quantization level is the proportionality factor for converting discrete numerics to values in units of physical parameters or measurements. Minimum sampling rate (or *sampling frequency*) depends on the *bandwidth* of a signal. From the *Nyquist theorem*, we can obtain the minimum *sampling frequency* for a band-limited signal, which is twice its bandwidth. We should note that all biomedical signals are band-limited signals. For example, consider a system of acquisition of digital ECG. The data may be generated at a rate ranging from 1.6 to 8 kbps (kilobits per second). The bandwidth of an ECG signal is approximately 100 Hz. Hence the minimum sampling rate for digitization is 200 samples per second. If the dynamic range of the amplitude of the signal is divided in 256 discrete levels, digital representation of the value of a sample requires 8 bits. From these two factors, we get the minimum data rate for acquiring an ECG signal of 1.6 kbps. If we consider simultaneous transmission of multiple-channel ECG signals, we can multiply this rate with the number of channels. For a 12 lead ECG, the minimum data rate is 19.2 kbps. The size of the file for storing these signals also depends upon this data rate and the required period of data acquisition. For example, to store a 12 lead ECG signal corresponding to three cardiac cycles, the file size could be on the average 6 kilo bytes (kB). The average heart rate of a human being is considered here as 72 cycles per minute. File size for an image would be quite high, if it is stored in the uncompressed or raw form. A color image of size 512×512 with 24 bits per pixel representation requires a data storage space of about 786 kB.

Videos require much larger data rate for real-time transfer of data. Typically, an echocardiogram video with a frame size of 480×640 pixels at a rate of 30 frames per second and 24 bits per pixel requires a data rate of about 221 Mbps for real-time transfer or more than 16.5 gigabytes (GB) of storage for a 10 min video. For real-time transmission, we need a very high speed communication link, whose technology is not only costly, but may also not be available always. Usually data compression is used in this scenario with an acceptable level of degradation in the visual quality. For example, for transmission of echocardiogram videos, Motion JPEG compression technique is approved by the DICOM committee, which may provide a compression ratio as high as 20:1 [7]. For 3D medical images from methods such as PET, CT, MR, etc., a much higher bandwidth is required. However, if the real-time transfer of data is not needed, the data may be transferred over a low-speed communication link for a longer period of transmission.

5.2.2 Information Technology

Information technology deals with representation, processing, storage, retrieval, communication, and presentation of data. High-end servers supported by back-end *relational database management systems* (*RDBMSs*) are being used for providing various healthcare solutions. Multimedia systems with advanced graphics hardware and high-resolution color display monitors are not only capable of displaying images and videos, but also are used for visualization of realistic images and animations enhancing human perception and understanding about a process or a phenomenon. There has been rapid advancement in the computing power of these systems as well as in computation techniques for processing multimedia data, which include text, audio, image, video, etc. Rapid advancement in web technology and data communication infrastructure enables these systems to exchange information and provide services across the globe. That is why this era of information technology is witnessing phenomenal growth in telemedicine services.

For exchanging multimedia data and information, it is desirable to follow an international standard or format of data representation, thus making the system integration simpler and effective. These standards specify the syntax and semantics of representation of data and its content. They also specify how the content should be delivered or visualized. For storage, transmission, and efficient representation, various standards for compressing data exist. These standards specify alternative description of the same content, specially required while compressing multimedia data. Typical examples of these standards are MPEG for audio; JPEG and JPEG-2000 for digital images; and MPEG-II, H.263, H.264, etc. for videos. These compression techniques could be *lossy* or *lossless*. In a lossy scheme, the decompressed data is not the exact copy of the original one, but it is very close to it. The degree of approximation determines the quality of decompressed data. In the lossless scheme, the recovered data exactly matches the original one. During telemedicine sessions, there are requirements for real-time transmission of audio and video for conferencing. They are compressed using lossy schemes such as MPEG, H.264, etc. But if digital images are required to be transmitted for diagnostic purposes, there should not be any loss of information. In such cases, lossless or semi-lossless (e.g., *JPEG-LS*) schemes are used. As lossy compression techniques provide higher compression compared to lossless techniques, under a resource-constrained environment, they may be used in providing services. For example, for transmission of data in a low-bandwidth channel, pragmatic uses of lossy compression schemes (such as JPEG) are observed. In Table 5.1, data rates required to transmit different compressed multimedia data are shown.

TABLE 5.1

Data Rates for Real-Time Transmission of Different Compressed Multimedia Data

Multimedia Data	Data Rates
Usual data	100 bps–2 kbps
Image	40–150 kbps
Voice	4–80 kbps
Stereo audio	125–700 kbps
VCR quality video	1.5–4 Mbps
3D medical images	6–120 Mbps
HDTV	110–800 Mbps
Scientific visualization	200–1,000 Mbps

5.2.3 Telecommunication Technology

Telecommunication technology involves the transmission of signal through a communication channel, which could be a coaxial cable, free space, optical fiber, etc. In transmission of digital data, it is required to convert or encode the data into analog signal or symbols. At the receiving end, the encoded analog signal is decoded in the form of digital data. To bring the bandwidth of the transmitting signal within the bandwidth of the communication channel, it is required to *modulate* the input signal. At the receiving end, the inverse operation, called *demodulation*, is carried out to reconstruct the signal in its original form. Usually, in a modulation technique, a carrier sinusoidal signal is used whose amplitude, shifts in frequency, or phase vary in proportion to the input signal, which is referred to as *amplitude, frequency*, and *phase* modulation, respectively.

The voice signal transmitted through analog telephony has a bandwidth of approximately 4 kHz. Communication links in this system also operate in the same range. That is why no modulation–demodulation is required in local exchanges of analog telephony. But, in digital telephony, it is required to convert digital data to analog symbols during transmission, requiring modulation of signal in the transmitting end and demodulation in the receiving end. For duplex communication, the device interfacing with the communication link is required to perform both the tasks, and it is called *modem* (*modulator and demodulator*). A modem acts as an interfacing device between a computer and a communication link, such as dial-up lines, satellite links, etc. while communicating data.

If the bandwidth of a communication channel is insufficient for accommodating the signal bandwidth, the receiving end gets a distorted signal, leading to loss of information. The bandwidth of a communication link is a key technical feature determining the quality of data transmission. For example, a real-time television standard video signal has a bandwidth of about 7.5 MHz. It is not possible to transmit this signal through usual telephone lines. However low-quality video transmission is possible through communication links with much lower bandwidth. In the context of digital transmission, bandwidth is also specified by an equivalent data rate of transmission, expressed in bits per second (bps). For example, using a modem in an ordinary telephone line, on the average a data rate of 30 kbps (maximum 57.6 kbps) could be supported. It is not possible to transmit real-time MPEG-I video at a rate of 1.54 Mbps through this link. On the other hand, a T-1 telephone link running at 1.54 Mbps could be used for transmission of such a video. With the advancement of technology, more and more data communication services offering higher data rates are made available at affordable cost. In this regard, optical communication links play a major role. They support data rate in ranges of gigabits per second (Gbps). A few examples of data communication services available from a service provider are given in Table 5.2.

Apart from wired links, wireless communication is also used for this purpose. With the rapid advancement of mobile communication technology, mobile telephone services are increasingly used for data communication. In Table 5.3, the data rates supported by different wireless technologies are summarized.

5.2.3.1 Computer Networking

Computer networking connects a group of computers for exchanging information. They communicate data following some protocols, understood by all the participating nodes. Here, a computer connected to a network is called a *node*. A node plays different roles in this network. For example, it may act as a terminal host to run the service of telemedicine, or it may perform similar tasks of a telephone exchange for establishing communication

TABLE 5.2

Different Communication Links

Links	Bandwidth
POTS	2.4–57.6 kbps
ISDN (BRI)	128 kbps
T-1 and fractional T-1	384 kbps–1.54 Mbps
DSL	128 kbps–9 Mbps
Primary rate ISDN	1.54 Mbps
T-2 leased lines	6.312 Mbps
T-3 leased lines	46 Mbps
T-4 leased lines	273 Mbps
Broadband ISDN	150–1,200 Mbps
SONET	51.84–4,976 Mbps

TABLE 5.3

Different Wireless Communication Services

Links	Bandwidth
GSM	9.6–43.3 kbps
GPRS	171.2 kbps
2G	64 kbps
3G	2 Mbps
4G	50–100 Mbps
Satellite	2.4 kbps–155 Mbps

links between two terminal hosts. For data communication, International Standard Organization (ISO) has identified seven layers of Open System Interconnection (OSI) protocols, namely *physical, data link, network, transport, session, presentation,* and *application* layers. Higher layers are those which appear later in this sequence. They take care of communication of data directly related to users, applications, processes, etc., while lower layers implement the protocols for basic signaling, channel sharing, and routing, being more integrated with network architecture and topology. There are various types of topologies for connecting computers. They may be connected through a grid of point-to-point connections, as it is done in a telephone network. They may share a common channel for transmission and reception, such as a common coaxial cable. This topology is referred to as *bus topology.* In a *ring topology,* they may form a ring, where each node is connected to two nodes. They may also be connected to a *switch* in a *star topology.* The switch establishes a link between two communicating nodes like a local loop of a telephone exchange. Different networking technologies are available depending upon the area of coverage. For example, for connecting computers in a campus, organization, etc., a *local area networking* (LAN) is used. On the other hand, *wide area networking* (WAN) connects computers situated in distant geographic locations. For LAN, different options are available on bus architecture (Ethernet and Token Bus protocols), ring architecture (Token Ring protocol), and tree architecture (nodes acting as switches, e.g., ATM cell switching, Gigabit Ethernet protocol, etc.). In Table 5.4, data rates supported by different protocols are listed.

Wide area networking (WAN) technologies are available as different services provided by telecommunication service providers. For example, in conventional telephone lines,

TABLE 5.4

Data Rates Supported by Different LAN Technologies

LAN Technology	Data Rate
Ethernet	10 Mbps
Token Ring	4–16 Mbps
Token Bus	16 Mbps
ATM	155 Mbps
Gigabit Ethernet	1,000 Mbps

services such as local loop subscription, integrated services digital network (ISDN), digital subscriber line (DSL), etc. are available. In some countries, there may exist special data communication infrastructure such as a ring architecture linking different service stations in a large area for supporting higher data rates. There are protocols for communicating data at a very high rate for running services in this kind of specialized infrastructure, e.g., switched multimegabit data services (SMDS), Frame Relay, Synchronous Time Division Multiplexing (STDM in SONET), etc.

5.3 Medical Informatics

Medical informatics primarily focusses on representation and computation on medical and health information. These are needed for design and development of a medical information system providing healthcare services and business. Different international standards have been evolved for exchange and management of information related to these areas. These standards are developed so that the systems designed by different organizations and running on different platforms could be made interoperable. Let us consider a few example cases. If a hospital requires to send information and a query regarding a patient to another referral hospital, the systems at both ends should have a protocol for exchange of information. In another situation, for getting payment from an insurance company, the hospital may require to send relevant information of their services and their charges to them. All these different types of information related to healthcare services should be sent in a format following a standard of data representation. For exchanging messages and queries among multiple information systems operating independently, there should be a common language and a common format for representing patients' data. Two such standards are very widely used in medical business world, namely HL7, which is a messaging protocol specifically developed to exchange health/medical/patient information between information systems, and the other standard, DICOM, which is used for representing medical images, waveforms, etc. and exchange of information in radiological imaging systems such as Picture Archiving and Communication System (PACS). There are also other types of health standards, which focus on requirements of a particular department or domain. For example, International Statistical Classification of Diseases and Related Health Problems (ICD9, and ICD10) are meant for identifying and classifying diseases. The standard Logical Observation Identifiers Names and Codes (LOINC) is used for denoting laboratory observations, health measurements, observations, and documents. Another standard Systematized Nomenclature of Medicine – Clinical Terms (SNOMED CT) targets at defining multilingual vocabulary of clinical terms.

5.3.1 HL7

Health Level Seven International (HL7), one of several ANSI-certified standards operating in the healthcare arena, aims at providing standards for the exchange, management, and integration of data related to healthcare services. It aims at facilitating clinical patient care and the management, delivery, and evaluation of healthcare services. HL7 focusses on the interface requirements of the entire healthcare organization. It tries to model different events that take place in healthcare services. Consider an event of a patient's admission to a hospital. The Patient Administration System (PAS) logs the details about demographics of the patient, and the admission, such as the name of admitting doctor, ward of admission, financial transactions, etc. Several other systems in the hospital, such as pathology laboratory information system or pharmacy system, will require these details at certain stages of care and services. Hence, this event in PAS may trigger forwarding of a message to each of the interested systems about the patient. Otherwise, those systems may wait until the patient needs their service. In any case, the PAS requires to communicate information about a patient to these systems. Before the introduction of HL7, developers of various information systems that needed to communicate had to discuss among themselves and work out a mutually acceptable common format for information exchange for developing interfaces. This is not only an expensive process, but it is also very time consuming. HL7 provides a common language for all health/medical information systems for exchanging and sharing of information. Developers need to make their systems HL7 compliant to connect to other systems through requests and responses to queries.

5.3.1.1 Messages

An information system usually creates and sends a HL7 message in response to an *event*. Examples of such events are patient admission, discharge, ordering diagnostic tests, reporting of test results, etc. It may also be a response to a query from another system. Each *message* depicts information about that event following a predefined syntax. Each HL7 message consists of a set of *segments*, which are its building blocks. Each segment is uniquely identified by a three-character code followed by a predefined format of specific *fields*. The end of a segment is denoted by the *end of line* (or *carriage return*) character. Every message begins with a *message header segment* (identified by the code 'MSH'). In a segment, pieces of related information are kept together. For example, the Patient Identification (PID) segment identifies a single patient, by the patient's ID number, name, address, and date of birth. The fields within a segment are separated by the | (*pipe*) character, and *components* within a field by the ^ character. The components may further be divided into *subcomponents* (with the & character). For specifying a sequence of data values within a field, the character ~ is used as a separator. These symbols of separators are defined in the message header of the segment. In HL7 messages, all data are represented by a selected set of displayable characters. The default is ASCII displayable character set.

Different types of messages are triggered by different HL7 events. Each message type defines a sequence of segments, providing all the required information regarding that event. In a message, there are some segments, which are mandatory, and some are optional. For example, in the event of 'Admission and Visit Notification' (ADT message), there has to be a minimum of four segments, such as Message Header (MSH), Event Type (EVN), PID, and Patient's Visit (PV1). But many other optional segments may exist there, describing information about a patient's next of kin, disability, allergy, previous diagnosis, etc. Similarly, in a segment, which is constructed as a sequence of data *fields*, some of these fields are mandatory. For an optional but empty field, field separator symbols are simply repeated. However, the sequence is terminated by the last non-empty field, without denoting subsequent stream of empty fields. For example in the PID segment, there could be as many as 30 fields. However, only two of them, namely patient's internal ID and patient's name, are minimally required for forming the segment. As these fields occur in third and fifth places, respectively, a valid PID segment could be as given below:

```
PID|||2-647009||Ghosh^Kallol
```

An example of HL7 message related to the event of patient's admission/transfer is shown below:

```
MSH|^~\&|BPHC||SDH|Reg|||ADT^A01|MSG00005|P|2.3 EVN|A01|201605061725
PID|||2-647009||Ghosh^Kallol||19610416|M|||24, Bataram
Lane^^Kharagpur^West Bengal^721302
  ||(3222)277-2345|(3222)277-2346||S
PV1||E|Emergency||||1234^Hossain^Sirajuddin|||SUR
```

The message could be interpreted in the following format:

1. **Admit/Visit Notification**
   1. Message Header
      a. From: BPHC
      b. To: SDH
   2. Event
      a. Date: 2016-05-06
      b. Time: 17:25
   3. Patient Identification
      a. Internal Patient ID Number: 2-647009
      b. Family Name: Ghosh
      c. Given Name: Kallol
      d. Birth Date: 1961-04-16
      e. Sex: M
      f. Street Address: 24, Bataram Lane
      g. City: Kharagpur
      h. State or Province: West Bengal
      i. Zip or Postal Code: 721302
      j. Phone Number Home: (3222)277-2345
      k. Phone Number Business: (3222)277-2346
      l. Marital Status: S
   4. Patient Visit
      a. Patient Class: E
      b. Point of Care: Emergency
      c. Attending Doctor's ID Number: 1234
      d. Family Name: Hossain

e. Given Name: Sirajuddin
f. Hospital Service: SUR

5.3.2 DICOM

The DICOM standard came out of the collaboration between American College of Radiology (ACR), American College of Cardiology (ACC), American Society of Echocardiography (ASE), European Society of Cardiology (ESC), American Society of Nuclear Cardiology (ASNC), and the National Electrical Manufacturer's Association (NEMA). Previously the standard was called ACR/NEMA. Later, with several revisions, it was renamed DICOM in 1991. In its present release, there are 21 parts in DICOM conformance statement covering various issues such as information object definitions, service class definitions, data structures and encoding, data elements dictionary, message exchange protocol, network communication support for message exchange, media storage and file format, grayscale standard display function, security and system management profiles, web services, application hosting, imaging reports using HL7 Clinical Document Architecture, transformations between DICOM and other representations, etc.

A DICOM file has a preamble of 128 bytes, a prefix of 4 bytes denoting 'DICM' as identification to DICOM standard, and a list of data elements. Each data element encodes related information about a patient or metadata information about the representation and organization of data in the file. There is a special data element called 'Transfer Syntax', which is used for describing how the fields in a data element are to be interpreted. A data element, may either be represented by *explicit value representation* or by *implicit value representation*. In the first case, data types of the values are explicitly specified. In this case, a data element has four fields, namely *tag* (4 bytes), *value representation* (VR of 2 bytes), *value length* (VL of 2 bytes), and *value field* (VF of an even number of bytes as mentioned in VL). For implicit value representation, the default data type, as defined in the standard, is assumed, and it is not described in the data element. Hence in this case, it requires three fields, namely tag (4 bytes), VL (4 bytes) and VF. A tag denotes the identity of a variable, whose value is stored in the element. The tag is encoded by a 2 bytes group code and a 2 bytes element code. For example, the variable, a *patient's name*, is under the group *patient*, and it is encoded as (0x0010, 0x0010). In this case, the group, patient, is identified by its *group code* (0x0010), and the variable, *patient's name*, is identified by the *element code* of that group (0x0010). A few examples of some other groups are shown in Table 5.5.

TABLE 5.5
Typical Groups in DICOM Standard

Group Name	Hex Encoding
File meta information	0x0002
General series information	0x0008
Patient information	0x0010
General study information	0x0020
Image information	0x0028
Image pixel data	0x7FE0
Waveform information	0x5400
Equipment information	0x003A

TABLE 5.6

Some Vital Tags for Rendering Images in DICOM Standard

Tag Description	Hex Encoding (Group, Element)
Transfer syntax tag	(0x0002, 0x0010)
Samples per pixel	(0x0028, 0x0002)
Number of frames	(0x0028, 0x0008)
Number of rows	(0x0028, 0x0010)
Number of columns	(0x0028, 0x0011)
Number of bits allocated	(0x0028, 0x0100)
Number of bits stored	(0x0028, 0x0101)
Pixel data	(0x7FE0, 0x0010)

A few examples of elements relevant for rendering an image in DICOM format are also listed in Table 5.6.

Different types of services are also specified in DICOM standard. They are called *service classes*. These services could cater to storage, query, and retrieval of data. It may be for printing images. There are service classes for managing different schedules. These are known as SOP classes, and each of them has a unique identifier.

5.3.3 RIS and PACS

The *Radiological Information System* (RIS) and the PACS are two major components of digital radiology [8]. The RIS is responsible for coordinating activities of the radiological investigations, such as scheduling examinations, capturing relevant clinical information, providing the list of examinations to PACS, postimaging interpretation and reporting, billing, etc. The PACS takes care of image acquisition, storage, distribution, and reporting. It consists of three basic components, namely, image acquisition, PACS core, and interpretation workstation. The modules in PACS core are database manager, image archive, image manager, and interfaces with RIS. The communication with and within PACS core takes place following the DICOM protocol, whereas communication with RIS involves HL7 messaging. Image manager controls the workflow of the PACS, such as interacting with RIS, managing storage and distribution of images, routing of images, responding to other modules of PACS, checking integrity of data, etc. In Figure 5.1, interactions among these systems are shown in a schematic diagram.

5.4 Telemedicine Systems

From the early 20th century, there have been various initiatives on running telemedicine services using the state-of-the-art telecommunication technology of those years. Many of these systems were built up to provide the proof of concepts and deployed under various pilot projects. Even today, various such pilot projects are being reported. However, in many hospitals and healthcare organizations, systematic and regular use of telemedicine services has been taking place.

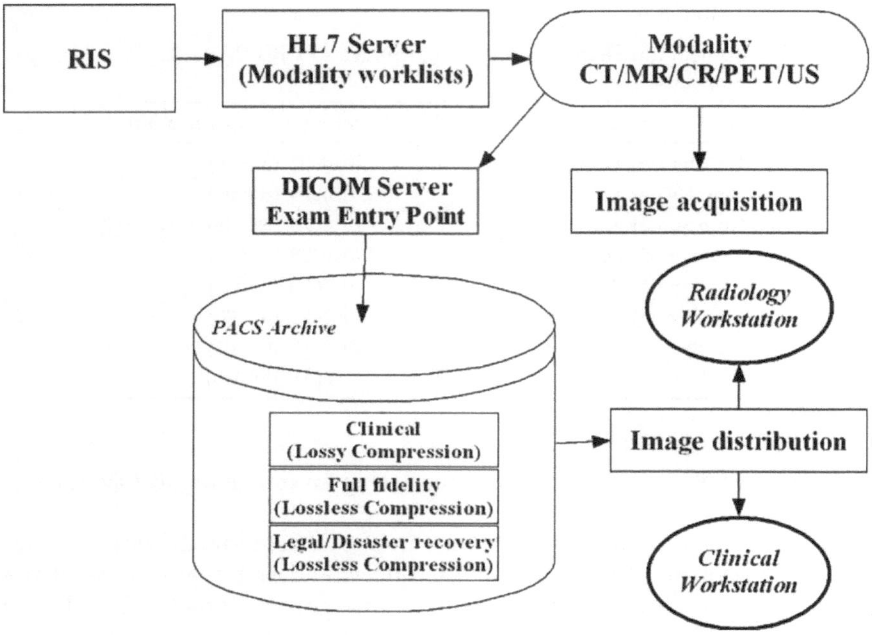

FIGURE 5.1
Integration of RIS and PACS.

5.4.1 Early Telemedicine Systems

One of the earliest reports of use of telecommunication for transmitting biomedical signal came from the work of the inventor of electrocardiography (ECG), Einthoven [9], in 1906. In his time, the cost of the galvanometer designed by him was much higher than telephone charges. That is why he used telephone lines for transmission of ECG signals. Even when medical instrumentation technology became relatively less expensive, in a country like USA, in 1920s, ECGs and EEGs were transmitted to distant places from the diagnostic laboratory using telephone lines. Some other types of services were also reported in those years. Maritime sailors used to get medical advice from offshore physicians using telegraphy and radio transmission of voices. There were also reports of various uses of telecommunication technology in USA for the purpose of medical consultation in subsequent decades. In 1950s, a state mental hospital and the Nebraska State Psychiatric Institution [10] were operating telepsychiatry using microwave communication links. During the same period, a joint telemedicine programme of NASA and US Public Health Services continued to serve the Papago Indian Reservation in the state of Arizona [10]. In 1960s, point-to-point conferencing between doctors used to take place using closed-circuit televisions. In 1970s, urban hospitals used to coordinate training and activities of paramedics in remote Alaskan and Canadian villages using ATS-6 satellite systems [11]. In Japan, pilot projects on telemedicine [2] were initiated as early as in 1970s using telephone lines and cable television system.

5.4.2 Telemedicine Systems at the Dawn of Internet Age

After 1990s, various developments and pilot deployment of telemedicine technology were reported due to the rapid advancement in digital technology and availability of this technology at a much lower cost in comparison to earlier systems. These are the years of preinternet or early internet era, when web technology was not yet matured. A few such examples are given below.

In Japan, a telemedicine network [12] was created for the purpose of education and training of medical professionals. The project started in 1994, when a system was built up by connecting National Cancer Centers in Tokyo and Chiba, separated by a distance of 30 km. The services were run over an optical leased line with 6 Mbps data rate. Later the services were catered through an upgraded link supporting 18 Mbps data transfer rate using ATM protocol. Additionally an experimental B-ISDN link with 156 Mbps data rate was used to enhance the quality of services. In later years, the network was expanded to include as many as 14 regional cancer centers using frame relay communication services [12]. The system was facilitated by TV conferencing. The participants could discuss sharing a still image of HDTV resolution. Regular conferences on various areas of healthcare, such as nursing, radiology, oncology, pathology, etc., were held using this network with an annual participation of more than 15,000 people.

In Lincolnshire, U.K., in 1996, telemedicine had been regularly used for handling accident and emergency cases [13]. In this system, specialists from a District General Hospital (DGH) at Boston were used to provide consultation to two Minor Injury Units (MIU) at Skegness and Johnson via an ISDN line supporting 128 kbps data rate. The system had a desktop videoconferencing unit on a Pentium PC and used the store and forward technology for image transfer. For digitizing patient's records, it used scanners, digital cameras, etc. The system mostly handled orthopedic emergency cases. Consultation and treatment of patients predominantly suffering from fractures, sprains, strains, and laceration were carried out, using online transmission of X-Ray images.

Demonstration of viability of telepathology using the store and forward technology [14] was reported from Korea. At the Samsung Medical Center, Seoul, pathological slides were captured by a microscope attached with a digital camera using a phototube adapter. These images were sent to two different places, namely Korea University Hospital, Seoul, Korea, and John Hunter Hospital, Newcastle, Australia. In addition, the slides were also sent to an independent pathologist. File sizes of images were reduced to a great extent by compressing them using the JPEG compression scheme at moderate quality. A high degree of concurrence of pathological reports with the independent reports by directly viewing the slides was reported in this experimentation. 95% concurrence in diagnosis was achieved with those from Korea University Hospital, and the reports from John Hunter Hospital concurred in 97% cases.

A pilot project on remote diagnosis of pediatric echocardiograms [15] using real-time transmission on an ISDN link was reported from the Duke University Medical Center, Durham, North Carolina. The center provided expert opinions to nine small centers equipped with a USG imaging system and a cardiologist (for adults) but without having any specialist for pediatric echocardiography. Distances of subcenters from the referral centers vary from 9 to 200 km (on the average 160 km). A videoconferencing unit was used for transmission of cardiograms at 15–18 frames per second. An independent study of the videotaped echocardiograms was carried out by a pediatric cardiologist during the period between January 1998 and January 2001. It was found that there was concurrence of reports in 383 cases out of 401. However, it was difficult to interpret colored Doppler images using the system.

In Taiwan [16], a telemedicine network was established for clinical consultation using T1 leased line and ISDN link. In this network, the National Taiwan University Hospital (NTUH) acted as the referral center, which was connected to several other remote sites, e.g., with Chinshan Health Station and National Chung Kung University Hospital. At NTUH, teleconferencing sessions were conducted. The participants could get access to a multimedia database. Images in different modalities such as CT, MR, and X-Ray represented in DICOM 3.0 standard were used in consultations. MPEG-compressed videos of USG, endoscopes, etc. were transmitted. The system was also capable of using the store and forward image transfer. A high degree of acceptance of the system was reported from the feedback from patients, consulting physicians, and technicians.

In 2001, at the Indian Institute of Technology, Kharagpur, a telemedicine system [17] was developed for the treatment of leprosy and other skin-related diseases. Later the system catered services to some other areas of medicine, such as teleradiology and hematology. The system was called TelemediK and designed for usage in a country like India, where basic healthcare facilities were poor and data communication infrastructure was also not developed during that period. The system was designed to run even on low-data-rate plain old telephone system (POTS) links. However, its features were scalable as and when higher data rate links were made available. It had a back-end MS-SQL RDBMS server and used the store and forward technology for performing consultation on patient records. The system provided interfaces for entering patient records, organizing them as a single unit, and transferring them to the remote end using the FTP. At the remote end, the experts used a browser to view the incoming patients' records and provided their opinions, which were again transmitted back to the nodal centers. It had also an online data conferencing module, which enabled consulting physicians to discuss and annotate on shared images. The conversation was carried out over a separate dedicated telephone line. The system could operate with low-cost peripheral equipment such as ordinary digital camera, scanner, webcams, etc. It was operational between the School of Tropical Medicine, Kolkata (which acted as the referral Institution) and Habra State General Hospital, 24 Parganas(N), and Cooch Bihar M J N Hospital, Cooch Bihar (which acted as nodal centers). About 1,500 patients were treated by this system for a period of 2 years.

5.4.3 Modern Day Telemedicine Systems

Due to the phenomenal growth of internet and rapid advancement of web and mobile technology, remote healthcare delivery has become a serious agenda to the health policy makers. This has led to the expansion of national health networks in various countries and also implementation of various telemedicine programs across the globe. A few examples are cited below.

In Denmark, due to the adoption of health informatics, interoperable health information systems connect all private doctors. All the hospitals in this country, connected by a central network, are using electronic medical records (EMRs), integrated ePrescriptions, RIS, and PACS [18].

In Israel, healthcare services are provided under insurance coverage through one of the four nationwide non-profit-making Health Maintenance Organizations (HMOs), namely, Clalit, Maccabi, Meuhedet, and Leumit. All the hospitals and HMOs use medical information systems [19] capable of sharing electronic medical records, information about healthcare services, etc.

In England, the Health and Social Care Network (HSCN)[2] is a broadband network that connects all health and care organizations of the English National Health Service (NHS).

[2] https://digital.nhs.uk/services/health-and-social-care-network.

It provides the infrastructure facilitating coordination of health and social care services through reliable exchange of medical information.

During this period, a good number of modern-day telemedicine systems are reportedly designed using internet protocols and web technology. Costs of these systems have significantly reduced, and they could be accessed by standard internet browsers from remote locations. With the development of cloud technology, cost of hosting such a system has further reduced. In addition, penetration of mobile technology in our society and availability of moderately high bandwidth wireless data services make the systems easy to access and easy to deploy. Different useful *apps* in mobile environment are increasingly being reported [20]. Many of these systems are designed and implemented targeting specialized areas of healthcare. A few typical initiatives during this period are discussed below.

In Brazil, healthcare services have been provided using the Santa Catarina State Integrated Telemedicine and Telehealth System (STT) [21] since 2005. The system was developed as a partnership between the Santa Catarina State Health Department (SES/SC) and the Federal University of Santa Catarina (UFSC). It uses a statewide network for healthcare services run by Brazilian universalized public healthcare system (SUS), which includes remote diagnosis on several specialties such as electrocardiography, dermatology, electroencephalography, radiology, etc. It was observed that there was a quantum jump in the utilization of the resources, when a PACS network was integrated with the system.

In the Department of Psychiatry, University of California, San Francisco (UCSF), USA, a program[3] has been initiated by the Young Adult and Family Center to help young adults and adolescents suffering from mental health using various internet-based technologies, videoconferencing, text messaging and chats, and social media platforms, such as Facebook.

In 2006, a door-to-door screening for patients suffering from diabetes in 42 villages in Southern India was carried out using a mobile van [22]. Further screening for complications, on the patients found to be suffering from the disease, was carried out by performing a series of tests on them. These included tests related to retinopathy using retinal photography and slit lamp, peripheral vascular diseases using Doppler ultrasound imaging, etc. The van was fully equipped with advanced equipment run by paramedical staff. Consultations for diabetic retinopathy and foot problems were carried out with a specialist in Chennai (capital of the Indian state Tamil Nadu) using the *very small aperture terminal* (VSAT) communication link of Indian Space Research Organization (ISRO). Retinal images and images of foot lesions were transmitted using this link. The other partner of the program was World Diabetes Foundation in Denmark. The program ran for 4 years, and it was reported that, during this period, only 2% of the patients needed to be sent to the hospital in Chennai.

In [23], implementation and testing of a system for teleophthalomogy has been reported. Medical videos of HDTV quality were transmitted to the ophthalmologists, whereas moderate- to low-quality videos were used in videoconferencing between a patient and a doctor. Requests for teleconsultation were sent a priori following a secured communication protocol between the hospital information systems at both ends. The system used private protocols based on TCP and UDP connections under OpenSSL frame work for exchange of information. The system was experimentally tested by running trials on 100 patients of Tan Tock Seng Hospital (TTSH, Singapore) with a clinic. Two ophthalmologists were

[3] http://psych.ucsf.edu/telemedicine-project.

involved in checking the patients in this trial. One of them consulted the patient in the clinic with the conventional face-to-face mode, and the other expert consulted the same patient with TeleOph independently. Based on their diagnosis reports, it was observed that both of them had agreement in all the cases.

An interesting application of telemedicine using wearable smart garment has been reported in [24]. The garment was fitted with devices monitoring ECG, respiration, and movement of a patient. It was used for the home monitoring of cardiac patients. These signals were stored and analyzed in a touchscreen computer with a dedicated software for data handling and a *universal mobile telecommunications system* (UMTS) dongle for data transmission, via email, to three cardiologists. In a pilot study, three patients participated daily in 3 min telemonitoring sessions for 30 days by using the platform. The system functioned reliably by transmitting good quality biomedical signals within this period. In another study, the same garment was used to evaluate the effect of high-altitude hypoxia on 30 healthy subjects living at a height between 3,500 and 5,400 m. In this case also, the study could be continued without much trouble. The subjects were found to be comfortable with the vest in most cases.

At the Indian Institute of Technology (IIT), Kharagpur, a web-based centralized telemedicine system [25] was developed in 2008, named iMediK, with Windows operating system with a back-end MS-SQL RDBMS. In addition to facilitating remote clinical consultations on different areas of medicine, the system had also a few specialized modules for treating pediatric HIV patients and patients suffering from drug-resistant tuberculosis. The system had interfaces with mobile devices and was operational in government hospitals of West Bengal for a period of 5 years. In 2011, iMediK became operational in government hospitals of Tripura, another northeastern state of India. Before using iMediK, these hospitals were using the peer-to-peer telemedicine system TelemediK2005, developed by the same Institute. In 2016, iMediK was replaced by a similar system, iMediX, which operates with Linux OS and mySQL as the back-end RDBMS. It is currently being used by more than 20 hospitals in Tripura with an annual patient turnover of more than 7,000.

In 2011, a system for managing neonatal patients in the NICU of IPGMER, Kolkata, had been installed [26]. This also used a web-based centralized server and performed teleconsulation with the special newborn care units (SNCU) of the hospitals of West Bengal, Chhattishgarh, Odisha, and Rajasthan in India.

5.5 iMediK: A Case Study

In this section, we discuss the design and implementation of iMediK, a web-based telemedicine system. The system implemented many of the features of TelemediK2005, designed previously for peer-to-peer teleconsultation. Further, as it was meant to be used in a public network, secure accessing of data was given prime importance while designing the system. It was also designed as a platform to host information systems of multiple hospitals which can operate independently, as well as refer patients to the other for teleconsultation. iMediK was primarily meant for managing outdoor patients and performing remote consultation on patient's data through offline and online data exchanges. It had a browser-based graphical user interface (GUI). It did not have any module for videoconferencing, which was assumed to be arranged independently over the network. Its major features are summarized below.

1. It was designed on a four-tier architecture for secured processing of a request through an internet browser.
2. It implemented the draft of Electronic Patient Record (EPR) standards proposed by National Task Force for Telemedicine Standards, Ministry of Communication and Information Technology, Govt. of India [27].
3. It had the provision for uploading and display of structured and unstructured text, multimedia, graphics, as well as vector data.
4. It incorporated functionalities for visualization and annotation of human profiles and medical images.
5. It provided an interface for performing data conferencing with electronic whiteboards for online consultation among multiple doctors.
6. It had specialized modules for treating HIV pediatric patients and for patients suffering from drug-resistant tuberculosis.
7. It included utilities for patient information backup and restoration.
8. It could generate and display patient statistics.
9. It had interfaces for entering and browsing patient records, annotating images by graphics and text, and providing medical advice from a mobile device.

5.5.1 System Architecture

A web-based telemedicine system should be secure, robust, flexible, and interoperable, so that it (a) protects the privacy, confidentiality, and integrity of the patient records from external as well as insider attacks; (b) accommodates a large number of simultaneous users without failing; (c) adapts easily to new changes, if required; and (d) is capable of exchanging information with other systems. In addition, it is desirable to keep the cost of development and maintenance low.

For making the system secure and reliable, a multi-tier application development architecture has been followed in the design [25]. In a typical multi-tier system, different segments of the application are implemented in separate layers. The modules in these layers communicate among themselves to serve a request generated from the client's end. In iMediK, there are four layers, namely the Database Layer, the Business Logic Layer, the Presentation Layer, and the Web Proxy Layer. A firewall is to be placed before the Presentation Layer, and the Web Proxy Layer accepts the request from the client and forwards it to the Presentation Layer for its processing. In Figure 5.2, a brief schematic diagram [25] of the four-tier system is shown. Descriptions of different layers are briefly presented below.

5.5.1.1 Database Layer

It is the lowermost layer. In the RDBMS, all the medical records of the patients as well as their personal details are stored. However, to restrict the access of a patient's identity by users other the attending doctors, the clinical records are kept separately from personal information. The medical information is stored in a structured form in a large number of relational tables of the database. There are different types of logins for different types of users in the database. The database can only be accessed from the Business Logic Layer. Depending upon the roles of users, appropriate logins are used to get access of the database.

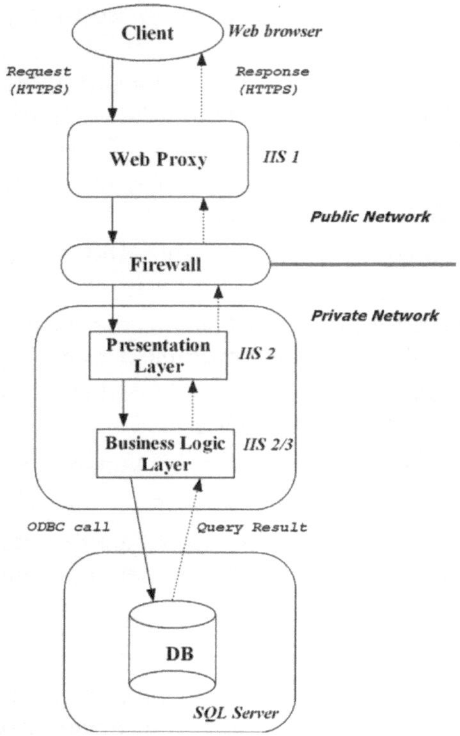

FIGURE 5.2
Four-layer architecture.

5.5.1.2 Business Logic Layer

The core of the application is the Business Logic Layer. It implements all the computational logics and transactions of the system. It is responsible for all communications with the database. In this layer, a set of objects or classes is implemented. They are invoked against any database operation or computation to be performed on the retrieved or submitted data. The Business Logic Layer (a) accepts all data requests from presentation layer, (b) verifies whether the current user is authorized to access the requested data, (c) determines the 'role' of the current user, and (d) then connects to the database using the database login of the specified 'role'. Once the data is retrieved from the database, this layer organizes the data in a format acceptable by the Presentation Layer and forwards it to that layer. In the same way, for storing data in the database, the request has to be processed by the Business Logic Layer.

5.5.1.3 Presentation Layer

In the Presentation Layer, the raw data is converted into user friendly HTML formats. This layer operates behind the firewall by accepting requests from the Web Proxy Layer, parsing it, and then forwarding it to the Business Logic Layer by invoking necessary modules in the Business Logic Layer to perform data insertion/retrieval operations. Once it gets the results of the processing from the Business Logic Layer, it generates response to the client according to its desirable format and forwards it to the Web Proxy Layer. The Presentation Layer is capable of handling flexibly different requirements of presentation format for generating responses. For example, for a mobile device, it generates HTML pages according to the display characteristics of the device.

5.5.1.4 Web Proxy Layer

This is the only portion of the system that operates in the Demilitarized Zone. It accepts all client requests and generates corresponding responses. This layer also takes part in managing the user sessions. When a client first connects to the system, two sessions are created for handling subsequent client requests. The first session, called the *external session*, identifies the connection between the client and the Web proxy, and the second session, termed *internal session*, denotes the connection between the Web Proxy and the Presentation Layer. The Web Proxy is required to map the external session with the corresponding internal session. Otherwise, every client request gets served by a new session leading to wastage of server resources. The Web Proxy Layer also translates the URL from its public identity to its internal representation as understood by the Presentation Layer, and subsequently the client request is transferred to the Presentation Layer.

Let us illustrate with an example the functioning of the telemedicine system by processing any external request through these layers. Suppose a doctor makes a request for viewing the list of patients under her treatment. The Web Proxy first processes the request by appropriate translation of URL and mapping sessions. Then the request is transferred to the Presentation Layer. In the Presentation Layer, the session status is verified, followed by invocation of the appropriate Business Logic Component for accessing the database. The Business Logic Layer determines the permission and the 'role' of the user, then connects to the database using the 'role' of the user, and processes the query in the database. After retrieving the list of patients as a result of database transaction, the Business Logic Layer returns the same to the Presentation Layer. The Presentation Layer organizes the output in the HTML format and sends the response to the Web Proxy, which in turn forwards the response to the client.

5.5.2 Implementation

Microsoft.Net Framework was used in the implementation of the system [28]. For the back-end database, Microsoft SQL server was used. All other layers were hosted in Windows server using Internet Information Services (IIS). The Web Proxy application was written in C#, handling all HTTP requests to the web site. It was a custom implementation of the HttpHandler class in .Net platform implementing several submodules, such as UrlManager, SessionMappingManager, CookieManager, AppLogManager, FormsManager, and ResponseToClientManager.

In the Presentation Layer, there were a set of HTML, ASP, and ASP.Net pages. The Business Logic Layer had .Net Remoting components, hosted in the IIS server. There were interfaces with these components in the Presentation Layer, and all communications between them were SOAP formatted. For a secured access by a client, the system was hosted by an HTTPS server.

5.5.3 The Data

There were different types of data used in the system, and they could be broadly categorized into five classes:

1. **Data related to a patient's personal information**: This category includes the information on a patient's name, address, sex, age, etc. The patient is also given a unique identity number (called as *Pat_Id*) by the system. Other than the attending doctors, the patient is identified only by this *Pat_Id* to other users of the system.
2. **Data related to a patient's medical information**: Medical information and health records of a patient are captured in various forms, such as ordinary text (e.g., prescriptions), structured text (e.g., test reports), digital images (e.g., X-Rays), graphics primitives (e.g., polygons, lines, etc.), graphs (e.g., ECGs), and video clips (e.g., ultrasonograms).

   Together with personal information, these records form the EPR of a patient. In iMedIK, the then existing standard of EPR as recommended by a standard committee set up by the Govt. of India was followed. These data were captured during patient's registration and consultation with the local physician. Prescriptions and advice were generated by a doctor locally or from a referral hospital.
3. **Data for patient management in telemedicine**: For maintaining the current list of patients waiting for telemedicine, the information about these patients were stored in a queue called *Tele_Pat_Queue*. There was also a queue of local patients requiring local consultation. This queue was referred to as *L_Pat_Queue*. Once a patient was served in a queue, she was deleted from it. The system administrators were also given the power to add and delete any patient in the patient queues as and when required.
4. **Data related to the physicians**: The physicians' personal information and unique identification keys were also used in the system.
5. **Data for system management**: For system management, the list of users at different levels had to be maintained. Their passwords also had to be managed by the system. Besides maintaining the valid users' list, the system also maintained the log files for data transfer, online telemedicine sessions, etc.

5.5.4 Data Conferencing

Apart from offline data uploading and downloading in the system, there were provisions for online chatting and data conferencing with multiple participants. During data conferencing, the participants would communicate by sending text and graphics primitives. The interface provides a common canvas for conferencing with graphic primitives (i.e., by drawing, annotating, etc.) of a patient's data. The canvas could be filled with various background images, such as white background or whiteboard, a medical test image, special anatomical images such as human profiles, etc. It is possible to do conferencing with multiple canvases. The interface implemented operations such as drawing lines, contours, circles, ellipses, text annotations, etc. Some of the image processing operations, such as zooming of a portion of image, edge extraction, enhancement, etc., could also be carried out using these modules. Annotations and drawing of different participants were identified by distinct colors of text and graphics.

5.5.5 Summarized Display of Medical Information

The system could generate a single-page summarized display of patient records. Following the links in the page, a doctor could get the details about different heads of the patient records. This report was generated following a user-defined XML template. Depending on requirements for a disease, this template could be designed. In Figure 5.3, a typical example of a summarized report generated for a patient suffering from pediatric HIV is shown.

5.5.6 Specialized Modules

iMediK had specialized modules for treating patients suffering from pediatric HIV [29] and drug resistant tuberculosis. There were special interfaces for data entry and visualization of data in charts and graphs. A few examples are shown in Figure 5.4. There were decision support systems for prescribing drugs following WHO guidelines (Figure 5.5). These modules also had specialized interfaces for mobile devices.

5.5.7 Interfaces with Mobile Devices

iMediK developed interfaces with mobile devices [30]. The responses from the Business Logic Layer were formatted in appropriate resolution and sizes for displaying in mobile devices in the Presentation Layer. The standard internet browser of the mobile device was able to display those pages within its displayable zone. It was convenient for a viewer to view those documents without repeated scrolling of those pages. For displaying images with higher resolution, interactive zooming mechanism was also in place. A few such examples of displays of various types of medical information in a mobile device are shown in Figure 5.6.

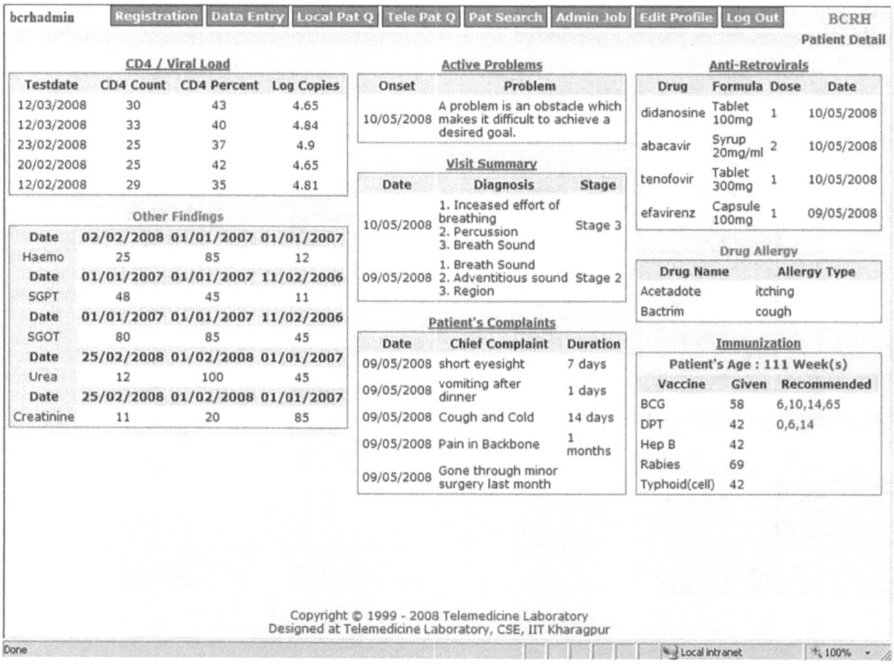
bcrhadmin | Registration | Data Entry | Local Pat Q | Tele Pat Q | Pat Search | Admin Job | Edit Profile | Log Out | BCRH

Patient Detail

CD4 / Viral Load

Testdate	CD4 Count	CD4 Percent	Log Copies
12/03/2008	30	43	4.65
12/03/2008	33	40	4.84
23/02/2008	25	37	4.9
20/02/2008	25	42	4.65
12/02/2008	29	35	4.81

Other Findings

Date	02/02/2008	01/01/2007	01/01/2007
Haemo	25	85	12
Date	01/01/2007	01/01/2007	11/02/2006
SGPT	48	45	11
Date	01/01/2007	01/01/2007	11/02/2006
SGOT	80	85	45
Date	25/02/2008	01/02/2008	01/01/2007
Urea	12	100	45
Date	25/02/2008	01/02/2008	01/01/2007
Creatinine	11	20	85

Active Problems

Onset	Problem
10/05/2008	A problem is an obstacle which makes it difficult to achieve a desired goal.

Visit Summary

Date	Diagnosis	Stage
10/05/2008	1. Inceased effort of breathing 2. Percussion 3. Breath Sound	Stage 3
09/05/2008	1. Breath Sound 2. Adventitious sound 3. Region	Stage 2

Patient's Complaints

Date	Chief Complaint	Duration
09/05/2008	short eyesight	7 days
09/05/2008	vomiting after dinner	1 days
09/05/2008	Cough and Cold	14 days
09/05/2008	Pain in Backbone	1 months
09/05/2008	Gone through minor surgery last month	

Anti-Retrovirals

Drug	Formula	Dose	Date
didanosine	Tablet 100mg	1	10/05/2008
abacavir	Syrup 20mg/ml	2	10/05/2008
tenofovir	Tablet 300mg	1	10/05/2008
efavirenz	Capsule 100mg	1	09/05/2008

Drug Allergy

Drug Name	Allergy Type
Acetadote	itching
Bactrim	cough

Immunization

Patient's Age : 111 Week(s)

Vaccine	Given	Recommended
BCG	58	6,10,14,65
DPT	42	0,6,14
Hep B	42	
Rabies	69	
Typhoid(cell)	42	

Copyright © 1999 - 2008 Telemedicine Laboratory
Designed at Telemedicine Laboratory, CSE, IIT Kharagpur

Done | Local intranet | 100%

FIGURE 5.3
Summarized report generated for a dummy pediatric HIV patient.

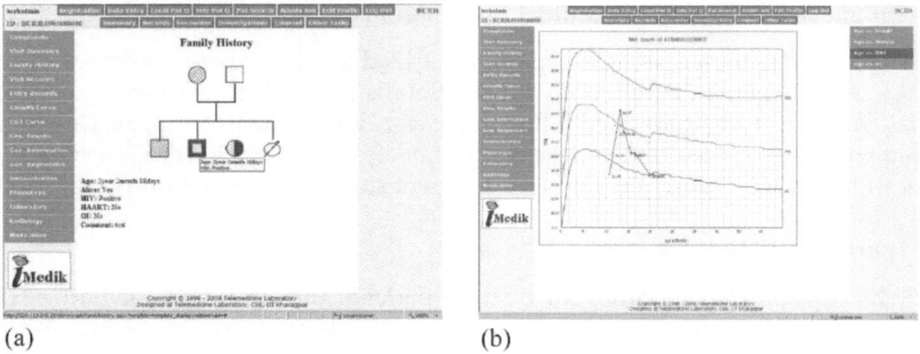

FIGURE 5.4
(a) Chart showing history of HIV infection in a family and (b) growth chart of a pediatric patient.

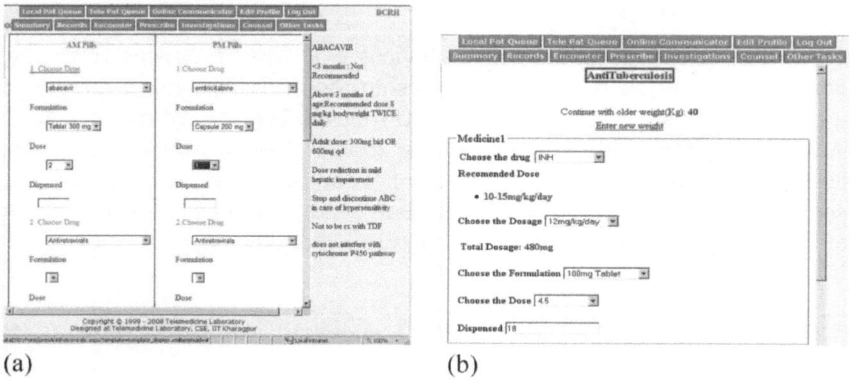

FIGURE 5.5
Prescribing drugs and doses for (a) pediatric HIV patients, and (b) drug-resistant tuberculosis.

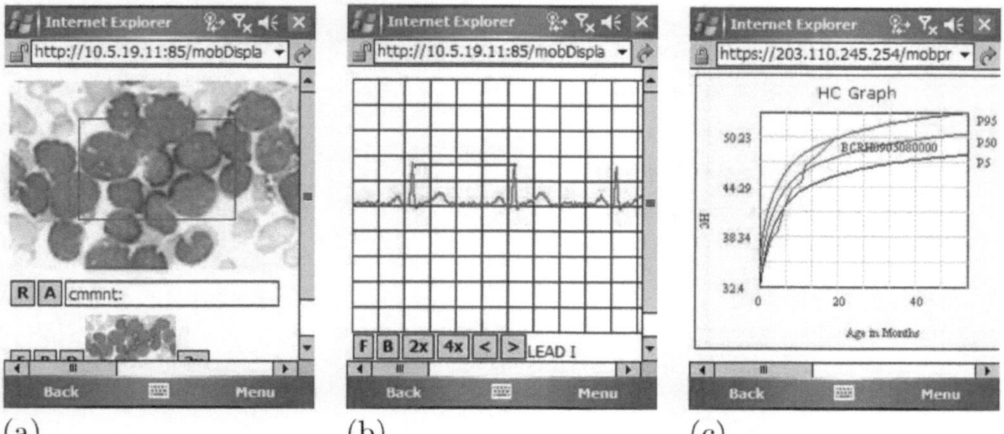

FIGURE 5.6
(a) Zoomed portion of an image of a pathology slide, (b) display of an ECG waveform, and (c) growth chart of a patient.

5.6 Distributed Telemedicine Systems

Centralized server model (such as iMediK) is cost effective and requires less communication overhead for exchanging information, compared to the peer-to-peer communication between a group of servers. But there are several drawbacks of using a centralized server. For any interruption in the connectivity to the central server, the activities of a telemedicine center would get stalled. Similarly, the failure of a single server would affect all the hospitals subscribed to it. To address these issues, various models of distributed telemedicine services have been considered. These models attempt to take advantages of both the peer-to-peer and centralized data exchanges.

In [31], three different models of distributed telemedicine have been proposed, namely, (a) two-tier model, (b) hybrid model, and (c) hierarchical model. These models are built to serve the patients locally and also to get referral advice from remote ends. Their objectives and topological and architectural construction are briefly discussed here.

5.6.1 Two-Tier Server Model

In this model, there are two tiers of servers. In the periphery, the lower tier consists of servers called the peripheral servers (PSs) that serves patients locally. Only when a referral of a patient is needed, data transfer takes place to the central server called the main server (MS). The MS acts as the communication hub of PSs.

5.6.2 Hybrid Server Model

This hybrid model keeps open the option for having an independent server in the periphery or not. The hospital may function only with the centralized server. In that case, any interruption of connectivity with the central server would affect the local services. But the hospital has the advantage of running with low resources.

5.6.3 Hierarchical Server Model

Hierarchical server model takes note of the fact that healthcare services are usually run through a hierarchy of primary, secondary, and tertiary care. At each level, a hospital functions independently, but a specific routing mechanism is in place for referring a patient to a higher level of care. In this model, several strategies have been proposed to minimize the physical transfer of patient records from one server to the other through proper delegation of rights to view records to authenticated users.

5.7 Privacy and Confidentiality

In telemedicine services, maintaining privacy and confidentiality of a patient is of great importance. As medical records contain various pieces of sensitive information, sufficient measures should be taken to avoid unauthorized access of data. The measures should be made transparent to the users, so that the system and services are trusted by both the patients and service providers. In USA, healthcare service providers and technology developers need to follow the principles and guidelines set by the Health Insurance Portability and Accountability Act (HIPAA) of 1996. The Govt. of India is considering passing a similar kind of law in the form of Digital Information Security in Healthcare Act (DISHA). However, these laws regulate the activities of hospitals and clinicians while acquiring patient's medical information and sharing them for consultation. There is no regulatory agency to check the use of patient's data by the technology solution providers, when their systems directly acquire data from a patient with a routine consent from them. For example, routine transmission of patient's medical data from an app or a medical device may be shared with an advertiser sponsoring the app [32]. Presently, it depends upon the technological companies how the information will be used and to what extent it would be disclosed to a patient. There is no such legal limit. There is always a risk of compromising patient's privacy in such cases.

In telemedicine system, sufficient care should be taken, so that even if an unauthorized person or process gets access to the medical records, the data should appear meaningless. Encryption of data at different levels can enforce this property. The data may be stored in encrypted form in the database. It needs encryption during transmission, and finally the recipient and the sender may use an end-to-end encryption–decryption scheme. The first two tasks are taken care of by the database system and the communication protocol followed by the browser and server (e.g., HTTPS). But the third measure may be required to be implemented in the telemedicine system.

5.8 Conclusion

The present trend in delivery of healthcare is to make it synonymous to home care. Under this context, telemedicine technology is becoming more and more relevant and has huge potential to grow. The emergence of health gadgets based on IoT (Internet of Things), expansion of cloud computing infrastructure, and large-scale penetration of smartphones empowered by powerful processors and graphics engines have brought another paradigm shift in the design and development of this technology. Next-generation telemedicine technology would be more focussed on a specific domain of healthcare and oriented to cater to needs related to personal health. On the other hand, more and more hospitals would be connected, and the systems would become interoperable following the international standards. This would also eliminate the boundary between a hospital management information system (HMIS) and a telemedicine system.

References

1. R.S.H. Istepanian. Telemedicine in the United Kingdom: Current status and future prospects, *IEEE Transactions on Information Technology in Biomedicine*. 3(2):158–159, 1999.
2. T. Takahashi. The present and future of telemedicine in Japan, *International Journal of Medical Informatics*. 61:131–137, 2001.

3. H.S. Chen, F.R. Guo, C.Y. Chen, J.H. Chen, and T.S. Kuo. Review of telemedicine projects in Taiwan, *International Journal of Medical Informatics*. 61:117–129, 2001.
4. B. Jankharia. Current status and history of teleradiology in India, *International Journal of Medical Informatics*. 61:163–166, 2001.
5. R.K.C. Hsieh, N.M. Hjelm, J.C.K. Lee, and J.W. Aldis. Telemedicine in china, *International Journal of Medical Informatics*. 61:139–146, 2001.
6. N. Kasitipradith. The ministry of public health telemedcine network of Thailand, *International Journal of Medical Informatics*, 61:113–116, 2001.
7. J.D. Thomas, D.B. Adams, S. DeVries, D. Ehler, N. Greenberg, M. Garcia, L. Ginzton, J. Gorcsan, A.S. Katz, A. Keller, and B. Khandheria. Guidelines and recommendations for digital echocardiography, *Journal of the American Society of Echocardiography*. 18(3):287–297, 2005.
8. K.J. Dreyer, D.S. Hirschorn, J.H. Thrall, and A. Mehta (Eds.), *PACS: A Guide to the Digital Revolution*, 2nd Ed., Springer, New York, 2006.
9. W. Einthoven. Het telecardiogram, *Ned T Geneesk*. 50:1517–1547, 1906.
10. T.L. Huston and J.L. Huston. Is telemedicine a practical reality? *Communications of the ACM*. 43(6):9–13, 2000.
11. C.W. Dohner, T.J. Cullen, and E.A. Zinster. ATS-6 satellite evaluation: The final report of the communications satellite demonstration in the WAMI Decentralized Medical Education Program at the University of Washington Prepared for Lister Hill National Center of Biomedical Communication, Seattle, University of Washington, 1975.
12. H. Mizushima, E. Uchiyama, H. Nagata, Y. Matsuno, R. Sekiguchi, H. Ohmatsu, F. Hojo, T. Shimoda, F. Wakao, T. Shinkai, and N. Yamaguchi. Japanese experience of telemedicine in oncology, *International Journal of Medical Informatics*. 61:207–215, 2001.
13. M. Beach, P. Miller, and I. Goodall. Evaluating telemedicine in an accident and emergency setting, *Computer Methods and Programs in Biomedicine*. 64:215–223, 2001.
14. E.S. Lee, I.S. Kim, J.S. Choi, B.W. Yeom, H.K. Kim, G.H. Ahn, and A.S.Y. Leong. Practical telepathology using a digital camera and the internet, *Telemedicine Journal and e-Health*. 8(2):159–165, 2002.
15. A.S. Milazzo Jr, J.R. Herlong, J.S. Li, S.P. Sanders, M. Barrington, and A.R. Bengur. Real time transmission of pediatric echocardiograms using a single isdn line, *Computers in Biology and Medicine*. 32:379–388, 2002.
16. C.C. Lin, H.S. Chen, C.Y. Chen, and S.M. Hou. Implementation and evaluation of a multifunctional telemedicine system in NTUH, *International Journal of Medical Informatics*. 61:175–187, 2001.
17. J. Mukherjee, A.K. Majumdar, A. Banerjee, B. Acharya, A. Nayak, and U.V. Reddy. Telemedicine for leprosy, *IETE Technical Review*, 18(4):243–252, 2001.
18. M. Birkemose. Evaluation of an innovative telemedicine project: Learning from the development and implementation process, Master's Thesis, Department of Clinical Medicine the Faculty of Health Sciences, The Arctic University of Norway, 2015.
19. B. Rosen, R. Waitzberg, and S. Merkur. Israel health system review, *Health Systems in Transition*, 17(6): 1-212, 2015.
20. L. Mertz. mHealth to the rescue: Growing use of wireless and mobile technologies improves community health, even in rural areas, *IEEE Pulse*. 7(6):16–24, 2016.
21. A. Savaris, A.A.G.M. Filho, R.R.P. de Mello, and G.B. Colonetti. Integrating a PACS network to a statewide telemedicine system, In *IEEE 30th International Symposium on Computer-Based Medical Systems* (*CBMS*), pp. 356–387, Greece, 2017.
22. V. Mohan and R. Pradeepa. Telemedicine in diabetes care, *IEEE Pulse*. 5(3):22–25, 2014.
23. H. Yao, Y. Wu, Z. Wei, Z. Zhao, L.H. Ngoh, R.H. Deng, and S. Yu. Teleoph: A secure real-time teleophthalmology system, *IEEE Transactions on Information Technology in Biomedicine*. 14(5):1259–1266, 2010.
24. M.D. Rienzo, P. Meriggi, F. Rizzo, P. Castiglioni, C. Lombardi, M. Ferratini, and G. Parati. Textile technology for the vital signs monitoring in telemedicine and extreme environments, *IEEE Transactions on Information Technology in Biomedicine*. 14(3):711–717, 2010.

25. A.K. Maji, A. Mukhoty, A. K. Majumdar, J. Mukhopadhyay, S. Sural, S. Paul, and B. Majumdar. Security analysis and implementation of web-based telemedicine services with a four-tier architecture, In *2nd International Conference on Pervasive Computing Technologies for Healthcare*, pp. 46–54, Finland, 30th January–1st February, 2008.
26. S. Ray, D.P. Dogra, S. Bhattacharya, B. Saha, A. Biswas, A.K. Majumdar, J. Mukherjee, B. Majumdar, A.K. Singh, A. Paria, S. Mukherjee, and S.D. Bhattacharya. A web enabled health information system for the neonatal intensive care unit (NICU), In *IEEE World Congress on Services*, pp. 451–458, Washington, DC, USA, 4–9 July, 2011.
27. DIT Technology Working Group for Telemedicine Standardization. Ministry of Information Technology, Government of India, 2003.
28. A. Maji. Vulnerability Analysis of a Multi-tier Architecture for Web-based Services with Application to Telehealthcare, M.S. Thesis, Department of Computer Science and Engineering, Indian Institute of Technology, Kharagpur, 2009.
29. S. Paul, S. Das Bhattacharya, D. Patra, A.K. Majumdar, J. Mukhopadhyay, B. Majumdar, and A. Sudar. A web-based electronic health care system for the treatment of pediatric HIV, In *11th International Conference on e-Health Networking, Applications and Services (Healthcom)*, pp. 175–180, Australia, 16–18 December, 2009.
30. S. Paul, J. Mukhopadhyay, A.K. Majumdar, B. Majumdar, and S. Das Bhattacharya. Methodology to visualize electronic health record for chronic diseases on small display screens, In *Proceedings of the International Conference on Advances in Computing, Communications and Informatics*, pp. 505–510, India, 3–5 August, 2012.
31. D. Patra, S. Ray, J. Mukhopadhyay, B. Majumdar, and A.K. Majumdar. Achieving e-health care in a distributed EHR system, In *11th International Conference on e-Health Networking, Applications and Services (Healthcom)*, pp. 101–107, Australia, 16–18 December, 2009.
32. J. L. Hall and D. McGraw. For telehealth to succeed, privacy and security risks must be identified and addressed, *Health Affairs*. 33(2):216–221, 2014.

6

Biomedical Signal Analysis[1]

Simone Benatti, Victor Kartsch, and Fabio Montagna
University of Bologna

Elisabetta Farella and Velu P. Kumaravel
Fondazione Bruno Kessler

CONTENTS

[1] Authors having contributed equally to this chapter.

6.1 Introduction

Traditionally, the monitoring of physiological signals has been conducted in labs and at medical premises. Reliable data used by clinicians to track our health status or smarten artificial intelligence methods help identifying important events and providing immediate feedback. Bringing such physiological signal monitoring practice to the final user in daily life has great potential to improve our quality of life. Furthermore, all these data can be used offline to deepen our understanding of the human body and its behavior. The traditional way of collecting biomedical signals (like electroencephalogram [EEG], exploratory data analysis [EDA], and electromyography [EMG]) using electrodes connected to a biomedical equipment through wire cables can be uncomfortable for the users. The current acquisition systems, such as EEG, EDA, EMG, and electrocardiogram (ECG), are typically fixed, wired and cumbersome. Consequently, to enable ambulatory monitoring, they must be substituted with smart, wireless, comfortable and wearable solutions. Some progress has been accomplished in this field [1].

This chapter introduces some of the typical signals used to monitor physiological information from the human body. We start with a brief overview of EEG, EMG, and EDA by presenting their nature and how they can be collected. Then, we provide an overview of common feature extraction techniques to process the acquired raw data, both in the time and frequency domain.

Finally, we describe some popular machine learning algorithms used for activity and gesture recognition with biosignals. Concerning, the described acquisition system and interface, we mainly introduce acquisition interfaces that enable the development of wearable devices, considering low-cost energy-efficient platforms. We choose computational methods that can be embedded on state-of-the-art advanced but low-power microcontrollers.

6.2 EMG Signal

6.2.1 Signal Description

The EMG signal measures the electrical activation of the muscular fibers. Similar to the nerves, the muscle tissue conducts electrical potentials, and these electrical potentials are called muscle action potentials (APs). Along the nerve cell membranes, the APs are generated by the passage of Na^+ ions and K^+ ions. These reactions are the electrical messages

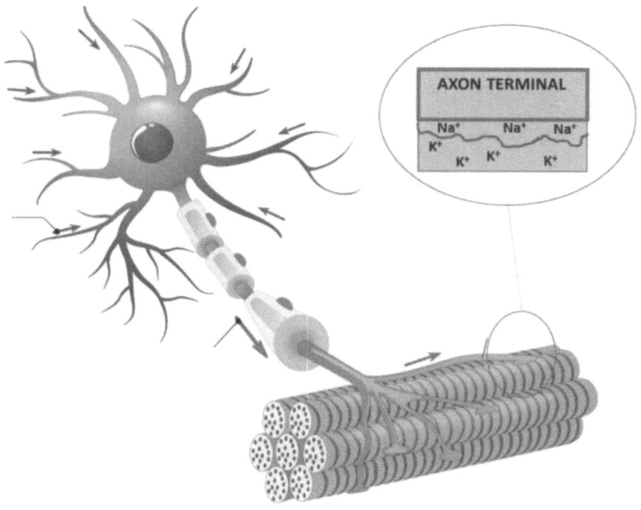

FIGURE 6.1
EMG and muscular contraction.

sent by the brain to the muscles to start contractions. The Na^+ ions flow through the cellular membrane via the Na channels. Due to the influx of Na^+ ions, the cell membrane depolarizes and the nerve impulses propagates toward the target muscle cells (Figure 6.1). The Na^+ ions' flow into the nerve cell is an essential step in the transmission of the AP to the nerve fibers and along the axons, and it causes a release of the Ca^{++} ions, causing cross-bridge binding and muscle sarcomere contraction [2].

The EMG signal, acquired by electrodes placed on the skin surface, is composed of all the AP of the cells underlying the electrodes that are aligned parallel to the muscle fibers. The surface EMG sensors consist of two conductive plates, and each one is connected to the inputs of a differential amplifier that can sense the AP of the muscular cells.

The maximum amplitude of such signal is 20 mV (−10 to +10), depending on several factors, such as the diameter of the muscle fiber, the distance between the active muscle fiber, and the detection site and the properties of the electrode. Even if the maximum bandwidth of the amplifier does not exceed 2 kHz, this signal is also very noisy and difficult to manage. The main causes for this are the noises caused by motion artifacts, fiber crosstalk, electrical equipment, and the floating ground of the human body, not referred to a solid ground potential [3,4].

6.2.2 Signal Acquisition

The typical EMG acquisition is based on active (high-end devices with integrated conditioning circuitry) or passive (low-cost passive patches that needs external conditioning circuitry) sensors (Figure 6.2).

Active sensors provide a high-quality preamplified EMG signal suitable for direct interface with microcontrollers. For instance, Ottobock sensors [5] represent the commercial solution for EMG acquisition for high-end prosthetics, both in research and industrial applications. These sensors perform a full-analog signal conditioning based on a bandpass discrete filter, an instrumentation amplifier with a high gain stage, and an offset cancellation feedback circuit that requires the use of a dedicated metal plate as the reference electrode

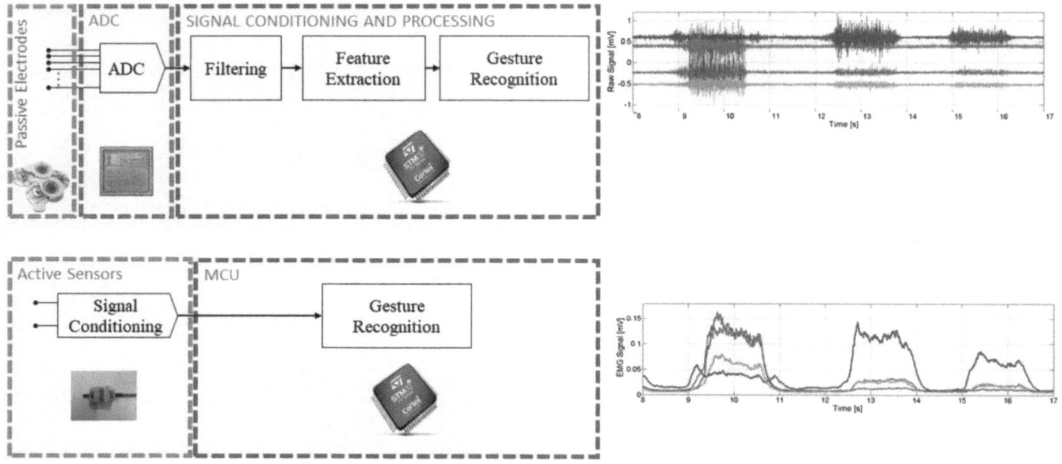

FIGURE 6.2
Acquisition setup for passive (high) and passive (low) EMG sensors. In the passive sensor setup, the ADC directly acquires the signal, while the filtering and the feature extraction are made in the digital domain. In the active electrode setup, filtering and feature extraction are made directly by the active sensor, while the digital platform performs only the last part of the processing [25].

for each sensor. In an Ottobock sensor, the EMG signal is amplified and integrated to reach an output span of 0–3.3 V, ideal for the single-ended stage of an embedded microcontroller analog to digital convertor (ADC). The bandwidth of the Ottobock sensor is 90 Hz to 450 Hz with a further Notch filter for the 50 Hz. This is because the sensors for the classification of the gestures do not need extensive frequency information but a clear low noise signal.

Passive sensors are conductive plates (Ag-AgCl or Au), which can be connected to the conditioning circuitry, such as a biopotential analog front end (AFE) or an operational amplifier with a high-common-mode rejection ratio (>90 dB) and a high-input impedance (>1 GΩ) amplifier. With passive sensors, the signal quality depends on the conditioning circuitry. It is possible to acquire raw EMG data with a frequency up to 20 kHz while the signal conditioning or filtering is performed in the digital domain.

6.3 EEG Signal

6.3.1 Signal Description

EEG measures the electric brain activity caused by electric currents that flow within the neurons of the brain. Neurons are the cells that constitute the brain tissues and their structure is depicted in Figure 6.3. Despite the apparent simplicity of the neural cell structure, the biophysics of neural current flow relies on complex models of ionic current generation and conduction. When a neuron is excited by other neurons through the burst of APs, excitatory postsynaptic potentials (EPSPs) are generated at its haptic dendritic branches. The dendritic membrane becomes depolarized and extracellularly electronegative with respect to the cell body. This potential difference causes a current, called primary current, to flow from the non-excited membrane of the body to the dendritic tree, sustaining the

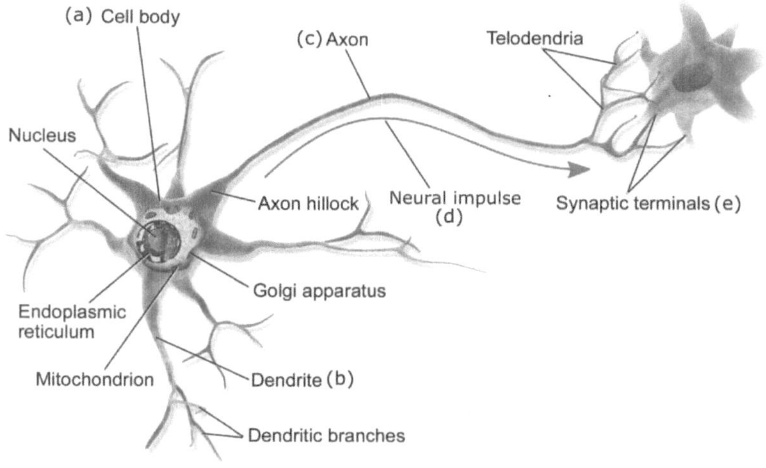

FIGURE 6.3
Structure of a neuron. The cell body (a) contains the cell nucleus and acts as the cell's life support center. The cell body gathers and aggregates the signals arriving from other cells though dendrites (b). The axon (c) allows the neuron to spread the signal away from the cell body to other neurons. The information is constituted by a neural impulse (d), flowing from the cell body to the peripheral synaptic terminals (e), which in turn communicate with another neuron.

EPSPs [6]. The principle of conservation of electric charges imposes that the flow is looped with extracellular currents, called secondary current. Both, the primary and secondary current contribute to generate the magnetic field measured by the electrodes on the scalp; however, the spatial cell arrangement is critical for the superposition neural currents to produce measurable fields. In fact, the measured EEG signal is the result the EPSPs of thousands of synchronously activated neurons because of the coherent distribution of their large dendritic trunks that are locally oriented in parallel and perpendicularly pointed to the cortical surface [6]. The signal must cross several layers before reaching the electrodes placed on the scalp, especially the skull, which attenuates the signal approximately 100 times more than the soft layers [7]. The noise within the brain and over the scalp contributes in the lowering of the signal-to-noise ratio (SNR); therefore, only a large amount of active neurons can generate enough potential to be recorded by the scalp electrodes [7].

6.3.2 Signal Acquisition

The EEG signal acquisition is an easy and non-invasive operation, which allows a great spread of this technique over others (magnetoencephalography, functional magnetic resonance imaging, and Electrocorticography). In fact, to acquire the EEG signal, it is sufficient to measure a set of electric potential differences between pairs of electrodes attached on the scalp. However, detecting EEG activity is not a trivial task, as the sensors and circuitry must cope with non-stable, skin-electrode interface, as well as with an intrinsically high-noise signal. Apart from brain activity, which are unrelated to the steady-state visually evoked potentials (SSVEPs), the additional sources of noise can come from the acquisition system like electrical noise and external interference. The most common source of EEG-signal degradation is the finite contact impedance at the interface between the electrode and the skin. A high value of contact impedance leads to a potential divider effect at the amplifier input.

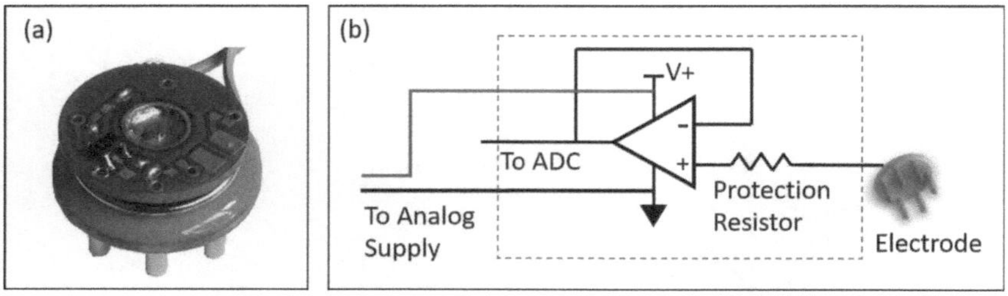

FIGURE 6.4
(a) Picture of the g.SAHARA dry electrode with custom amplifier PCB. (b) Electrical schematics of the custom amplifier PCB.

This causes a reduction of the capability to reject common-mode noise, such as that from mains. It increases the noise generated at the metal-skin interface and augments the effect of interference, coupling through capacitive effects to the cables or the artifacts because of cable movement and the microphone and piezoelectric effect. The contact impedance is minimized in clinical EEG protocols by removing the superficial skin layers by abrasion and inserting a conductive gel or paste in-between the two surfaces. Obviously, the skin preparation is unsuitable for non-clinical settings, where the system setup needs to be as quick and easy as possible for an untrained person, and the associated risk of infection is unacceptable.

To minimize the setup time and allow self-positioning of the system, zero-preparation electrodes can be adopted as an interface between the system and subject. The two options available are dry and wet electrodes. The dry electrodes are recognized as the best option for zero-preparation time. However, they present contact impedance up to three orders of magnitude higher than wet electrodes with skin preparation; hence, to mitigate such high contact impedance, an amplification stage is directly placed on the electrode.

Figure 6.4a and b respectively show a picture and the schematic of the active sensor custom printed circuit board (PCB) designed in Ref. [8]. As the single-ended amplification stages with a gain higher than one reduce the rejection of common mode noise, only a signal buffering is performed on the active electrode by a low-power, low-noise, and rail-to-rail operational amplifier, connected as a unity-gain buffer. Protection resistors with 68 kΩ are used to limit patient auxiliary current in cases of single fault condition below the applicable limit of 50 μA. The operational amplifier is an AD8603 from analog devices, which has a quiescent current of 50 μA and a low-voltage noise (2.3 μV peak-to-peak in the 0.1–10 Hz band and 25 nV / $\sqrt{\text{Hz}}$ at 1 kHz). While the total input capacitance is below 5 pF, the input leakage current is below 1 pA at room temperature, which translates into an input impedance in excess of 500 MΩ in the EEG band.

6.4 Electrodermal Activity

6.4.1 Signal Description

The changes in the electrical properties of the skin, that is, the skin conductance, are commonly referred to as electrodermal activity. Since the second half of 19th century, when psychological factors related to electrodermal phenomena were first discovered, electrodermal recording has become a huge topic of interest in the field of psychophysiology.

Psychophysiology is a multidisciplinary field (neurology, neuroscience, psychology, and physiology), which is concerned with the mind, brain, and body interaction. Typically, it relies on several kinds of physiological signals, such as the heart rate variability; blood pressure; respiration rate and volume; facial EMG; gastrointestinal activity; pupil diameter; levels of certain hormones, such as cortisol; and measurements related to skin proprieties, that is, electrodermal activity (EDA) and skin temperature.

To understand the principle behind the EDA measurements, it is important to know the physiological aspects of the skin. The skin is a complex layered structure that contains several sensory receptors, hair follicles, blood vessels, and sweat glands. The density of the sweat glands varies with the location of the skin, and they are under the control of the sympathetic nervous system (SNS). During stressful situations, the SNS elicits the sweating activity, thus, increasing the skin conductance. More details and descriptions of various skin layers can be found in Ref. [9].

Sweat is a relatively good conductor; it is equivalent to 0.3% NaCl salt solution. As the sweat ducts from underlying cells penetrate the corneum, that is, the superficial layer of epidermis, these sweat-filled ducts produce many low-resistance parallel pathways. The skin conductance varies depending on the present and past eccrine activity, which is caused both by the need to regulate body temperature and SNS. The electrodermal measurement techniques include the following:

a) Endosomatic recording without the application of external current to the skin.
b) Exosomatic recording with direct current in which a small constant direct current voltage (say, 0.5 V) is applied to the skin surface, using silver-silver chloride electrodes and a sodium or potassium chloride electrolyte solution (the most commonly used technique).
c) Exosomatic recording with alternating current.

To measure non-thermoregulatory electrodermal phenomena, it is recommended to place the active electrodes in the smooth and hairless skin regions, such as the palms and soles, of the non-dominant hand. Especially in case of ambulatory applications, fingers are also preferred. Usually no pretreatment of electrode sites is needed. However, prior to the attachment of electrodes, the sites may be washed with lukewarm water. Washing them with soap is not recommended as this causes swelling of the epidermis.

The electrolyte solution should contain a chloride salt like NaCl with molar concentrations in the range of 0.050 to 0.075. The pregelled commercial ECG electrodes are unsuitable for EDA recording, as they contain higher salt concentrations, which can be altered by the sweat at the interface of electrodes. In general, dry electrodes are not recommended as the increased contact of metal electrodes with the skin causes a slow humidity buildup, resulting in a drift toward increased conductance [10].

Any EDA signal can be broadly classified into two components: tonic and phasic. Usually, the tonic component is a slow-changing component, which is denoted by the skin conductance level. It is related to the general level of autonomic arousal. The phasic component, also known as the skin conductance response (SCR), consists of a rapid increase in skin conductance often associated with distinct stimuli, such as a deep breath, followed by a gradual decrease in skin conductance, returning to the baseline. In literature, SCR is sometimes also termed as electrodermal response. The SCRs can be further classified into two types, namely, event-related SCR (associated with stimuli) and non-specific SCR (responses unassociated to any stimuli).

The internal variables that could affect the EDA measurements include disruptions of the skin-electrode interface (mechanical pressures on the electrodes, loose contact, and/or changes of the skin caused by the electrolyte solution) and artifacts, such as body movements, speech, and irregular breathing. In addition, there are wide individual differences in both tonic and phasic EDA related to demographic variables, such as age, gender, and ethnicity [10]. The external variables that influence electrodermal recording include the environmental variables, such as temperature, humidity, and ambient noise. Long-term ambulatory recording of skin conductance outside of a temperature-controlled laboratory has also found EDA measures to be positively correlated with changes in ambient temperature (in Ref. [7] recordings duration was about 7h continuously; in Ref. [11] EDA was recorded for up to 28h continuously). Medication is also considered an artifact that significantly influences the EDA values, especially the psychiatric medicines, such as anti-depressants.

6.4.2 Signal Acquisition

Biosignals have very specific requirements, and many projects end up heavily bounded by high-cost hardware materials. Physiological computing poses different challenges related to biosignal acquisition like the need for high SNR or greater accuracy in the sampling rate. We explore EDA signal acquisition using BITalino [12], a commercial platform that is a good compromise between different requirements, such as matching a low-cost, form-factored support for easy prototyping and obtaining sufficient signal quality for a wearable EDA platform.

BITalino is a low-cost modular biosensor kit. It consists of different independent functional blocks, such as sensor chips (EEG, ECG, EDA, EMG, accelerometer, and light sensor); a Bluetooth chip; a rechargeable battery supply (500mA 3.7 lithium polymer); a power management unit, which acts as a voltage regulator and provides information on battery status; an 8-bit Atmega microcontroller; and input–output units, such as buttons and LEDs.

The EDA sensor (Figure 6.5a) has an inbuilt analog circuitry to improve the SNR. The bandwidth of the filter is designed to pass frequencies between 0 and 2.8Hz. So, a sampling rate of 10Hz is used. The operating voltage is 3.3V, and the range of skin conductance that can be measured is between 0 and 25 μS.

The transfer function that translates the ADC units to the equivalent skin conductance value is as follows:

$$\text{EDA } (\mu\text{S}) = \frac{\left(\frac{\text{ADC}}{2^n}\right)\cdot \text{VCC}}{0.132} \tag{6.1}$$

where EDA is the skin conductance value in microsiemens; ADC is the value sampled from the input channel; n is the ADC (number of bits) resolution; and VCC is the regulated voltage from the power supply.

Dry electrodes are used to overcome the disadvantages of gel electrodes, such as lack of reusability, longevity, and comfort. In dry electrodes, the metal electrodes are directly placed on the skin surface. Since no gels are used, they establish a good contact after perspiration, where the sweat acts as the electrolyte at the skin-electrode interface. A dry electrode has similar characteristics to a polarizable electrode, considered as a leaky capacitor. This imposes the need for analog conditioning circuitry to have a very high input impedance in the range of GΩ. Moreover, the circuitry should be placed very close to the electrode to prevent an electromagnetic interference. BITalino is interfaced with

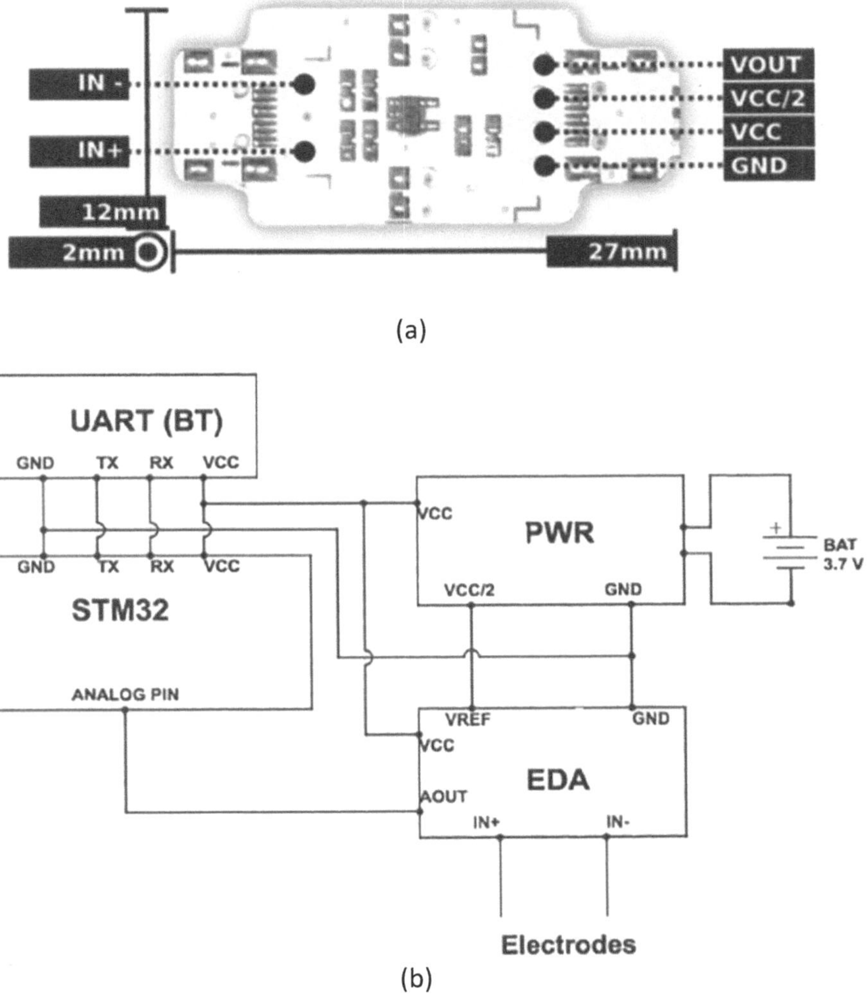

FIGURE 6.5
(a) BITalino EDA sensors, pins and dimensions. (b) Hardware building blocks of the acquisition setup.

a Cortex-M4 microcontroller (i.e., STM32F4 integrating the ADS1298 AFE). Figure 6.5b shows the hardware acquisition building blocks.

STM32 is programmed to receive the analog signal from the BITalino EDA sensor unit. One of the 12-bit ADCs of the microcontroller is configured to sample the data at a predefined interval (in this case, 100 ms). The sampled ADC bits are then transferred to the personal computer (PC) by configuring the universal asynchronous receiver/transmitter communication protocol. The baud rate used is 921,600 kbps. The EDA sensor is supplied with a reference voltage of VCC/2 (i.e., 1.65 V) by the power management unit. The EDA being a slowly changing signal and the 0–2.8 Hz bandpass filter used in the sensor warranted a low sampling rate of 10 Hz. Furthermore, with such a sampling rate, the motion artifacts are almost avoided during the acquisition. However, it is necessary to use a moving average filter to avoid the false identification of the cognitive event. To perform an acquisition test,

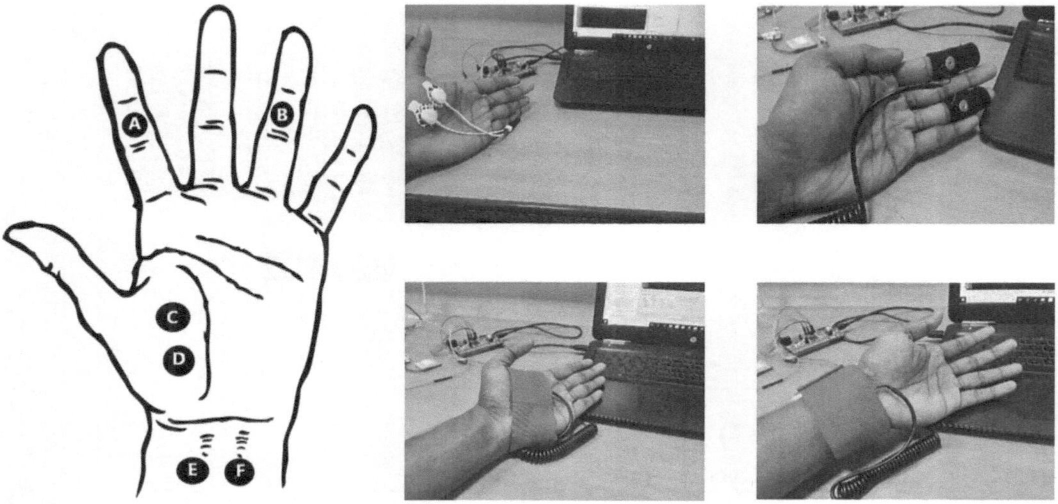

FIGURE 6.6
Electrodes placements for the EDA acquisition - scheme and real: A, B = dry electrodes for the fingers; C, D = both dry and gel electrodes for the palm; E, F = dry electrodes for the wrist.

some shrewdness must be applied. Since the EDA signals are strongly dependent on the environment, it is better to maintain a stable room temperature. We acquired the signal with both gel and dry electrodes on the palm, and only with dry electrodes in the other position (Figure 6.6). Furthermore, to obtain the subject's EDA reaction and repeatable experiments, we applied the following rules. Firstly, the subjects were asked to limit as much physical movements as possible to relax completely and sit in a posture they were most comfortable with. Secondly, the subject's signals were recorded for 60 s while they relaxed to find the baseline of each subject. Then, they were asked to perform the Valsalva maneuver [7,13,14] for about 10 s. The total experiment was limited to 120 s, so the remaining 50s were used to further relax. The Valsalva maneuver was chosen as it stimulates the SNS [15]. We repeated the protocol before and after asking the subjects to wash the areas where the electrodes were applied. Figure 6.7 shows the signal before and after the Valsalva maneuver, performed at 60 s in the plot. The rise time indicates the number of seconds it takes to reach the SCR peak, while the decay time is during which the signal falls to 50% of the peak amplitude. Figure 6.8 shows the signal acquired from different positions with the described setup. The raw EDA signal is smoothed using an exponential smoothing function.

From our results, it has been demonstrated that palm is the most sensitive part of the body to acquire a strong EDA signal. Signal captured from fingers also proved to sufficiently reflect responses to psychological stimuli. The EDA signals acquired from the wrist were the most inconsistent with a corresponding low amplitude (in the range of μS) and therefore difficult to distinguish from noise. Furthermore, according to the literature and our observed acquisitions, each subject has a different skin-conductance baseline. In addition, signal amplitude changes after washing hands, that is, skin conductance decreases after washing hands. All these factors determine high signal variability among subjects and in time. This along with the difficulty of preserving adequate conditions to minimize noise due to temperature changes and body movements makes the EDA an unreliable signal to be used independently to evaluate psychophysiology conditions of a subject, and thus it is generally used in combination with EEG, ECG, etc. [11].

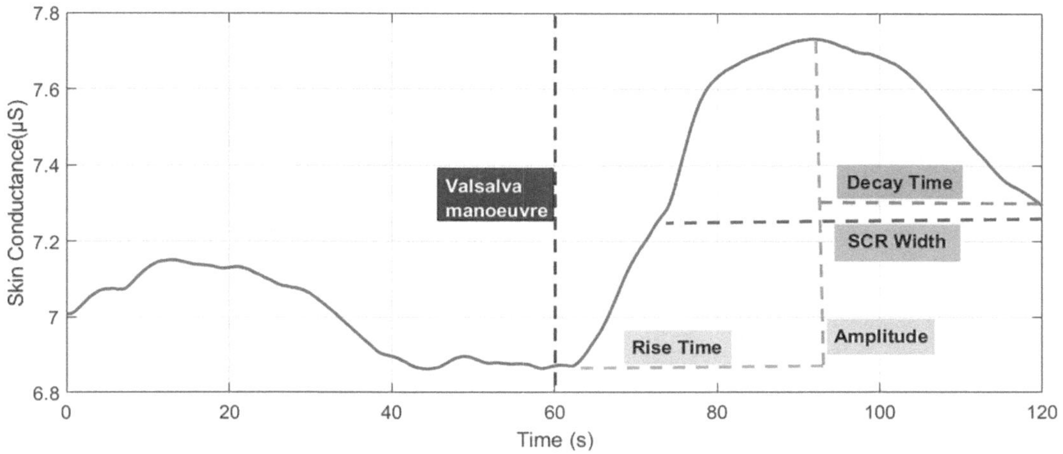

FIGURE 6.7
Characteristics of the acquired EDA Signal.

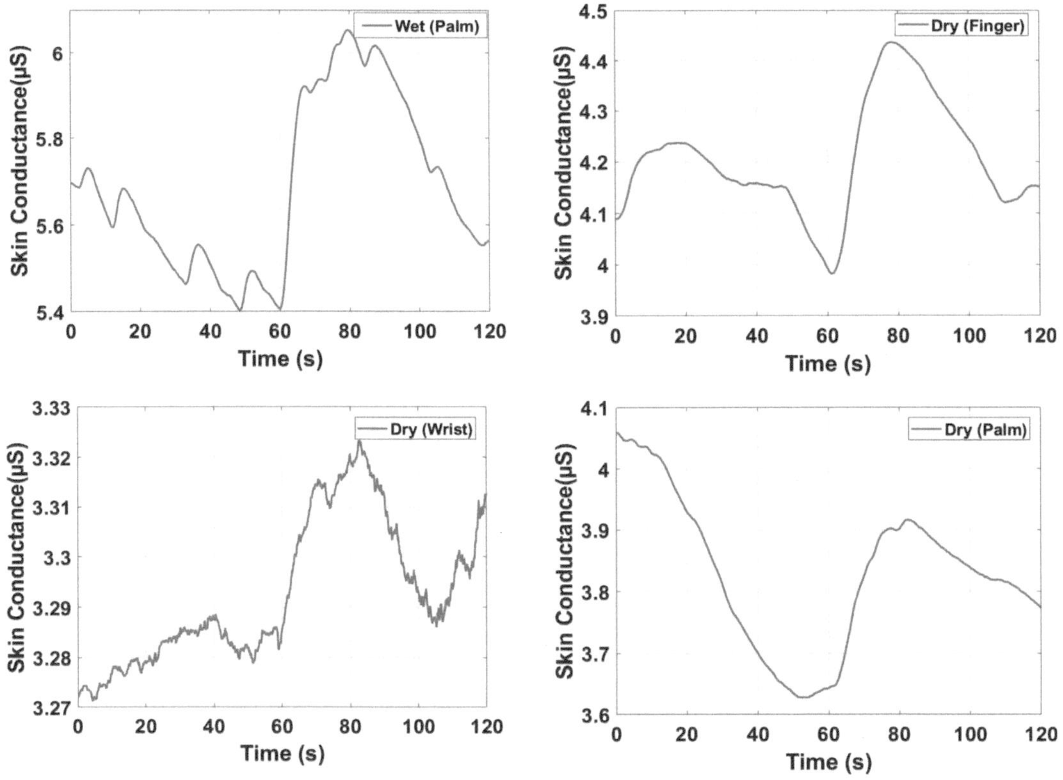

FIGURE 6.8
EDA Signal acquired at different positions using different electrodes - from top left to bottom right: wet electrodes on palm, dry electrodes on fingers, dry electrodes on wrist, dry electrodes on palm.

6.5 Signal Analysis

Biosignals contain essential qualitative and quantitative information about a given event that, in most cases, is not visible to the naked eye. In most cases, this is due to, among numerous reasons, the superposition of several signals from different sources that do not allow recognizing the significant patterns of a given event. Using several mathematical transformation and filtering techniques, it is possible to extract only relevant information. These, so-called features of the signal, can be roughly divided into three different categories, namely, time, frequency, and time frequency and are employed depending on the nature of the phenomena that is studied. In this section, we will discuss the extraction of these features while delineating general guidelines for their use.

6.5.1 Time-Domain Analysis

A signal in correspondence to a given event could be retrieved in the form of consecutive time-series raw samples that indicate the evolution of the phenomena. This signal might also include information about other events out of interest and noise. Thus, it is essential to transform the signals in different domains to better represent and separate the clinically significant components. These features are time-locked, mostly with some latency, with respect to the original event. In the following section, we discuss the most common feature extraction mechanisms used for biosignal processing (EMG and EEG) and in Figure 6.9 we show some examples of the extracted features.

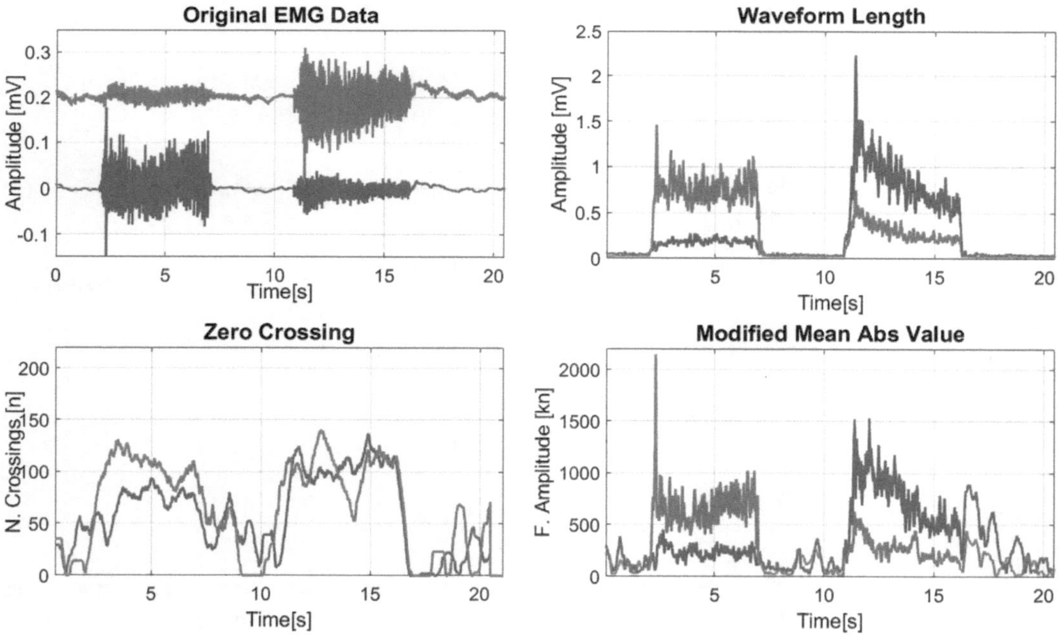

FIGURE 6.9
Example of the feature extraction output from an EMG signal. The original signal contains information about two channels with two different consecutive contractions.

6.5.1.1 Integrated EMG (IEMG)

The IEMG is calculated by summing the absolute values associated with the amplitude of the time series discrete samples of the EMG signal. This feature is regularly employed to detect the muscular activation onset [16,17]. It is evaluated as follows:

$$\mathrm{IEMG}_k = \sum_{n=1}^{m} |x_n| \tag{6.2}$$

where k is the single output from the evaluation window, m is the evaluation window size, and X_n is a single point in the evaluation window.

6.5.1.2 Mean Absolute Value

The mean absolute value of the time series samples is mostly used to recognize the contraction level EMG signals. The data can also be multiplied by a window function to provide smoothness between consecutive windows, producing the so-called modified mean absolute value (MMAV) [16,17]:

$$\mathrm{MMAV}_k = \sum_{n=1}^{m} W_n \, |X_n| \tag{6.3}$$

where W_n is the window function.

6.5.1.3 Variance (VAR)

The VAR includes information on how the samples vary from mean, thereby providing information on the signal spread. It can be calculated as follows:

$$\mathrm{VAR}_k = \frac{1}{m-1} \sum_{n=1}^{m} X_n^2 \tag{6.4}$$

6.5.1.4 Root Mean Square (RMS)

Traditionally used on EMG signals, RMS provides information on the signal envelope [18]. It helps in presenting quantitative information about the muscle activation patterns, converting the highly variable components of the signal to a stable and smooth set of points used to separate complex gesture sets. The RMS is expressed as follows:

$$\mathrm{RMS}_k = \frac{1}{m} \sqrt{\sum_{n=1}^{m} X_n^2} \tag{6.5}$$

6.5.1.5 Waveform Length (WL)

WL is used to extract the cumulative length of the waveform along the time [15,17]. It is related to wave width, frequency, and time and extracted using the following equation:

$$\mathrm{WL}_k = \sum_{n=1}^{m-1} |X_{n+1} - X_n| \tag{6.6}$$

It can be applied in EMG gesture recognition and EEG classification.

6.5.1.6 Zero Crossing (ZC)

ZC is the total number of times the signal crosses a fixed value (zero of the signal). This value is set upon evaluating the output signal, and its purpose is to remove background noise from the window. A starting point can be set at the mean value of the evaluation window. It is calculated as:

$$ZC_k = \sum_{n=1}^{m-1} \operatorname{sgn}(X_n \quad X_{n+1}) \cap |X_n \quad X_{n+1}| \geq \text{threshold} \tag{6.7}$$

where sgn(x) is equal to one if the signal is above the threshold and zero otherwise.

6.5.1.7 Willison Amplitude (WAMP)

WAMP represents the number of times the difference between two (or more) consecutive samples exceeds a certain threshold [20], and it is expressed as follows:

$$\text{WAMP}_k = \sum_{n=1}^{N-1} f(|X_{n+1} - X_n|) \tag{6.8}$$

where $f(x)$ is equal to one if $x \geq$ threshold and zero otherwise. The WAMP provides information about the firing of the motor unit APs on EMG signals.

With a few exceptions, the features presented up to now are more applicable to EMG signals. In the following section, we introduce the frequency-domain features that are more recurrent in EEG signals. When possible, a time-domain feature might be selected since the complexity of the calculation could be less computationally expensive. Nevertheless, this is only possible when the signal of interest has a larger SNR than other components.

6.5.2 Frequency-Domain Analysis

These features are based on the spectral content of the signal. This process can itself constitute feature extraction and, indeed, has been used largely in EEG [19,20]. Nevertheless, further information can also be extracted after this transformation. Figure 6.10 shows a few examples of the extracted features.

6.5.2.1 Fourier Transform

The Fourier transform allows signal conversion from the time domain to the frequency domain. When working with discrete signals, the discrete Fourier transform (DFT) offers the only workable solution that can provide computationally efficient calculations deployable using low-power microcontrollers. The DFT is denoted by:

$$X_k = \sum_{n=1}^{m-1} X_n \; e^{\frac{i2n}{m}kn} \tag{6.9}$$

Computing the spectrum directly from the definition is not practical. Instead, the preferred method employs the fast Fourier transform, reducing the complexity of the computation from $O(m^2)$ to $O(m \log m)$. This modification allows the computation of even large windows

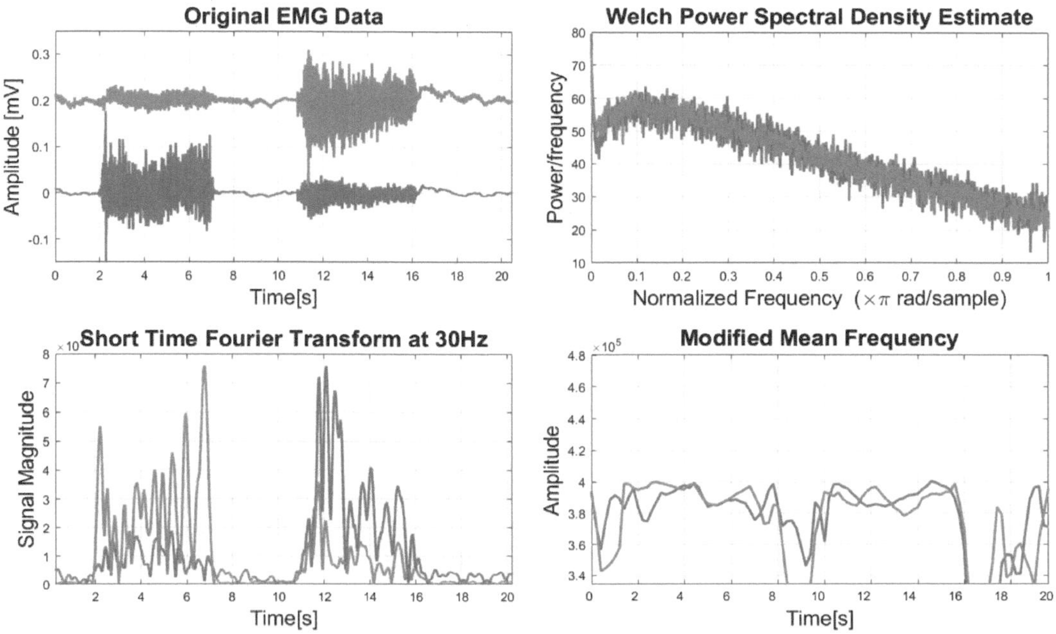

FIGURE 6.10
Example of the feature extraction considering the frequency components of the signal.

in a relatively short time, much desired in real-time and embedded applications. There are several algorithms to compute the DFT, the Cooley–Tukey being the most widely used. The DFT provides information about the phase and magnitude of the signal. The magnitude is squared to estimate the power of the spectra. The features presented below start from this calculation and aim to provide more information about the studied phenomena.

6.5.2.2 Short-Time Fourier Transform

Biosignals are intrinsically non-stationary, and thus, the processing and feature extraction involves more sophisticated methods than in the case of stationary signals. In particular, Fourier analysis does not perform well with random variations that occur in a large window of data [21]. To solve this issue, the frequency components of the signal are calculated using chunks of data, also called windows. The size of this window has a significant impact on the feature extraction process and has to be set carefully.

The resulting process provides information about how the spectral components are evolving in time, where the number of fast Fourier transform points and window overlap are the parameters to be determined to improve the output of the transformation in regard to the expected feature behavior.

The short-time Fourier transform is denoted as:

$$X_k = \sum_{n=1}^{N-1} X_n \; W_n \; e^{\frac{i2n}{m}kn} \tag{6.10}$$

where W_n is a window function (e.g., Hann, Hamming, rectangular).

6.5.2.3 Modified Mean Frequency (MMNF)

MMNF is the sum of the products between the signal amplitude and relative frequency, divided by the total sum of the amplitudes of the spectrum [19], expressed as:

$$\mathrm{MMNF}_k = \frac{\sum_{j=1}^{M} f_j A_j}{\sum_{j=1}^{M} A_j} \tag{6.11}$$

where A_j corresponds to the spectrum power of the signal at the j frequency component.

6.5.2.4 Canonical Correlation Analysis (CCA)

CCA normally deals with multivariable systems and can measure the linear dependency between two multidimensional variables by finding a couple of linear combinations, one for each multidimensional variable, which maximizes their correlation. In other words, it allows to determine the correlation of a set of signals with a reference. The concept of CCA will be further exposed when presenting some use cases of the feature extraction techniques in this section.

6.5.3 Time-Frequency Analysis

Discrete wavelet transform (DWT) is the most popular signal processing tool in the time-frequency domain and performs well with EMG signal pattern recognition [22]. The wavelet transform is based on the multiresolution analysis of a given signal through a scalable modulated sliding window.

6.5.3.1 Wavelet Transform Overview

The wavelet transform provides a time-frequency representation of a non-stationary signal, with information on the frequency content of a signal in the time domain. Formally, the continuous wavelet analysis (CWT) of a given function $f(t)$ of the scale factor s, and the translation τ is given by:

$$\gamma(s,\tau) = \int f(t)\, \Psi_{s,\tau}(t)\, dt \tag{6.12}$$

The wavelets are a set of basic functions $\Psi_{s,\tau}(t)$ obtained by scaling and translating a function $\Psi(t)$ called *mother wavelet*:

$$\Psi_{s,\tau}(t) = \frac{1}{\sqrt{s}} \Psi\left(\frac{t-\tau}{s}\right) \tag{6.13}$$

As the value of s is increased, the wavelet is dilated, ranging the analysis from the higher to lower frequencies. The result is the transformation value. For each value of s, the wavelet is shifted, increasing the value of τ. The implementation of this analysis requires using a wavelet that can be translated and scaled not continuously but in discrete steps, as showed below:

$$\Psi_{j,k}(t) = \frac{1}{\sqrt{s_0^j}} \Psi\left(\frac{t - k\tau_0 s_0^j}{s_0^j} \right) \tag{6.14}$$

with j and k integer values. The results of the transformation performed with discrete wavelets is a series of coefficients, named wavelet decompositions.

Although a computer can compute this CWT, it is highly redundant and computationally demanding because of all the combinations of s and τ. The CWT provides correlation between a given signal and a test function (the wavelet) at different scales. In the discrete case (the DWT), the signal is passed through a cascade of both low pass and high pass filters to obtain information on high and low frequency bands, remembering that filtering a signal is equal to performing the convolution of the signal with the impulse response of the filter. To calculate both the components we introduce the scaling function φ:

$$\varphi(st) = \begin{cases} 1 & \text{if } 0 < st < 1 \\ 0 & \text{otherwise} \end{cases} \tag{6.15}$$

The wavelet functions $\Psi()$ are associated to the high pass components while the scaling functions $\varphi()$ are related to low pass filtering. The filtering changes the signal resolution, and the scale is changed by downsampling the original signal.

The wavelet transform can be considered a bandpass filter and the series of scaled wavelet a filter bank. Using a finite number of values for the decomposition, we extract information in time-frequency domain by the recursive filtering of the given signal. Figure 6.11 shows the sequences of high and low pass filtering sequence performed to obtain the DWT coefficients. For a DWT decomposition of level n, the coefficients of the low pass filters are named *detail coefficients* (D_n) while the coefficients of the high pass filters are named *approximation coefficients* (A_n).

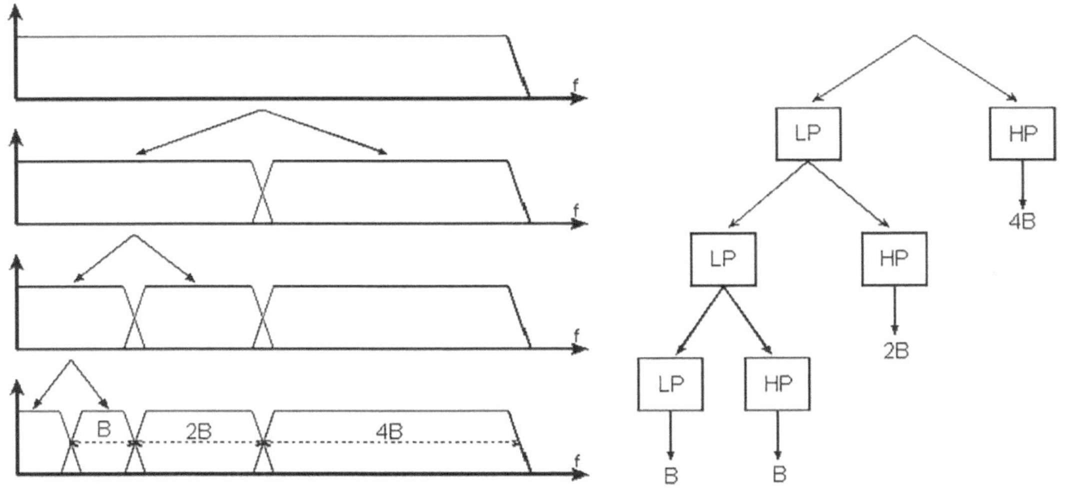

FIGURE 6.11
Wavelet decomposition scheme.

6.5.4 Pattern Recognition Analysis

The recognition and decoding of the user's intended hand movement are major challenges in the development of a human–machine interface (HMI) system for prosthetics. The classification of hand gesture with pattern recognition methods is a well-studied field of research, and there are many contributions from signal processing and machine learning communities [23]. Literature analysis demonstrates that pattern recognition approach is a good way to restore a natural gesture control overcoming the limitations of the interfaces used in state-of-the-art prosthetics control. Support vector machine (SVM) [24] is one of the most accurate algorithms for pattern recognition with this kind of signals. Typically, before applying pattern recognition algorithms, it is preferable to extract meaningful information for the input data. This step is called feature extraction. Practically, the algorithms presented in sections 6.5.2 and 6.5.3 are used to create a feature vector on which the classifier works. The classifier outputs are named classes and are associated to the gesture recognized, during the training stage of the algorithm. In the following section, we give an overview of the supervised learning approaches and describe the intuitions and the key concepts of the classical logistic regression approach and its main differences with the SVM.

6.5.5 Hidden Markov Model (HMM)

The HMM is a statistical algorithm used to model generative sequences characterized by an underlying process generating an observable sequence. The HMM is a framework of the Markov processes that are used to describe the temporal evolution of a system S that can be in one of N states $S_1, S_2, \ldots, S_N$. The system changes its state depending on the probability associated with each state even though the states in many cases are not directly observable. Hence, they must be considered as hidden random variables that generate an observable output. The HMM describes observations $o = \{o_1, o_2, \ldots, o_t\}$ with their states $Q = \{q_1, q_2, \ldots, q_t\}$. The two fundamentals hypothesis for the HMM are:

1. The state of the system at any given instant t depends only on the state at instant $t-1$.

$$p\left(q_t \mid q_{t-1}, o_{t-1}, q_{t-2}, o_{t-2}, \ldots, q_1, o_1\right) = p\left(q_t \mid q_{t-1}\right) \tag{6.16}$$

2. The observations o at any instant t depends only on the state at the instant t.

$$p\left(o_t \mid q_t, q_{t-1}, o_{t-2}, q_{t-2}, o_{t-2}, \ldots, q_1, o_1\right) = p\left(o_t \mid o_t\right) \tag{6.17}$$

The basic elements to define the HMM problem are:

- A set of N hidden states, $S = \{s_1, s_2, \ldots, s_N\}$. Often these states are related to gesture recognition.
- A set of M distinct observation symbols, $V = \{v_1, v_2, \ldots, v_M\}$. These represent the values of the observations as output of the system.
- The state transition matrix, $A = \{a_{ij}\}$, where a_{ij} is the probability of making a transition from state s_i to state s_j: $a_{ij} = P\left(q_t = S_j \mid q_{t-1} = S_i\right)$, with $1 \le i, j \le N$ and where q_t denotes the state at time t.

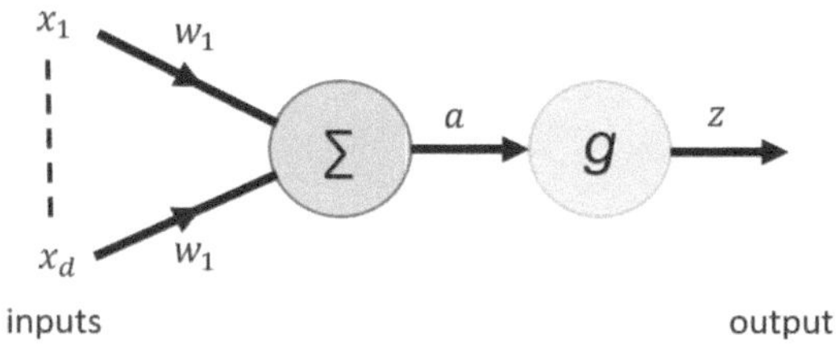

FIGURE 6.12
Example of a processing unit (neuron), where inputs (x_d) and weights (w_d) are used for a weighted sum (Σ), and the sum passes through an activation function, producing an output (z).

- The observation symbol probability distribution matrix, $B = \{b_j(k)\}$, where $b_j(k)$ is the probability of emitting v_k in state s_j at time t: $b_j(k) = P(o_t = v_k \mid q_t = s_j)$, with $1 \le j \le M$; $1 \le k \le M$, and where o_t denotes the observation at time t.
- The initial state distribution matrix, $\Pi = (\pi_i)$ where π_i is the probability that s_i is the initial state: $\pi_i = P(q_1 = s_i)$, with $1 \le N$.

Following the convention, an HMM can be expressed as:

$$\lambda = (A, B, \pi) \tag{6.18}$$

The transition probability matrix (A) and the observation symbol probability distribution matrix (B) that describe the HMMs are computed using the MATLAB® function *hmmestimate*. Before training the HMM, each training set is quantized into 12 levels; the same is done with the test sets before the classification. With 4 states and 12 possible emissions for each state, matrix A results to be 4×4 and matrix B to be 4×12. The definition of the matrices is performed offline with MATLAB, while the online recognition is calculated by multiplying A and B matrices and does not affect the real-time computing requirements.

6.5.6 Overview of the Artificial Neural Networks (NNs)

NNs, a computation paradigm where a general solution to a problem is learned starting from a given set of samples, are among the various approaches to pattern recognition analysis. The study of information processing inside the human brain inspires the theory of NNs. The human brain is composed of millions of neurons, which can be of a large variety of forms. Most of them have common features, such as *dendrites* as inputs and the *axon* as output. Each neuron communicates with other neurons through synapses, computing a weighted sum of the inputs and firing if the results exceed a given threshold [25,26].

An NN can be considered a non-linear mathematical function that transforms inputs into output variables through a set of weights. The process on which these weights are mapped is computationally intensive (learning process) and is based on the elaboration of a specific set of examples. After the learning process, the weights are fixed and new samples are classified using the trained network. Figure 6.12 shows an example of an artificial neuron. This is the basic processing unit on which the ith input (x_i) is multiplied by the ith weight (w_i) and summed to all the others plus a bias (w_0):

$$a = \sum_{i=1}^{d} w_i x_i + w_0 \tag{6.19}$$

We can rewrite the previous equation considering the bias as an extra output (x_0) set permanently to +1. The equation becomes as follows:

$$a = \sum_{i=0}^{d} w_i x_i \tag{6.20}$$

The weights and the bias can be negative or positive. The final output z is obtained by passing a through an activation function $g()$

$$z = g(a) \tag{6.21}$$

The non-linear activation function is of different types, such as linear, threshold, threshold linear, and sigmoidal.

An NN is composed of a set of processing units (*neurons*) and weighted connections between neurons. Of the several classes of NN, two fundamental classes are identifiable. The first one is the *feedforward* network (Figure 6.13a) in which the neurons are grouped in layers – one input layer, n hidden processing layers, and one output layer. Each neuron of each layer can be connected with the neurons of the next layer (in the output direction). A feedforward network is fully connected if each neuron is connected with all the neurons of the next layer. In addition, each input–output set is independent of other states. The second class is known as the *recurrent* network (Figure 6.13b) with at least one feedback from the kth output to the $(k+1)$th input. In some NNs, neurons can be connected to themselves (direct recurrences), to neurons on the preceding layers (indirect recurrences), and/or to neurons within the same layer (lateral recurrences).

The learning process of an NN can be performed in several ways. In unsupervised learning, the network tries to find similarity in the training set (a set of input patterns), generating pattern classes. Another possibility is the reinforcement learning, where the network receives a value that indicates if the result is correct or not and possibly how correct it is. The last one is supervised learning, the most popular way of training, where the training set is composed of input patterns and labels, which indicate the true class of each input [25]. Among all the possible NN topology and learning methods, the most common are

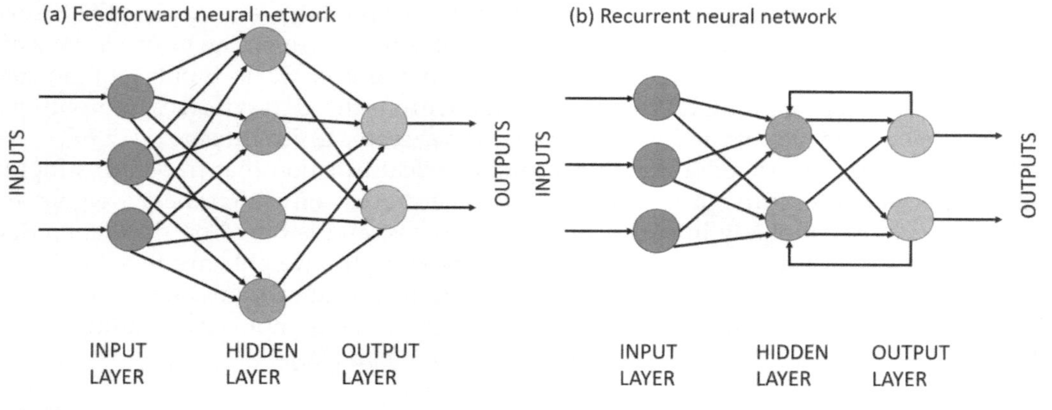

FIGURE 6.13
(a) An example of a fully connected feedforward NN with one hidden layer. (b) An example of a recurrent NN with one hidden layer.

the multilayer feedforward NNs, performing supervised learning through *backpropagation* technique. This method implies three main steps:

- the feedforward of the input patterns contained in the training set;
- the evaluation of the error and its backpropagation;
- the adjustment of the parameters (weights).

The learning process begins with random weights, and the three steps are reiterated until the error is minimal. The optimal weights are computed by minimizing the following function through *gradient descent*:

$$\Delta w_i = -\alpha \frac{\partial E}{\partial w_i} \tag{6.22}$$

where E is obtained by subtracting the real output to the one in output to the NN, and α is the learning rate (a constant, in each iteration). The Δw_i is the adjustment of each weight and depends on the amount of error generated by the previous weights.

6.5.7 Overview of the Support Vector Machine

The supervised learning approaches define a model from a dataset of given input $\left(x^{(i)}, y^{(i)}\right)$, where $x^{(i)}$ are the vectors of the input data and $y^{(i)}$ are the labels. The model is defined by tuning a vector of parameters $\boldsymbol{\theta}$ through the minimization of an opportune cost function (training). The model is represented by a boundary that separates the data in two or more classes (classification or prediction). We can define the hypothesis function as:

$$h_\theta(x) = g\left(\theta^{\mathrm{T}} x\right) \tag{6.23}$$

where

$$g(z) = \frac{1}{1 + \mathrm{e}^{-z}} \tag{6.24}$$

This is called the sigmoid or logistic function and gives the estimated probability that the output $y = 1$ with a given new input x is:

$$h_\theta(x) = P\left(y = 1 \mid x; \theta\right) \tag{6.25}$$

Observing the sigmoid function, if $z = \theta^{\mathrm{T}} x$ is greater than 0 the output of $g(z)$ is greater than 0.5, otherwise the output is –0.5. During the classification stage, if we have an input x for which $h_\theta(x) > 0.5$ (i.e., $\theta^{\mathrm{T}} x > 0$) the corresponding y will be equal to 1, otherwise 0. The tuning of the parameter $\boldsymbol{\theta}$ is obtained by solving an optimization problem by minimizing the following cost function with the input data of the training dataset:

$$\operatorname{cost}\left(h_\theta(x,y)\right) = \frac{1}{m}\sum_{i=1}^{m}\left[-y^{(i)}\log\left(h_\theta\left(x^{(i)}\right)\right) - \left(1 - y^{(i)}\right)\log\left(1 - h_\theta\left(x^{(i)}\right)\right)\right] \tag{6.26}$$

The computer science community developed the SVM in the 1990s in the framework of the statistical learning theory. The basic idea of this supervised learning algorithm is to

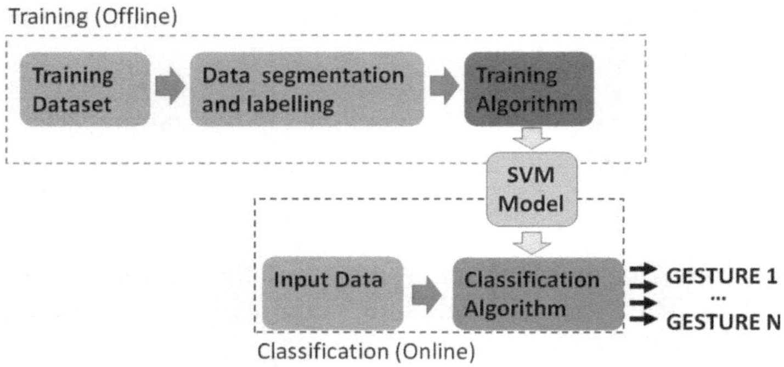

FIGURE 6.14
SVM algorithm diagram.

improve the performance of the classification obtained with logistic regression methods through the application of *large margin* classification [24]. The offline training phase of the algorithm uses labeled instances of data to calculate the optimal separation hyperplane (*maximum margin* hyperplane) between two classes of data through the solution of a convex optimization problem. Such separation plane is represented by a set of data vectors, namely the support vectors, which belong to the borders of the two classes and are used to classify new data instances. A diagram of SVM training and recognition is illustrated in Figure 6.14.

When the two classes are not linearly separable, the input data space can be mapped to a higher-dimensional space through a kernel function [27] to allow an effective separation [16,28]. The kernel is based on a radial basis function (RBF) and use a Gaussian function expressed by:

$$\exp\left(-\frac{\left\|x - l^{(n)}\right\|}{2\sigma^2}\right) \tag{6.27}$$

where x is the input vector, l^i is the i support vector, and σ is the variance.

Having two possible classes, denoted as Cl_1 and Cl_2, the formula of the decision function to classify a new input instance is:

$$f(x) = \sum_{i=1}^{N_{SV}} y_i \alpha_i K\langle x, s_i \rangle - \rho \quad f(x) > 0,\ x \in Cl_1, \quad f(x) < 0,\ x \in Cl_2 \tag{6.28}$$

where $x \in \mathbb{R}^{N_F}$ is the input features vector, $S_i \in \mathbb{R}^{N_F}$, $i = 1,\ldots, N_{SV}$ are the support vectors, α_i are the support values, with y_i depending on the class of reference ($y_i = +1$ for Cl_1, $y_i = -1$ for Cl_2), and $K\langle .,. \rangle$ denotes the kernel function.

The SVM is a binary classification algorithm, but its application can be extended for a multiclass scenario by reducing the problem to multiple binary classification problems for each class.

6.5.8 Overview of Hyper-Dimensional Computing (HDC)

Among other approaches based on the emulation of brain functionality, HDC [29] is a brain-inspired algorithm that uses point in a very high dimension (i.e., 10,000 space as an alternative to numbers). This high dimension allows a large number of nearly orthogonal hyper-vectors, which can be combined into a new hyper-vector through well-defined vector space operations. Multiplication, addition, and permutation (MAP) are the three main operations used in HDC. The hyper-vectors can be binary vectors composed of an equal number of randomly placed 0s and 1s, $\{0,1\}^D$. A measure of similarity between two hyper-vectors can be obtained by using the cosine similarity, which can be substituted by the Hamming distance in the binary case. The Hamming distance computes the number of bits at which two hyper-vectors differ. All the MAP operations can be replaced by simpler operations when computing binary hyper-vectors; in fact, addition (+) becomes a component-wise majority, multiplication a component-wise XOR (⊕), and permutation (ρ) a 1-bit rotation. The addition operation produces a new hyper-vector similar to the hyper-vectors in input (useful to represent sets), multiplication produces a dissimilar one (for binding two hyper-vectors), while permutation is well suited to store a sequence as it produces a dissimilar pseudo-orthogonal hyper-vector.

The HDC algorithm is composed of three main modules (Figure 6.15). In the first one, we map the samples into a hyper-dimensional space using the *item memory* matrix, which is composed of (pseudo)random orthogonal ($\perp$) hyper-vectors assigned to each acquisition channel (i.e., $E_1 \perp E_2 \ldots \perp E_i$). In the *continuous item memory* matrix, another matrix used for mapping, hyper-vectors are assigned to the possible analog values that the signal can assume. Orthogonal endpoint hyper-vectors are generated for the maximum and minimum signal level, and the other values between these two are derived from a linear interpolation. For instance, discretizing the features in K levels leads to K hyper-vectors $(V_1, V_2, \ldots, V_K)$, where $V_1 \perp V_K$. These two matrices are computed at the beginning of the training phase and remain constant throughout the lifetime of the application. In the second kernel, HDC encodes the features into a hyper-vector using the MAP operations to represent the event of interest. The *spatial encoder* captures the

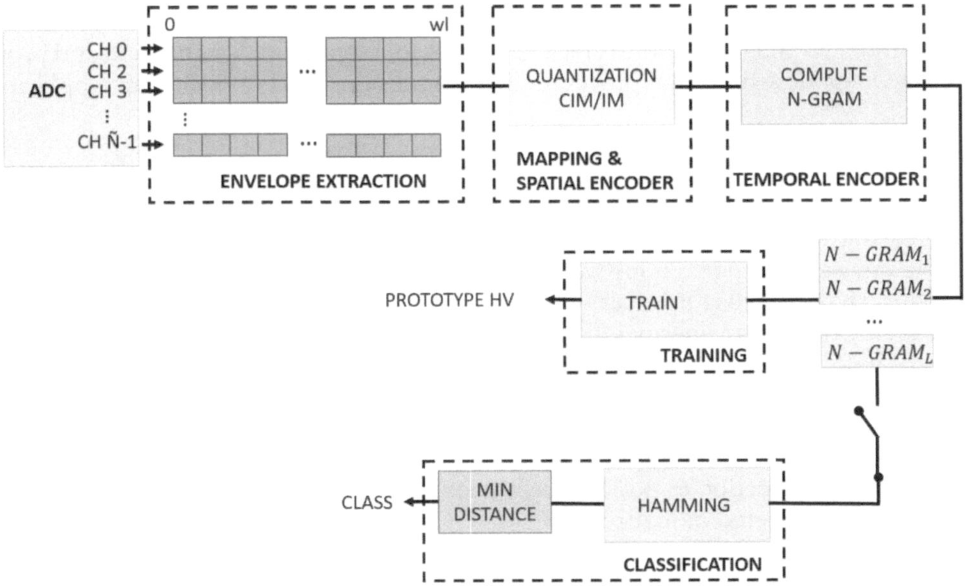

FIGURE 6.15
HDC processing chain.

spatial information contained in the signal using the component-wise XOR and the component-wise majority:

$$S^t = \left[\left(E_1 \oplus V_l(1)^t\right) + \cdots + \left(E_i \oplus V_l(i)^t\right)\right] \quad (6.29)$$

Sometimes, having the spatial information contained in S^t is not enough, we also need to include the temporal information. The *temporal encoder* uses permutations and multiplications between sequences of N hyper-vectors to create an N-gram, where N is the length of the temporal window we are considering.

The N-gram is created with the following equation:

$$S^t \oplus \rho^1 S^{t+1} \oplus \rho^2 S^{t+2} \oplus \cdots \oplus \rho^{n-1} S^{t+n-1} \quad (6.30)$$

where ρ^k indicates a rotation of k positions. These kernels are common for both training and classification. During training, the N-grams derived from features of the same class are added to form a prototype hyper-vector, which is stored on the associative memory matrix with the prototypes of the other classes involved in the elaboration. During classification, the N-grams in the output to the encoders are called *query* hyper-vectors and derived from unseen features. Hence, the query is classified computing the Hamming distances between the query and all the prototype hyper-vectors contained in the associative memory matrix. The label associated to the minimum distance is assigned to the query.

The HDC algorithm is very robust and the dimensions of the structures required to perform all the elaboration are fixed. The most remarkable advantage of this approach is the capability of performing the training 'one shot', which means that there is no need of time-consuming back propagation (like in the NN) or other convergence methods. For this reason, HDC is feasible for an embedded implementation to perform on-chip training.

6.5.9 Case Study and the Deployment of Biomedical Signal Analysis

In this section, we illustrate some typical use cases for biopotential acquisition and processing to showcase how biosignal processing techniques are used in healthcare applications.

6.5.9.1 A Minimally Invasive Low-Power Platform for Real-Time Brain–Computer Interaction Based on Canonical Correlation Analysis

An increasing trend in HMI is to add more sophisticated digital signal processing into wearable devices aiming to improve the natural interaction with the environment. This task is not trivial because it involves the design of an Internet of Things node that must deliver unobtrusiveness, energy efficiency, and computational power at a low cost.

Authors in Ref. [8] presented a wearable Internet of Things node designed for brain computer interface (BCI) spelling. The system targets the detection of visual evoked potentials through a multichannel CCA. The system is based on the detection of the SSVEP. These signals are evoked by varying the luminosity (contrast) of black and white 10×10 square checkerboards in function to periodic reference signals at different frequencies. The processing chain is depicted in Figure 6.16.

The HMI system works as follows. The subject fixes the sight at a given checkerboard from the layout, each one flickering at a different frequency. This elicits an EEG response, most at the visual cortex region, that is captured using an array of dry electrodes and the wearable node. The resulting EEG signals, thus, have the same frequency response of the

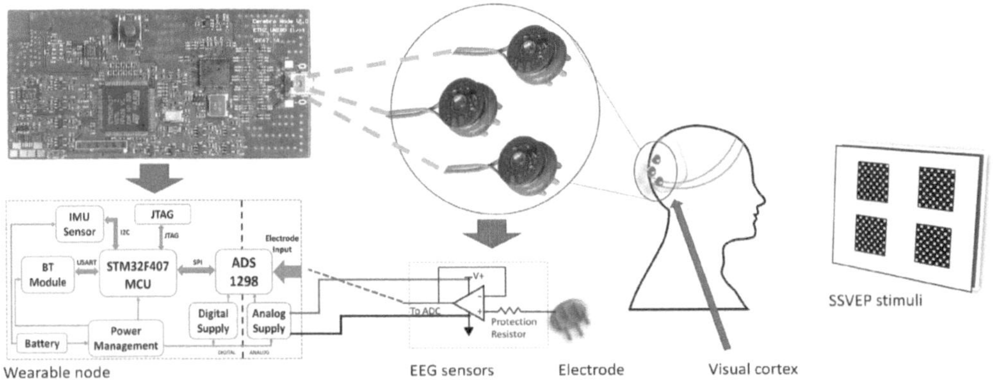

FIGURE 6.16
Experimental setup for the SSVEP classification © 2019 IEEE. Reprinted, with permission, from *IEEE Internet of Things Journal* (Volume: 6 , Issue: 1 , Feb. 2019), Salvaro et al., "A Minimally Invasive Low-Power Platform for Real-Time Brain Computer Interaction Based on Canonical Correlation Analysis".

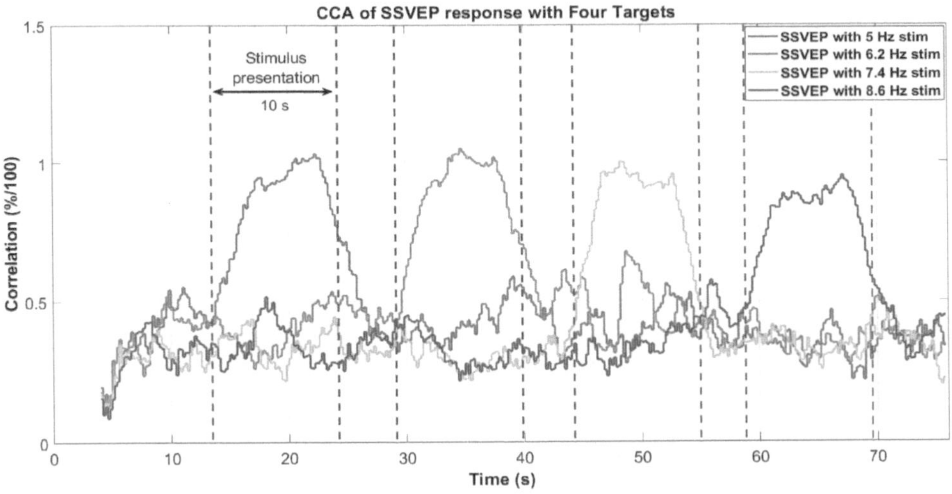

FIGURE 6.17
CCA analysis of the SSVEP signals.

stimulus and can be extracted from the array of electrodes using the CCA. Figure 6.17 shows an example of the CCA of the EEG signals from the three channels used while using four target frequencies. The pre-processing includes a high pass filter to eliminate the signal offset and drift. Finally, the class is deduced from the correlation values, where the highest correlation will be associated with the checkerboard that the subject is seeing.

While the checkerboards are flickering, the subject fixes the sight at a given target. The SSVEP response is later captured by the Internet of Things node, and the results of the classification are transmitted wirelessly to a host device.

In this test, the stimulus is presented for 10 s. The subject is asked to fix the sight at the different checkerboards, sequentially.

The CCA is calculated in three main steps that include a QR decomposition, singular value decomposition, and the extraction of the canonical coefficients. Finally, the Euclidian

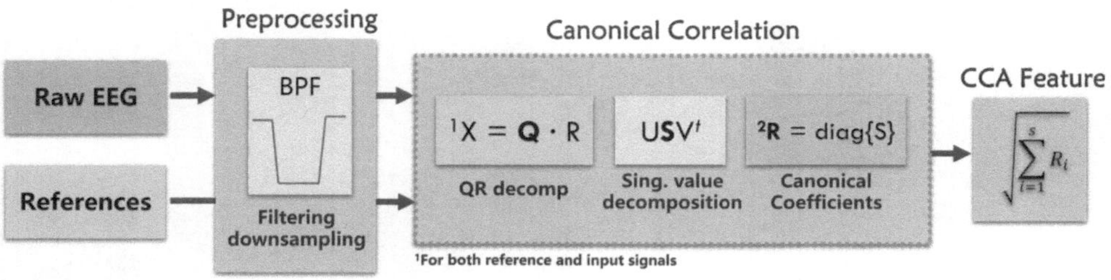

FIGURE 6.18
Real-time CCA classification chain.

norm is used as a feature extraction for the classification. Figure 6.18 illustrates the complete processing chain for the extraction and classification of the EEG signals.

The complete system runs in real time on a low-power microcontroller and provides a peak information transfer rate of 1.57 bps with a power envelope below 30 mW, providing up to 122 h of continuous operation. The system has several applications including robotic control and closed-loop HMI.

Here, we have combined different DSP techniques passing from offline experimentation and profiling to finally produce a device that can handle data processing while being energy-efficient. This achievement denotes that the efforts for practical systems must always be extended up to the architectural issues (i.e., computational power and power consumption).

6.5.9.2 EMG Prosthetic Control

Commercial EMG-based prosthetic controllers are based on threshold control, where a burst of contraction is decoded and mapped on a gesture. This approach is highly reliable but requires a high level of concentration from the users' end and a long learning curve. One of the open challenges in this field is to restore natural gesture control strategy, mapping the muscular activation patterns onto hand gestures. In this vein, supervised pattern recognition approaches are gaining traction [23,28,30].

Figure 6.19 shows the typical workflow of a pattern recognition application based on SVM. A dataset is collected and labeled to obtain an SVM model. Once the model is trained, it is used to classify new data. To design an intuitive controller, it is very important to implement a reliable control strategy since the classifier output is not sufficient to control the prosthesis. For instance, contraction spikes must be considered as artifacts if their duration is lower than 100 ms. Furthermore, since most prosthetic hands do not have position control encoders for the fingers, all gestures must start from the same position. A typical control strategy leverages the classification algorithm and implements a higher level finite-state machine. Figure 6.20 shows a typical SVM-based prosthetic controller. The data are acquired and sampled at 500 Hz by a 16-bit ADC. If the EMG level stays over a given threshold for more than 100 ms, EMG samples are classified by the SVM. To make the classification stable, the output of the classifier is low-pass filtered, using a majority voting strategy (e.g., using the majority on 20 samples). If the prosthetic hand is in the "rest position" (i.e., hand open), the controller drives the prosthesis to execute the movement; if the hand is already in a grasp, the controller drives the prosthetic hand to return to the "rest position". After the movement execution, it is necessary to retrigger the controller, waiting for muscular de-contraction. Figure 6.21 shows the control strategy on the EMG

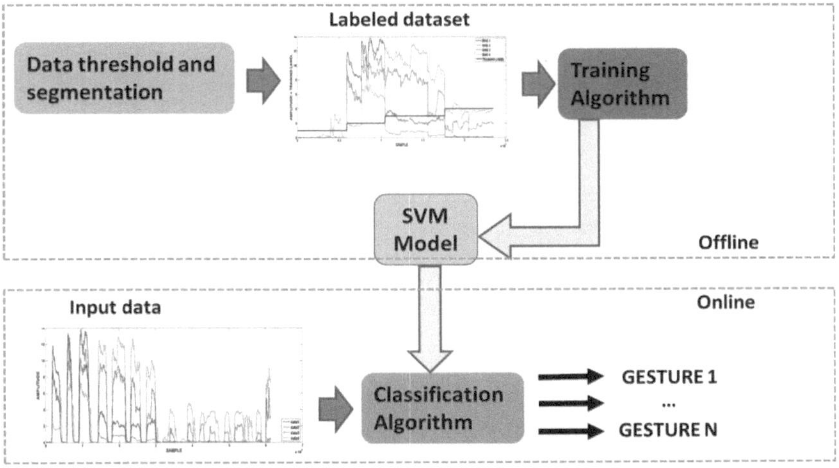

FIGURE 6.19
Algorithm workflow for SVM gesture classification.

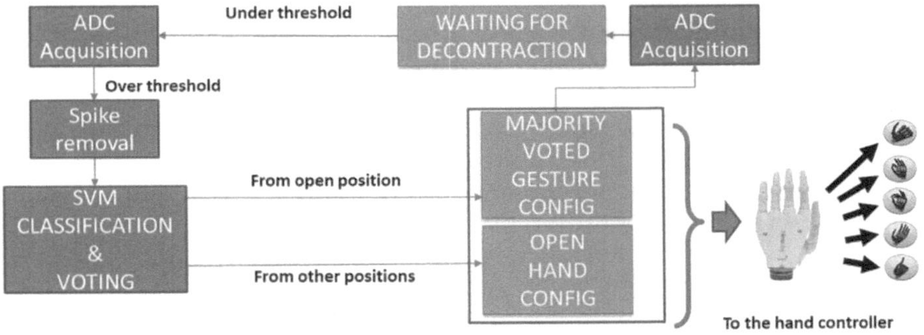

FIGURE 6.20
FSM for prosthetic control.

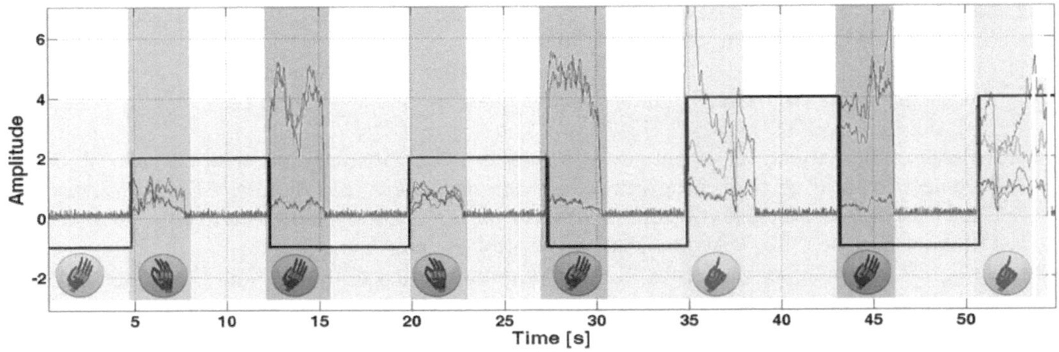

FIGURE 6.21
EMG control strategy.

plot. The red solid line shows the output of the controller. Once the gesture is recognized, the controller stays in the same state and, to perform new movements, it always returns to the "rest position" (open hand).

6.6 Conclusions

In this chapter, we have introduced some fundamental concepts regarding biopotentials, acquisition methods, and processing that take advantage of the ability of the human body to provide complementary information through biosignals. The current evolution of these methods, coupled with the new trend in wearable and ultra-low power architectures, allows reaching more ambitious application scenarios such as industry, gaming, and learning, while improving medical applications toward more useful and practical devices.

In particular, we explored the nature of EEG, EMG, and EDA signals that are the source of current HMI systems, and we described the state-of-the-art algorithms for the extraction of useful information. We also glanced upon the new low-power embedded platforms, which can process and classify different HMI paradigms in real time, also offering some insights on recent technology for dry signal acquisition.

To wrap all the presented information, we have provided two case studies for the evaluation and implementation of HMI systems, passing from the signal characterization and experimentation, digital signal processing, and embedded deployment. We believe that all this information presents a general picture of the challenges on the design and implementation of these devices and some possible practical solutions that will potentiate future applications.

References

1. Mihajlovic, V., Grundlehner, B., Vullers, R., & Penders, J. Wearable, wireless EEG solutions in daily life applications: What are we missing? *IEEE J. Biomed. Heal. Inf.* **19**, 6–21 (2015).
2. Basmajian, J. V. & De Luca, C. J. *Muscles Alive : Their Functions Revealed by Electromyography.* (Philadelphia, PA: Williams & Wilkins, 1985).
3. Milosevic, B., Benatti, S., & Farella, E. Design challenges for wearable EMG applications. In *Proceedings of the 2017 Design, Automation and Test in Europe* (*DATE*), Lausanne, Switzerland, 2017. doi:10.23919/DATE.2017.7927217.
4. Milosevic, B., Farella, E., & Benaui, S. Exploring arm posture and temporal variability in myoelectric hand gesture recognition. In *Proceedings of the IEEE RAS and EMBS International Conference on Biomedical Robotics and Biomechatronics*, Pisa, Italy, 2018–August, pp. 1032–1037.
5. Ottobock Sensor 13E200. Available at: www.ottobock.com/.
6. Baillet, S., Mosher, J. C., & Leahy, R. M. Electromagnetic brain mapping. *IEEE Signal Process. Mag.* **18**, 14–30 (2001).
7. Turpin, G., Shine, P., & Lader, M. Ambulatory electrodermal monitoring: Effects of ambient temperature, general activity, electrolyte media, and length of recording. *Psychophysiology* **20**, 219–224 (1983).

8. Salvaro, M., Benatti, S., Kartsch, V., Guermandi, M., & Benini, L. A Minimally invasive low-power platform for real-time brain computer interaction based on canonical correlation analysis. *IEEE Internet of Things Journal* **6**(1), 967–966 (2018). doi:10.1109/JIOT.2018.2866341.
9. Malmivuo, J. & Plonsey, R. *Bioelectromagnetism: Principles and Applications of Bioelectric and Biomagnetic Fields.* (Oxford: Oxford University Press, 2012). doi:10.1093/acprof:oso/9780195058239.001.0001.
10. Boucsein, W., Fowles, D. C., Grimnes, S., Ben-Shakhar, G., Roth, W. T., Dawson, M. E., & Filion, D. L. Publication recommendations for electrodermal measurements. *Psychophysiology* **49**, 1017–1034 (2012).
11. Doberenz, S., Roth, W. T., Wollburg, E., Maslowski, N. I., & Kim, S. Methodological considerations in ambulatory skin conductance monitoring. *Int. J. Psychophysiol.* **80**, 87–95 (2011).
12. Bitalino. Available at: https://bitalino.com/en/.
13. Opfer-Gehrking, T. L., Low, P. A., Huck, C. A., Novak, V., & Sandroni, P. Mechanisms of blood pressure alterations in response to the Valsalva maneuver in postural tachycardia syndrome. *Clin. Auton. Res.* **10**, 1–5 (2006).
14. Henderson, L. A., Macey, P.M., Macey, K. E., Frysinger, R. C., Woo, M. A., Harper, R. K., Alger, J. R., Yan-Go, F. L., & Harper, R. M. Brain responses associated with the Valsalva maneuver revealed by functional magnetic resonance imaging. *J. Neurophysiol.* **88**(6), 3477–3486 (2002). doi: 10.1152/jn.00107.2002.
15. Levin, A. B. A. simple test of cardiac function based upon the heart rate changes induced by the valsalva maneuver. *Am. J. Cardiol.* (1966). doi:10.1016/0002-9149(66)90200-1.
16. Taylor, J. & Cristianini, N. *Kernel Methods for Pattern Analysis.* (Cambridge: Cambridge University Press, 2004).
17. Phinyomark, A., Limsakul, C. & Phukpattaranont, P. A novel feature extraction for robust EMG pattern recognition. arXiv preprint arXiv:0912.3973, 2009.
18. Benatti, S., Rovere, G., Bösser, J., Montagna, F., Farella, E., Glaser, H., Schönle, P., Burger, T., Fateh, S., Huang, Q., Benini, L. A sub-10mW real-Time implementation for EMG hand gesture recognition based on a multi-core biomedical SoC. In *Proceedings of the 2017 7th International Workshop on Advances in Sensors and Interfaces (IWASI)*, Vieste, Foggia, Italy, 2017, pp. 139–144. doi:10.1109/IWASI.2017.7974234.
19. Sanei, S. & Chambers, J. A. *EEG Signal Processing.* (John Wiley & Sons Ltd., 2013). doi:10.1002/9780470511923.
20. Montagna, F., Buiatti, M., Benatti, S., Rossi, D., Farella, E., & Benini, L. A. Machine learning approach for automated wide-range frequency tagging analysis in embedded neuromonitoring systems. *Methods* **129**, 96–107 (2017).
21. Flandrin, P. Some aspects of non-stationary signal processing with emphasis on time-frequency and time-scale methods. In *Wavelets.* (Berlin, Heidelberg: Springer, 2012, pp. 68–98). doi:10.1007/978-3-642-97177-8_4.
22. Phinyomark, A., Limsakul, C., & Phukpattaranont, P. Application of wavelet analysis in EMG feature extraction for pattern classification. *Meas. Sci. Rev.* **11**, 45–52 (2011).
23. Meattini, R., Benatti, S., Scarcia, U., De Gregorio, D., Benini, L., & Melchiorri, C. An sEMG-based human-robot interface for robotic hands using machine learning and synergies. *IEEE Trans. Compon Packag. Manuf. Technol.* **8**, 1149–1158 (2018).
24. Boser, B. E., Guyon, I. M., & Vapnik, V. N. A training algorithm for optimal margin classifiers. In *Proceedings of the Annual Conference on Computational Learning Theory*, Pittsburgh, PA, 1992, pp. 144–152. doi:10.1145/130385.130401.
25. Kriesel, D. 2007, A brief introduction to neural networks. (dkrissel.com).
26. Bishop, C. M. Neural networks and their applications. *Rev. Sci. Instrum.* **65**(6), 1803–1832 (1994).
27. Breneman, J. Kernel methods for pattern analysis. *Technometrics* **47**, 237–237 (2009).
28. Benatti, S., Milosevic, B., Farella, E., Gruppioni, E., & Benini, L. A prosthetic hand body area controller based on efficient pattern recognition control strategies. *Sensors (Switzerland)* **17** (2017). doi:10.3390/s17040869.

29. Kanerva, P. Hyperdimensional computing: An introduction to computing in distributed representation with high-dimensional random vectors. *Cognit. Comput.* **1**, 139–159 (2009).
30. Kartsch, V., Benatti, S., Mancini, M., Magno, M., and Benini, L. Smart wearable wristband for EMG based gesture recognition powered by solar energy harvester. In *2018 IEEE International Symposium on Circuits and Systems* (*ISCAS*), Firenze, Italy, 2018, pp. 1–5. IEEE.

7

Pervasive Computing in Cardiovascular Healthcare

Chandrajit Pal, Naresh Vemishetty, and Amit Acharyya
Indian Institute of Technology (IIT) Hyderabad

CONTENTS

7.1 Introduction

Remote areas, generally, encounter scarcity of cardiologists and progressive facilities for the treatment of cardiovascular diseases (CVDs). However, same constraints will be self-addressed by the exploitation of recent advancements in wireless technology and its ubiquity. Clinical CVD identification is, generally, applied exploitation either of ordinary 12-Lead (S12) or Mason–Likar 12-Lead (ML12) system [1]. These systems encompass eight freelance leads/signals which limit their usage in telemonitoring applications with personalized remote health observation, home observation, etc. to name a few [1]. Credits go to high information measure and storage needs, knowledge, and low compression **quantitative relation** from signal compression techniques. Moreover, the reduced lead (RL) systems with 2–3 leads are generally utilized in telemonitoring application, which is not decent for diagnosis. Diagnosis from 12-lead electrocardiography (ECG) is cumbersome and time-consuming. Therefore, it is required to form a reduced 3-lead (R3L) system from S12 and ML12 systems at the transmission end, which reduces the number of signals to 3 and then reconstructs the S12 and ML12 systems at the receiver end through signal reconstruction. This can eradicate the aforementioned limitations. It can also be the reconstruction of standard 12-lead (S12) system from Frank vectorcardiographic (FV). A personalized reduced 2/3 lead system is required which can offer equivalent information as contained in S12 system, so as to accurately reconstruct S12 system from RL system for diagnosis. Among the 12-lead system, selection of such RL systems suitable for personalized remote health-monitoring applications is also a challenge, followed by its subsequent successful reconstruction of S12-lead. The accuracy and real-time performance of traditional methods needs improvement. Also if accurate Frank's vectorcardiogram (VCG) system is made available avoiding the limitations of proper electrode settings and hardware complexity, VCG will complement S12 system in diagnosis of CVDs.

There exists some literature that provides various solutions as described below.

- The first remedy proposed a personalized reconstruction methodology for the reconstruction of missing precordial leads of S12 and ML12 systems [1].
- Second solution proposed a robust and accurate method for the reconstruction of S12-lead system from FV system using personalized transformation (PT) matrices targeting personalized remote health-monitoring applications [2].
- Third strategy proposed a personalized R3L system formation methodology, which employs principal component analysis (PCA), thereby reducing redundancy and increasing signal to noise ratio (SNR) ratio, hence making it suitable for wireless transmission [3].
- Fourth, a technical methodology to the medical practitioners has been proposed for selection of RL systems suitable for personalized remote health-monitoring applications [4]. Subsequently, a novel S12-lead ECG reconstruction methodology is also reported which is shown to be more reliable than the state-of-the-art lead reconstruction methodologies.
- Subsequently, VCG-based substitution to complement S12 system in diagnosis of CVDs has been proposed.
- This is followed by convolutional neural network (CNN)-based method which proved to be more accurate and time-saving for deployment in nonhospital situations to synthesize a S12 ECG from a RL-set ECG recording [5,6].

7.2 Reported Solutions at a Glance

A. A R3L system from S12 and ML12 systems is formed at the transmission end reducing the number of signals to 3 and then reconstructs the S12 and ML12 systems at the receiver end using a personalized reconstruction methodology, thereby allaying aforementioned limitations. Among others, I, II, and V2 form the basis leads and the precordial leads form the target leads. Least square fit and heart vector projection (HVP) theory have been used to obtain PT coefficients [1]. Accuracy of reconstruction has been evaluated on PhysioNet PTBDB (Physikalisch-Technische Bundesanstalt database) and INCARTDB (Institute of Cardiological Technics 12-Lead Arrhythmia Database) [7,8], after wavelet-based preprocessing, using R^2 statistics, correlation (r_x), and regression (b_x) coefficients. Reusability of personalized coefficients has also been investigated. Mean R^2 values obtained from the reconstruction of target leads are 91.87% (PTBDB) and 83.75% (INCARTDB). R3L system reduces the number of leads/signals from 8 to 3, and as the results indicate the possibility of reusing the transformation coefficients, the number of electrodes can be reduced from 10 to 5. This increases the comfort of patients and caregivers [1].

B. The technique in Ref. [2] proposed a robust and accurate method for the reconstruction of S12-lead system from FV system using PT matrices targeting personalized remote health-monitoring applications. FV system is used in the 3D visualization of heart and diagnosis and prognosis of many cardiologic disorders including myocardial infarction, Brugada syndrome, etc. However, cardiologists are accustomed to S12 system pertaining to its old classical usage and widespread acceptability and hence, it is generally used as primary ECG acquisition system and state-of-the-art inverse Dower transform (DT) and affine transform (AT) are used to obtain FV system from S12 system. Here, authors [2] have proposed the acquisition of FV system and use PT to reconstruct S12 system from FV system. PhysioNet's PTBDB after wavelet-based preprocessing to remove baseline wandering and noise has been used in this investigation. The personalized coefficients have been obtained using least-square (LS) fit method and HVP theory and evaluation metrics used are R^2 statistics, correlation and regression coefficients. The proposed PT methodology has outperformed AT and DT by mean R^2 values of 16.36% and 26.53%, respectively. For the practical application of the proposed system, the reusability of personalized coefficients have been investigated which has been shown to outperform state-of-the-art AT and DT [2].

C. A personalized R3L system formation methodology has been proposed which employs PCA [3], reducing redundancy and increasing SNR ratio. This, makes it suitable for wireless transmission. Accurate S12 system is made available using personalized lead reconstruction methodology, thereby addressing medical constraints. Mean R^2 statistics values obtained for reconstruction of S12 system from the proposed R3L system using PhysioNet's PTB and T-wave alternans (TWA) databases were 95.63% and 96.37% respectively. To substantiate the superior diagnostic quality of reconstructed leads, root mean square error (RMSE) metrics obtained upon comparing the ECG features extracted from the original and reconstructed leads, using a recently proposed Time Domain Morphology and Gradient (TDMG) algorithm [9], have been discussed. The proposed system does not require any extra electrode or modification in placement positions and hence, can readily find application in computerized ECG machines.

D. Authors in Ref. [4] made an attempt to provide a technical methodology to the medical practitioners for the selection of appropriate RL system targeting personalized remote health-monitoring applications for ECG acquisition by selecting two popular RL systems and by comparing the performance extensively in terms of reconstructed S12-leads obtained from both of these RL systems using our proposed methodology.

E. In the study cited in Ref. [10], authors proposed a methodology to construct Frank's VCG from S12 system using PCA [10]. They have compared the work with state-of-the-art inverse Dower transform (IDT) and Kors transform (KT). Mean R^2 statistics and correlation coefficient values, obtained upon comparison of reconstructed and originally measured Frank's leads, for CSE multilead database and PhysioNet's PTBDB databases using our proposed method, IDT [11,12] and KT [13] were found to be (73.7%, 0.869), (57.6%, 0.788), and (56.2%, 0.781), respectively. From remote healthcare perspective, a reduced 2–3 lead system is desired and Frank lead system seems to be promising as shown by previous works. However, cardiologists are accustomed to S12 system due to its widespread usage and derived Frank lead system may not be sufficient. Therefore, to bridge this gap, authors have presented the results of personalized reconstruction of S12 system from derived VCG, obtained using proposed PCA-based method and compared it with results obtained when originally measured Frank leads were used.

F. In the study presented in Ref. [5], authors have presented a methodology based on CNNs for the synthesis of missing precordial leads [5]. The results in Ref. [5] show that the proposed method receives better similarity and consumes less time using the PTB database. Particularly, the performance of the methodology shows cent percent accuracy in reconstructing the pathological ECG signal, which is crucial for cardiac diagnosis. The CNN-based method is shown to be more accurate and time-saving for deployment in nonhospital situations to synthesize a S12-lead ECG from a RL-set ECG recording. This is promising for real cardiac care.

G. The authors in Ref. [6] have presented a novel, enhanced method for deriving 12-lead ECGs from a pseudoorthogonal 3-lead subset. Designers investigated how well the 12-lead ECG is reconstructed using both generic and patient-specific nonlinear reconstruction methods that are based on the use of artificial neural networks (ANNs) [6].

7.3 Explanation of the State-of-the-Art Methodologies

A. As shown in Figure 7.1, the envisaged remote health-monitoring scenario involves acquisition (raw ECG), preprocessing, coefficient generation, transmission, and lead reconstruction. In this paper [1], designers assume that the acquired raw ECG is available, and the main focus is on preprocessing, coefficient generation, and reconstruction of the missing precordial leads. The overall methodology can be summarized as follows:

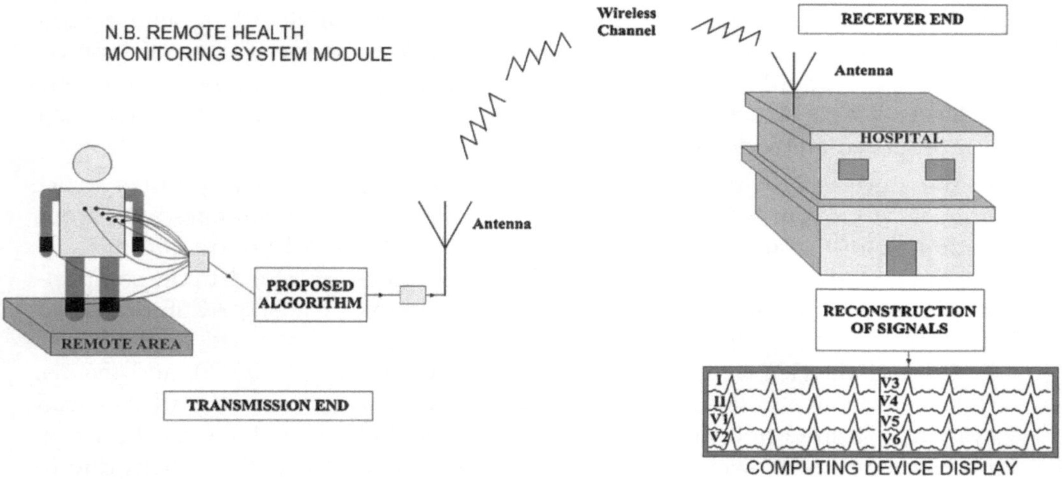

FIGURE 7.1
Proposed remote health-monitoring scenario [1].

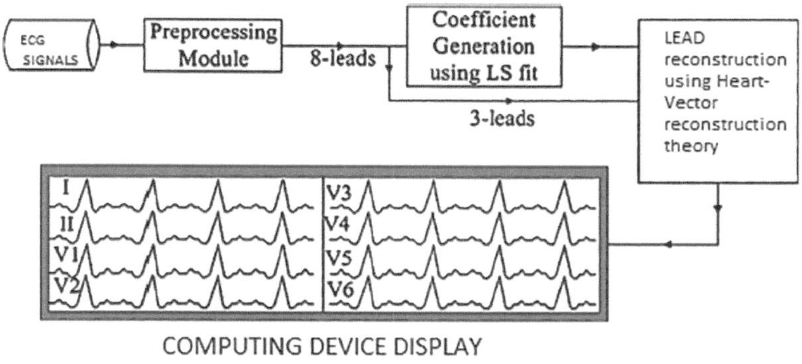

FIGURE 7.2
Functionalities of the remote health-monitoring scenario.

Raw ECG signal acquired is passed through a wavelet-based preprocessing module to remove baseline wandering and noise. This preprocessed signal is then used for generation of personalized coefficients using HVP theory and LS fit method [14]. The R3L system along with the transformation coefficients are then used to reconstruct the missing precordial leads using HVP theory [15,16]. Finally, the reconstruction results are evaluated using R^2 statistics, correlation (r_x), and regression coefficients (b_x) [1].

Figure 7.2 shows the summary of the methodology followed in the paper [1]. PTBDB and St. Petersburg INCARTDB sampled at frequencies 1 kHz and 257 Hz, respectively, have been used in this investigation. These are open-source ECG databases for academic research.

Table 7.1 contains the results of the evaluation metrics.

R^2 statistics has been used to evaluate the degree of association between the measured and the reconstructed signal. Perfect retracing of the measured wave by the reconstructed wave will be indicated by a value 100%. Correlation coefficient (r_x) is a metric to estimate the similarity between two signals and the regression (b_x) fairly estimates the amplitude differences between the measured and reconstructed signals.

Evaluation metrics are defined below:

$$R^2 = \left\{1 - \frac{\sum\left[\text{Derived(sample } k) - \text{Measured(sample } k)\right]^2}{\sum\left[\text{Measured(sample } k)\right]^2}\right\} \times 100 \tag{7.1}$$

$$r_x = \left\{\frac{\sum(\text{Measured sample } i) \times (\text{Derived sample } i)}{\left(\sum(\text{Measured sample } i)^2 \times (\text{Derived sample } i)^2\right)^{1/2}}\right\}, \tag{7.2}$$

$$b_x = \left\{\frac{\sum(\text{Measured sample } i) \times (\text{Derived sample } i)}{\sum(\text{Measured sample } i)^2}\right\} \tag{7.3}$$

TABLE 7.1

Mean R^2, r_x, and b_x [16,17] Values of Various Categories for the Transformation of R3L to S12 System

	V1			V3			V4			V5			V6			Average
	R^2	r_x	b_x	R^2	r_x	b_x	R^2	r_x	b_x	R^2	r_x	b_x	R^2	r_x	b_x	R^2
BB	90.40	0.947	0.899	95.78	0.979	0.975	87.39	0.937	0.913	88.47	0.944	0.919	92.41	0.962	0.952	90.89
HC	94.69	0.973	0.952	96.52	0.983	0.973	91.83	0.959	0.939	93.53	0.968	0.957	94.69	0.973	0.967	94.25
HY	97.44	0.987	0.979	96.23	0.981	0.968	87.45	0.934	0.897	89.90	0.945	0.899	95.74	0.978	0.957	93.35
MI	94.34	0.971	0.947	95.92	0.979	0.959	89.52	0.945	0.895	89.22	0.944	0.889	91.52	0.956	0.911	92.10
VA	93.73	0.967	0.942	94.25	0.970	0.945	89.97	0.948	0.906	91.89	0.959	0.921	93.78	0.968	0.934	92.72
ND	90.92	0.949	0.913	93.54	0.967	0.931	83.72	0.910	0.838	84.61	0.916	0.847	87.28	0.920	0.876	88.01

Note: Average value shown has been taken over the reconstructed precordial leads. It should be noted that the rest of the leads, namely, I, II, III, aVR, aVL, and aVF, are obtained with approximately [1].

Table 7.1 presents mean R^2, correlation (r_x), and regression (b_x) values of various categories of the working PTBDB [16,17].

The proximity effect can be observed from the table. Leads in close proximity of the basis lead have better reconstruction compared to those away from it; this is evident from the R^2 values of leads V1 and V3, which are higher than other precordial leads owing to their close proximity to V2, one of the chosen basis leads in this investigation. The lowest values can be found for lead V4.

R3L system was formed using leads I, II and one of the six precordial leads, that is, V1–V6 resulting in a total of six such combinations and along with FV system they all were used to reconstruct S12 system.

Table 7.2 presents the mean values of evaluation metrics for INCARTDB. It is seen that reconstruction results in the case of S12 system are superior to those of ML12 system. This difference is probably because of the electrode positions in ML12, which are on the torso and not on arms and legs as in S12 system.

Table 7.3 provides the mean values of PT coefficients for PTBDB (healthy control [HC] and unhealthy [UH]) and INCARTDB [1].

N.B. Patients in PTBDB were divided in healthy control (HC) and unhealthy (UH) subjects [1].

Figure 7.3a,b corresponds to patients with mean R^2 values of 94.28% (HC) and 90.06% (UH), respectively (mean taken over 5 reconstructed missing precordial

TABLE 7.2

Lead-Wise Mean R^2, Correlation (r_x), and Regression (b_x) Values of Recordings in INCARTDB [1]

	INCARTDB		
	R^2	r_x	b_x
V1	86.38	0.936	0.923
V3	86.91	0.940	0.926
V4	83.61	0.922	0.901
V5	83.74	0.921	0.911
V6	78.11	0.893	0.872
Average	83.75	0.922	0.907
Minimum	78.11	0.893	0.872

Note: INCARTDB, Institute of Cardiological Technics 12-Lead Arrhythmia Database.

TABLE 7.3

Mean Values of Personalized Transformation Coefficients Obtained [1]

	PTBDB [1]						INCARTDB [1]		
	Healthy Control			Unhealthy					
V1	−0.6767	−0.0561	0.4926	−0.6250	−0.0965	0.5006	−0.7450	−0.1431	0.5434
V3	0.7297	0.3298	0.8506	0.4449	0.3588	0.9637	−0.2267	0.5070	1.1181
V4	1.4101	0.5072	0.4149	0.9036	0.5792	0.5915	0.1140	0.7191	0.5214
V5	1.5883	0.4610	0.0314	1.0042	0.5933	0.1774	0.7173	0.6890	−0.0455
V6	1.1729	0.3618	−0.1173	0.7330	0.4710	−0.0442	0.6792	0.6080	−0.3074

Note: INCARTDB, Institute of Cardiological Technics 12-Lead Arrhythmia Database.

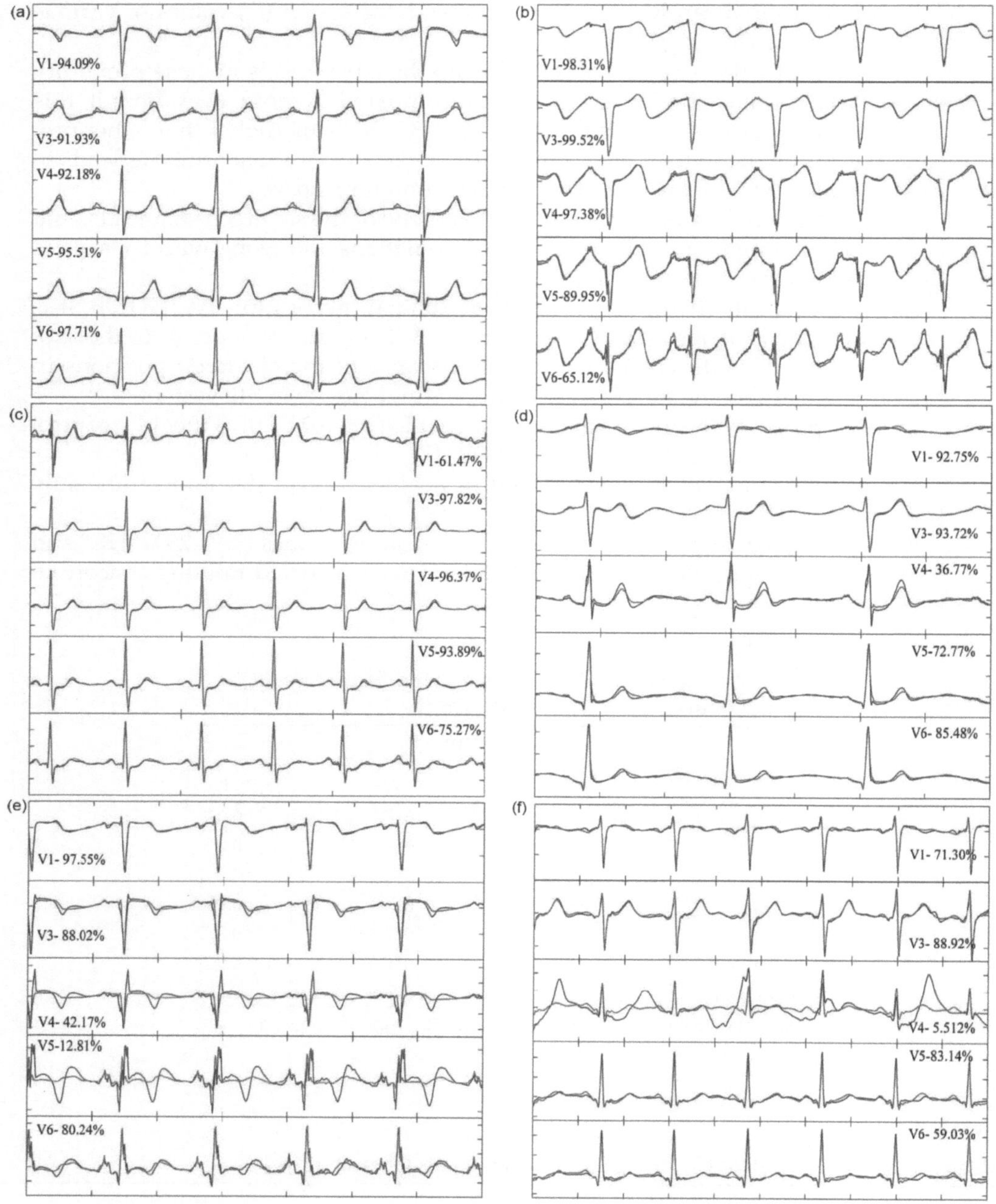

FIGURE 7.3
The comparative study between original (blue) and derived (red) leads of the subjects from PTBDB and INCARTDB. (a and b) The comparison of subjects from HC and UH category, respectively, with the mean R^2 values near to actual means provided in Table 7.1. (c) The corresponding mean case from INCARTDB. (d–f) Minimum case of mean R^2 values for the subjects from HC, UH, and INCARTDB, respectively. The corresponding R^2 values of the reconstructed lead have been provided with the figure. (Image appears in color in eBook version of this book.)

leads), which are nearly equal to the mean values given in Table 7.2. Figure 7.3c is a similar mean case subject from INCARTDB with mean R^2 value being 84.97%. Similarly, Figure 7.3d–f provides the reconstruction results of subjects with minimum mean R^2 values from HC, UH, and INCARTDB, respectively. The range of R^2 values for reconstructed leads shown in Figure 7.3 is from 5.512% to 99.52%. It should be noted that R^2 values of 80% and above can be considered to have high diagnostic value and R^2 values of 90% and above can be considered to be an accurate retrace of the original 80% (51 × 5) value and 66% leads had more than 90% R^2 value. For the reconstruction of 340 (68 × 5) precordial leads of ML12 system, about 73.82% had more than 80% R^2 value and 50% leads had more than 90% R^2 value. In Figure 7.3e,f, relative smoothing and stabilization of probably badly positioned leads V4 and V5 can be seen. This is one of the advantages of lead reconstruction as badly acquired leads can be reconstructed from the well-acquired basis leads.

In this section of study, authors have attempted to address the requirements of cardiologists and technological constraints encountered in telemonitoring applications [1]. A personalized reconstruction methodology for reconstruction of missing precordial leads of S12 and ML12 systems have been proposed and the results show that using this methodology reliable and accurate reconstruction is possible. R3L system used in this paper requires no change in the standard electrode positions of S12 or ML12 system and hence it is a practical option for clinical usage. Reconstruction of missing precordial leads of S12 outperforms ML12 system. Here, we have presented an in-depth analysis of reusability of transformation coefficients and analyzed the reconstruction results when the same coefficients are used for reconstruction. It has been inferred that if care is taken while placement of electrodes is at the standard positions, then reconstruction results are less likely to vary over time [1].

B. The methodology described in summary of the study cited in Ref. [2] is to reconstruct the eight independent leads of S12 system from FV's leads, that is, X, Y, and Z. Brief description of methodology is as follows: first, raw ECG and VCG leads are processed in a preprocessing module followed by coefficient generation using HVP theory and LS fit method.

The coefficients generated are then used to reconstruct the leads from Frank's leads using HVP theory.

A wavelet-based preprocessing module has been proposed. The preprocessing module consists of two steps, namely, removal of BW, and second, denoising, as shown in Figure 7.4 [2]. Denoising has been performed by using three variants, as shown in Figure 7.4, of wavelet transforms, namely, discrete wavelet transform, undecimated wavelet transform [18], and translation invariant wavelet transform [19]. BW has been removed using DWT [9]. The preprocessing module consists of four different approaches, all of which consisted of two steps. The first step involves the removal of BW using DWT following the procedure stated in Zhang [19] and is same in all the approaches, while the second step is denoising and performed in the second, third, and fourth approach.

On comparing the proposed PT with the previously proposed state-of-the-art transformations, namely, AT and DT [17,20], with the help of Table 7.4, it is found

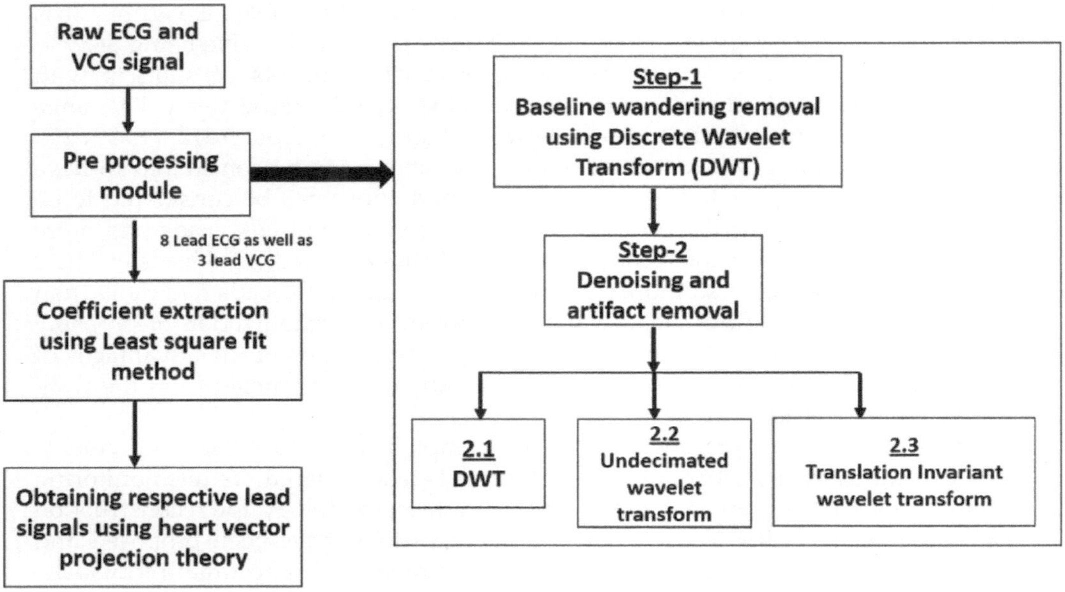

FIGURE 7.4
An overview of the complete methodology.

that there is a significant mean improvement of 12.06% over AT and 20.73% over DT for HC subjects, while for UH subjects it was found to be 20.67% and 32.33% over AT and DT, respectively.

Figures 7.5 and 7.6 provide an insight into the correspondence between the quality of reconstruction and R^2 values. With careful observations, it was found that the R^2 values of 80% or more can be considered to have high diagnostic value and reconstructed leads with an R^2 value of 90% or more, thereby producing a complete retrace of the originally measured signal.

In this work, designers [2] have proposed a personalized lead reconstruction methodology to reconstruct S12 system from FV leads. A wavelet-based preprocessing module has been introduced to remove BW and noise. Three variants of wavelet transform have been evaluated, and it has been found that translation invariant wavelet transform (TIWT) outperforms others, however not with significant differences. The proposed methodology was compared with the state-of-the-art DT and AT.

C. Figure 7.7 shows summary of the proposed methodology. Here, authors envisage [3] a scenario where S12-lead ECG is being acquired in a remote or hospital-based environment. Generally, in a remote health-monitoring scenario, these leads are needed to be transmitted to nearby state-of-the-art health center for diagnosis, storage, and updating of patient's health record. At the transmission end, the conventional S12 acquisition system is used to capture the ECG. Eight independent leads of the S12 system are then passed through the lead component (LC) system formation module. Using LC and S12 system together, the transformation coefficients are obtained by employing LS fit method. Then, LC leads along with the

TABLE 7.4

Mean R^2, r_x, and b_x Values Obtained on Applying the Proposed PT (Personalized Transformation), AT (Affine Transform), and DT (Dower Transform) on HC and UH Subjects [2]

	A1			A2			A3			A4			AT			DT		
	R^2	r_x	b_x	R^2	r_x	b_x	R^2	r_x	b_x	R^2	r_x	b_x	R^2	r_x	b_x	R^2	r_x	b_x
Healthy Control																		
I	92.89	0.963	0.926	93.01	0.963	0.927	93.05	0.964	0.928	93.00	0.963	0.927	76.58	0.899	0.812	66.91	0.893	0.808
II	97.36	0.987	0.975	97.43	0.987	0.976	97.46	0.987	0.976	97.38	0.987	0.975	93.84	0.976	0.927	92.28	0.973	1.028
V1	95.70	0.978	0.956	95.75	0.978	0.957	95.77	0.979	0.957	95.77	0.979	0.957	81.76	0.942	0.908	73.46	0.910	0.858
V2	93.90	0.969	0.943	93.94	0.969	0.944	93.95	0.969	0.944	93.96	0.969	0.943	75.79	0.921	0.929	63.87	0.806	0.651
V3	96.46	0.982	0.966	96.48	0.982	0.966	96.49	0.982	0.966	96.63	0.983	0.965	83.10	0.927	0.897	57.55	0.827	0.880
V4	97.42	0.987	0.974	97.44	0.987	0.974	97.46	0.987	0.974	97.65	0.988	0.976	82.34	0.937	0.925	72.56	0.921	0.972
V5	98.94	0.995	0.991	98.95	0.995	0.991	98.96	0.995	0.991	98.96	0.995	0.991	89.71	0.962	0.963	89.27	0.953	0.891
V6	98.98	0.995	0.991	99.00	0.995	0.991	99.01	0.995	0.991	99.00	0.995	0.991	92.70	0.970	0.997	90.58	0.958	0.844
Average	**96.46**	0.982	0.965	**96.50**	0.982	0.966	**96.52**	0.982	0.966	**96.54**	0.982	0.966	**84.48**	0.942	0.920	**75.81**	0.905	0.867
Unhealthy																		
I	90.04	0.948	0.909	90.34	0.950	0.913	90.39	0.950	0.913	90.40	0.950	0.913	48.32	0.866	0.884	60.23	0.867	0.681
II	92.79	0.963	0.936	92.99	0.964	0.939	93.04	0.964	0.939	93.07	0.964	0.939	85.54	0.932	0.894	78.64	0.921	0.999
V1	92.83	0.963	0.936	92.94	0.964	0.937	92.96	0.964	0.938	92.97	0.963	0.938	73.58	0.923	0.912	60.42	0.865	0.745
V2	91.06	0.953	0.916	91.14	0.953	0.917	91.13	0.953	0.917	91.16	0.953	0.918	67.18	0.893	0.867	55.62	0.807	0.612
V3	93.44	0.966	0.939	93.49	0.967	0.941	93.50	0.967	0.941	93.51	0.967	0.941	71.46	0.905	0.909	47.86	0.772	0.661
V4	94.51	0.972	0.951	94.56	0.972	0.952	94.57	0.972	0.952	94.57	0.972	0.952	72.67	0.909	0.890	46.56	0.779	0.711
V5	96.08	0.980	0.965	96.14	0.980	0.965	96.15	0.980	0.966	96.15	0.980	0.966	80.46	0.914	0.877	70.06	0.872	0.818
V6	95.45	0.976	0.957	95.55	0.977	0.958	95.57	0.977	0.958	95.57	0.977	0.959	82.81	0.943	0.917	69.36	0.903	0.832
Average	**93.28**	0.965	0.939	**93.39**	0.966	0.940	**93.41**	0.966	0.940	**93.42**	0.966	0.941	**72.75**	0.911	0.894	**61.09**	0.848	0.757

Note: PT was performed using various preprocessing approaches denoted in the table by A1–A4. The values denoted in 'bold' are mean R^2 values taken over the eight reconstructed independent leads of standard 12-lead system.

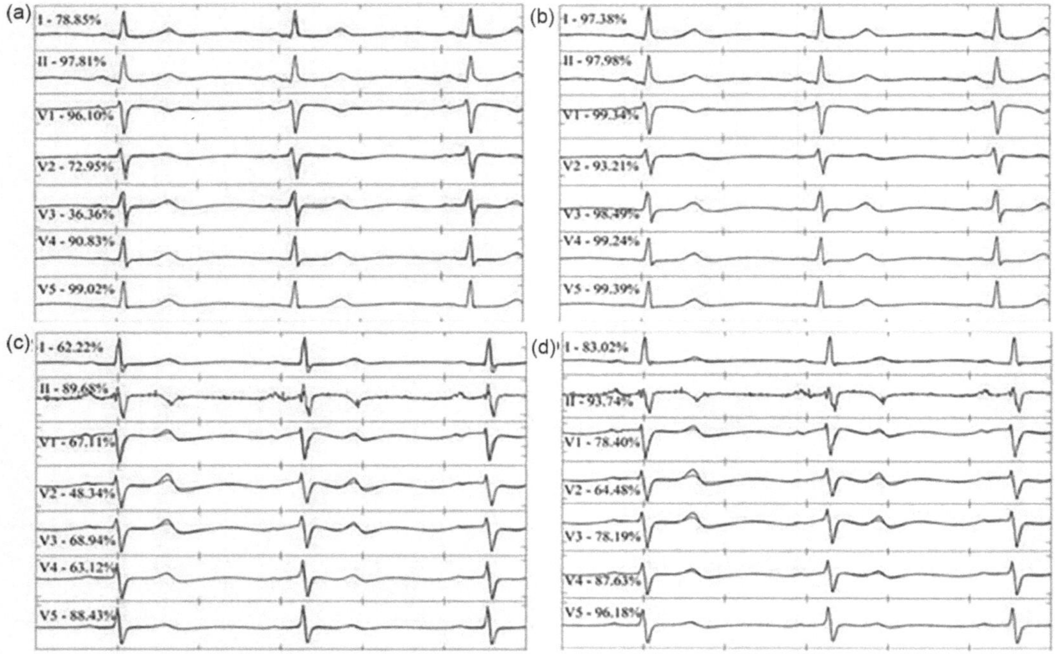

FIGURE 7.5
A comparative study between the reconstructed (red) and original (blue) signal. One patient each has been taken from HC and UH categories, and the reconstruction has been performed using AT and PT methodologies. The patients chosen have mean R^2 values of 83.89% (HC) and 72.82% (UH), obtained using AT, which is close to the overall mean using AT, i.e. 84.48% (HC) and 72.75% (UH) as can be seen from Table 7.4: (a) reconstruction of HC subject using AT, (b) reconstruction of HC subject using PT, (c) reconstruction of UH subject using AT, and (d) reconstruction of UH subject using PT. (Image appears in color in eBook version of this book.)

transformation coefficients can be transmitted and eventually eight independent leads of the S12 system can be reconstructed for diagnosis at the receiver's end in an optimum way.

Table 7.5 presents the mean R^2, r_x, and b_x values for the reconstruction of S12 system from LC and S12S systems using both PTBDB and TWADB. The average values have been computed over eight independent leads of S12 system. The R^2 values are higher for the precordial leads compared to bipolar leads I and II. The transformation of proposed LC system to S12 has outperformed S12S system. This confirms that the information obtained from standard leads is better represented in LC system than that of a subset [3].

Figure 7.8 presents the range of R^2 values with the mean denoted by the circle with green dot and minimum and maximum values denoted by the whiskers. It should be noted that Figure 7.8 shows minimum R^2 values being negative, which denoted that the reconstructed signal has failed to trace the original and is completely out of phase with it. From the range of values, it can be seen that the proposed LC system is considerably better and reliable for the reconstruction of precordial leads.

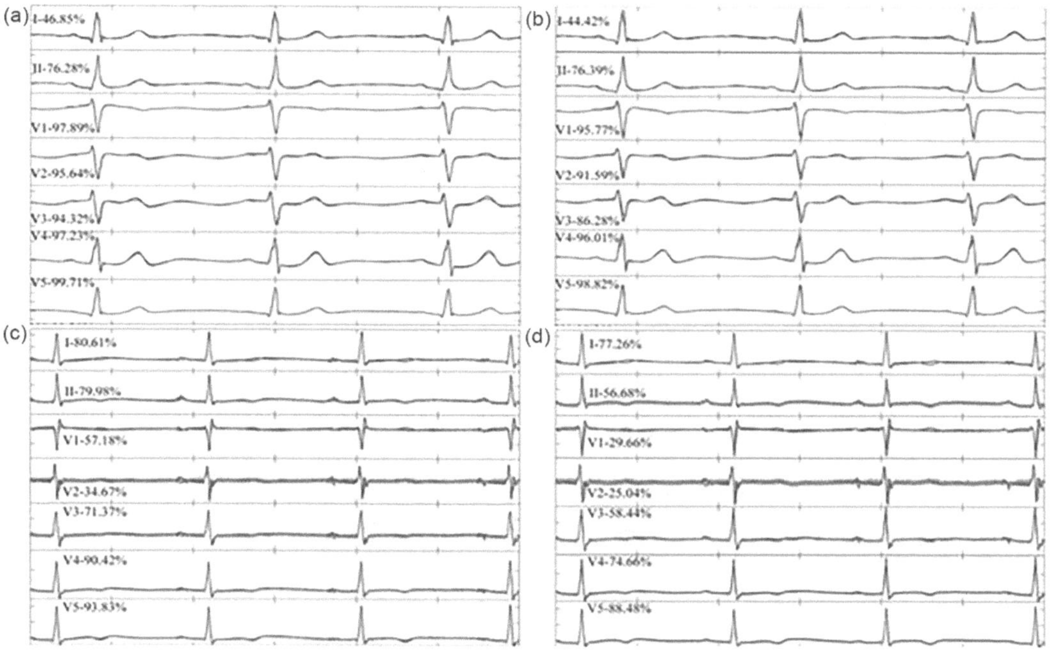

FIGURE 7.6
A comparative study between the reconstructed (red) and original (blue) signal. One patient each has been taken from HC and UH categories, and the reconstruction has been performed using AT and PT methodologies. The patients chosen have minimum mean R^2 values of 88.41% (HC) and 75.22% (UH), obtained using proposed PT, and the corresponding values when AT was used on the same subjects were 85.94% (HC) and 61.72% (UH): (a) reconstruction of HC subject using PT, (b) reconstruction of HC subject using AT, (c) reconstruction of UH subject using PT, and (d) reconstruction of UH subject using AT. (Image appears in color in eBook version of this book.)

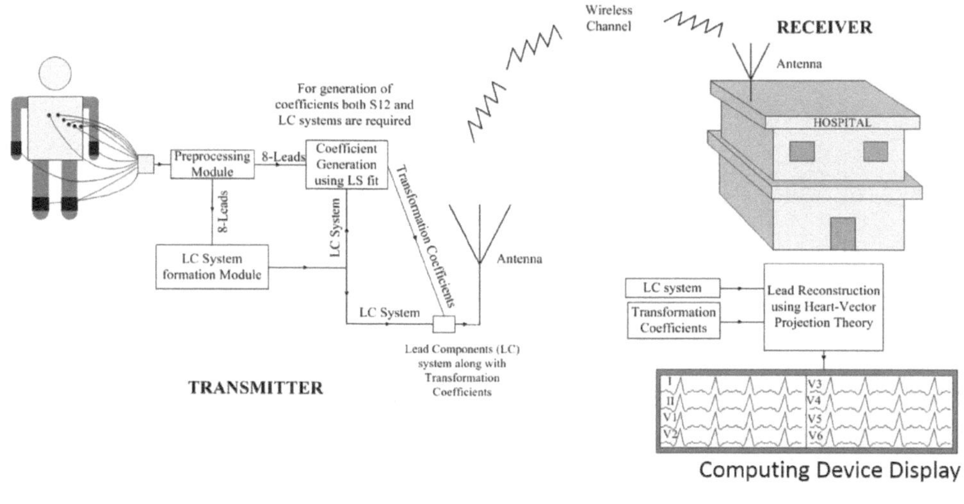

FIGURE 7.7
A summarized version of the proposed methodology. (Adopted from Ref. [1] and modified.)

TABLE 7.5

Mean R^2, r_x, and b_x Values of the Transformation of LC to S12 (1) and S12S to S12 (2) Systems for Both TWADB (72 Patients) and PTBDB (First Recording of 290 Patients) [3]

		TWADB									PTBDB								
		I	II	V1	V2	V3	V4	V5	V6	Average	I	II	V1	V2	V3	V4	V5	V6	Average
(1)	R^2	93.24	91.72	96.19	97.33	98.77	98.52	98.11	97.07	96.37	93.32	93.25	95.33	96.76	97.46	96.96	97.17	94.77	95.63
	r_x	0.974	0.956	0.984	0.989	0.994	0.993	0.992	0.987	0.984	0.970	0.968	0.978	0.985	0.989	0.987	0.987	0.975	0.979
	b_x	0.969	0.951	0.987	0.996	0.997	0.994	0.992	0.994	0.985	0.956	0.969	0.974	0.986	0.993	0.995	0.987	0.964	0.978
(2)	R^2	100	100	85.25	100	96.17	88.93	82.51	79.71	91.57	100	100	88.04	100	90.34	77.25	73.99	75.46	88.14
	rx	1	1	0.954	1	0.981	0.942	0.926	0.918	0.965	1	1	0.938	1	0.955	0.902	0.886	0.887	0.946
	b_x	1	1	0.958	1	0.969	0.928	0.931	0.935	0.965	1	1	0.918	1	0.947	0.890	0.876	0.877	0.939

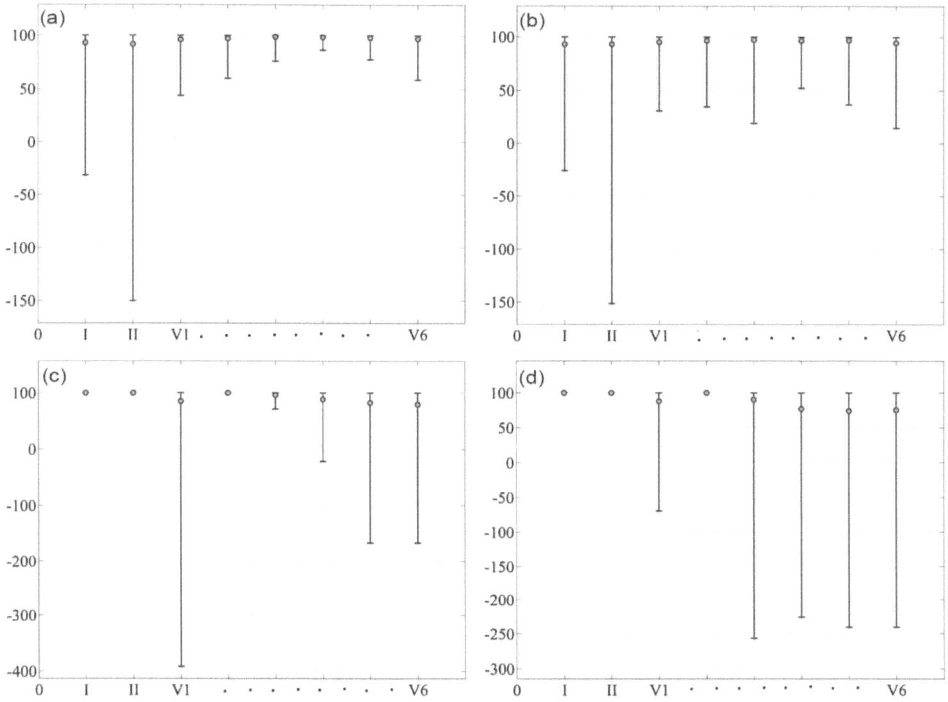

FIGURE 7.8
The mean (circle with green inside) and the range (whiskers) of R^2 values for both LC to S12 (a-TWADB, b-PTBDB) and S12S to S12 (c-TWADB, d-PTBDB) transformations [3].

Figure 7.9 shows the comparison between the originally measured (blue) and reconstructed (red) signal. One subject is each from TWADB (Figure 7.9a,b) and PTBDB (Figure 7.9c,d). (Image appears in color in eBook version of this book.) The two subjects have mean R^2 values of 91.69% (TWADB) and 88.11% (PTBDB), which is very close to the overall mean values stated in Table 7.5 for S12S to S12 transformation and the corresponding values for LC to S12 transformation were 96.71% and 92.64%. The R^2 values of respective leads have been indicated in the figure. The R^2 values in Figure 7.9 range from 54.61% to 100% and provide insight into the correspondence between the quality of reconstruction and R^2 values. Table 7.6 presents the RMSE values for the features extracted from eight independent leads of S12 system between the originally measured and reconstructed signal. Thirteen different features were extracted for both the methodologies. Mean improvement of over 45% was observed in the proposed LC to S12 compared to S12S to S12 transformation.

From a wide range of R^2 values in Figure 7.9 i.e. 54.61%–100%, we have found that R^2 value of 80% or above can be considered to have significant diagnostic value and a value of 90% or above is practically retracing the original waveform. For PTBDB and TWADB using the proposed methodology, approximately, 96% of patients were found to have mean R^2 value of 80% and above. Similarly, 91% and 93% of patients were found to have mean R^2 value of 90% and above for PTBDB and TWADB, respectively. For S12S to S12 transformation, the fraction of patients above with mean R^2 values ≥80% were 86% (PTBDB) and 92% (TWADB). The fraction of patients with mean R^2 values ≥90% were 76% (PTBDB) and 83% (TWADB) [3].

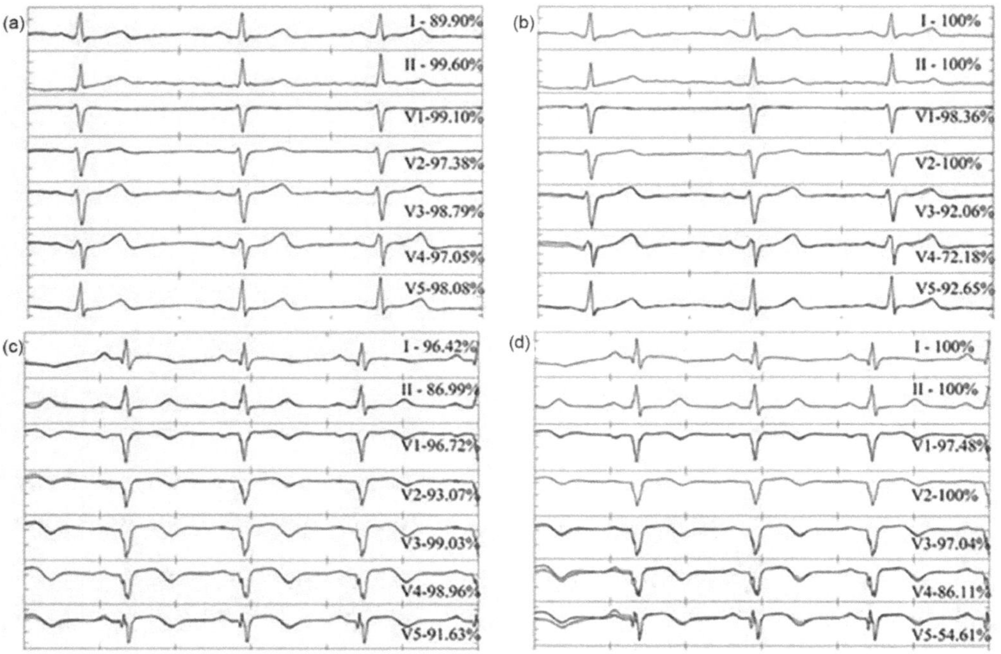

FIGURE 7.9
A comparative study between the two systems (LC and S12S) for the mean case S12S patient one each from TWADB and PTBDB. (a and b) Comparison of eight independent leadsi.e. I, II, V1–V6 for the TWADB and PTBDB. (c and d) A comparison of a mean case patient from PTBDB. The reconstructed lead (red) has been plotted over original lead (blue). Corresponding R^2 values have been shown on respective plots [3]. (Image appears in color in eBook version of this book.)

D. Figure 7.10 shows the two possible ways in which a patient can be registered at the health center. During the registration process, the transformation coefficients are generated which can be eventually used to reconstruct the S12 system. The first scenario shows the online registration process and second shows the offline registration process. 'Online' means that the patient needs not to be present in the hospital/health-center physically [4]. In this case, the RL or Frank leads can be captured and transmitted to the hospital/health-center for reconstruction of S12-leads. 'Offline' means that the patient is present in the hospital/heath-center physically. In this case, S12-lead reconstruction is done at the hospital/health-center itself. The aforementioned remote health-monitoring service can benefit both ambulatory patients and the patients living in rural or remotely accessed areas. The following section discusses the registration process in the context of RL system used.

1. R3L systems to S12 system [21–24]

   If the reconstruction methodology being adopted is the transformation of R3L systems to S12 system, then for coefficient generation the acquisition of only S12 system is required. These coefficients can then be used to reconstruct S12 system on eventual readings.

2. FV system to S12 system

   If the reconstruction methodology being adopted is the transformation of FV to S12 system, then for the coefficient generation a simultaneous acquisition

TABLE 7.6

Mean Root Mean Square Error (RMSE) Values for Various Features Extracted Using TDMG [9] for the Reconstruction Methodology

				LC to S12						S12S to S12				
Sl. no	**Feature (unit)**	**I**	**II**	**V1**	**V2**	**V3**	**V4**	**V5**	**V6**	**V1**	**V3**	**V4**	**V5**	**V6**
1	P duration (ms)	9.755	13.18	13.43	10.12	7.306	9.51	9.632	7.755	16.86	12.37	18	18.82	19.63
2	P height (μV)	52.89	63.59	92.53	84.82	50.52	81.96	37.58	37.03	133.3	110.1	177.7	113.4	89.27
3	PR interval (ms)	11.39	11.55	15.27	12.65	6	7.225	9.020	9.102	19.59	11.47	12.98	12.61	17.06
4	PR segment (ms)	10.78	6.449	15.14	8.163	5.551	9.388	5.265	5.020	18.33	11.35	16.78	12	16.04
5	Q peak (μV)	19.58	14.33	47.73	60.85	54.27	52.51	26.20	16.27	77.22	111.6	353.4	62.59	22.27
6	QRS length (ms)	10.65	7.184	7.633	3.755	4	6.816	5.184	6.122	9.306	9.143	11.63	10.61	10.37
7	QT interval (ms)	16.08	15.8	26.98	7.551	18.61	10	5.469	8.531	30.61	21.59	20.33	11.10	18.65
8	R height (μV)	47.26	51.03	47.12	99.95	60.18	88.54	48.41	31.72	78.21	182.4	480.4	158.9	90.75
9	S peak (μV)	9.125	10.36	19.27	0.035	0.036	2.825	9.714	10.20	28.90	2.689	4.220	19.85	17.29
10	ST interval (ms)	9.184	8.857	24.25	7.143	6.612	6.122	4.367	7.796	31.59	7.225	15.06	11.76	18.16
11	ST segment (ms)	14.16	11.92	18.61	10.37	7.184	11.27	6.490	10.90	28.86	7.633	17.76	18.08	20.37
12	T duration (ms)	43.32	64.35	107.5	114.5	139.0	105.2	73.27	43.96	161.2	199	202.5	135.7	98.76
13	T height (μV)	22.18	32.95	55.05	58.62	71.18	53.84	37.51	22.51	82.53	101.9	103.7	69.50	50.57

Note: Mean was taken over all the 49 patients (please see Section IV D). The leads I, II, and V2 formed the basis leads of S12S, so the RMSE values for them were zero and therefore have not been reported [3].

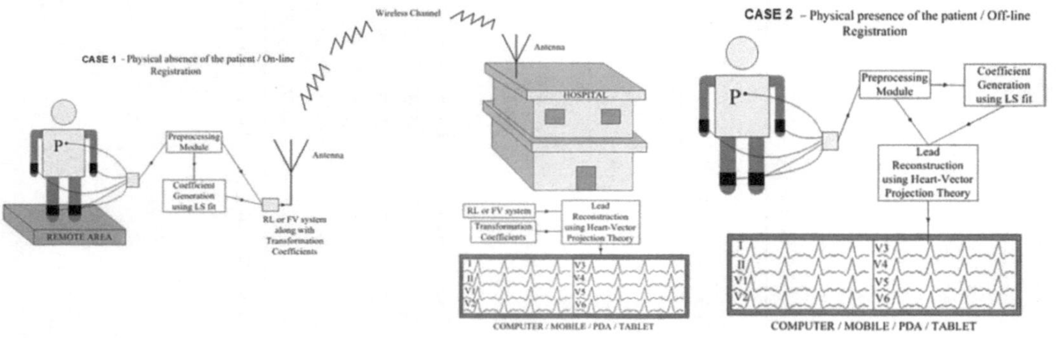

FIGURE 7.10
Proposed remote health-monitoring scenarios: (1) when patient may not be physically present for registration (online) and (2) when patient is available for registration [4].

TABLE 7.7
Denotations of R3L Systems and Various ECG Features Extracted Using TDMG Algorithm [4]

Basis Leads of RL (Reduced Lead) System	Denotation	ECG Features (Represents Unit)	Denotation
I, II, V1	I	P duration (ms)	**1**
I, II, V2	II	P height (mV)	**2**
I, II, V3	III	PR interval (ms)	**3**
I, II, V4	IV	PR segment (ms)	**4**
I, II, V5	V	Q peak (mV)	**5**
I, II, V6	VI	QRS length (ms)	**6**
Frank leads X, Y, and Z	**FV**	QT interval (ms)	**7**
		R height (mV)	**8**
		S peak (mV)	**9**
		ST interval (ms)	**10**
		ST segment (ms)	**11**
		T duration (ms)	**12**
		T height (mV)	**13**

of both the systems is required. On eventual monitoring, only FV system's acquisition is required which can be then transformed to S12 system.

Authors have used 'I', 'II', etc. in bold letters to denote R3L systems with precordial leads V1, V2, etc. as the basis leads in the R3L systems, respectively (Table 7.7). FV system has been denoted by FV in bold. Table 7.7 also shows the denotations of 13 different features that were extracted using TDMG for comparison starting from 1 to 13 in bold letters [4].

Table 7.8 presents the mean R^2, correlation and regression coefficient values of the FRM i.e. R3L systems to S12 system for the reconstruction of lead set 2. The last column of Table 7.8 shows the mean R^2 values over all the 12-leads of patients in the corresponding categories using FRM.

Table 7.9 shows the reconstruction results from SRM, i.e. FV to S12 system with average values mentioned in the last row [4].

Table 7.10 provides the mean RMSE values for all the seven methodologies. It can be seen that higher values of error have mainly occurred when height or depth of a

TABLE 7.8

Mean of R^2, r_x, and b_x Values of Various Categories for the Transformation of RL Systems (V1–V6: I–VI) to S12 Systems [4]

		V1			V2			V3			V4			V5			V6			
		R^2	r_x	b_x	R^2	r_x	b_x	R^2	r_x	b_x	R^2	r_x	b_x	R^2	r_x	b_x	R^2	r_x	b_x	Average
BB	I	100	1	1	83.36	0.91	0.872	79.93	0.9	0.865	80.54	0.91	0.886	87.64	0.938	0.909	92.59	0.963	0.945	93.07
	II	90.4	0.947	0.899	100	1	1	95.78	0.979	0.975	87.39	0.937	0.913	88.47	0.944	0.919	92.41	0.962	0.952	95.6
	III	87.05	0.929	0.864	95.07	0.975	0.942	100	1	1	94.16	0.971	0.965	89.36	0.949	0.93	92.25	0.962	0.952	95.89
	IV	84.68	0.919	0.849	77.53	0.873	0.779	89.47	0.943	0.896	100	1	1	91.6	0.963	0.954	92.03	0.962	0.951	94
	V	78.24	0.889	0.815	64.08	0.798	0.688	67.28	0.813	0.728	74.2	0.848	0.805	100	1	1	94.24	0.971	0.963	89.23
	VI	83.18	0.911	0.832	71.52	0.847	0.757	73.5	0.867	0.804	76.13	0.885	0.839	91.15	0.954	0.913	100	1	1	90.69
HC	I	100	1	1	94.57	0.972	0.947	93.51	0.967	0.935	93.14	0.965	0.936	95.29	0.976	0.957	96.82	0.984	0.972	97.76
	II	95.79	0.978	0.959	100	1	1	97.3	0.986	0.975	94.29	0.971	0.948	95.54	0.978	0.959	97.01	0.985	0.974	98.31
	III	92.17	0.958	0.924	95.55	0.978	0.961	100	1	1	96.77	0.984	0.97	96.11	0.98	0.963	96.81	0.984	0.972	98.1
	IV	82.23	0.901	0.827	80.59	0.888	0.816	92.57	0.962	0.93	100	1	1	97.66	0.988	0.978	96.6	0.983	0.969	95.79
	V	73.18	0.846	0.74	66.38	0.794	0.678	77.04	0.868	0.777	92.18	0.959	0.924	100	1	1	97.78	0.989	0.981	92.2
	VI	78.49	0.878	0.795	74.56	0.857	0.764	79.04	0.885	0.801	87.66	0.935	0.883	96.92	0.984	0.969	100	1	1	93.04
HY	I	100	1	1	96.15	0.98	0.962	90.83	0.953	0.921	85.62	0.925	0.883	89.13	0.939	0.893	95.39	0.976	0.953	96.41
	II	97.44	0.987	0.979	100	1	1	96.23	0.981	0.968	87.45	0.934	0.897	89.9	0.945	0.899	95.74	0.978	0.957	97.22
	III	95.58	0.978	0.961	97.3	0.987	0.972	100	1	1	93.52	0.966	0.949	91.61	0.955	0.918	95.98	0.979	0.958	97.82
	IV	91.88	0.958	0.925	88.94	0.941	0.888	93.48	0.967	0.93	100	1	1	94.45	0.971	0.941	96.63	0.983	0.962	97.1
	V	85.25	0.921	0.861	78.51	0.878	0.797	77.9	0.883	0.796	86.81	0.929	0.887	96.91	0.984	0.967	100	1	1	93.77
	VI	84.73	0.916	0.855	77.05	0.861	0.778	70.86	0.83	0.721	74.65	0.842	0.76	90.32	0.947	0.906	100	1	1	91.45
MI	I	100	1	1	92.1	0.958	0.923	87.73	0.935	0.88	84.07	0.914	0.844	87.56	0.934	0.875	91.19	0.954	0.907	95.2
	II	94.34	0.971	0.947	100	1	1	95.92	0.979	0.959	89.52	0.945	0.895	89.22	0.944	0.889	91.52	0.956	0.911	96.69
	III	89.82	0.946	0.907	95.35	0.976	0.959	100	1	1	95.55	0.977	0.954	91.49	0.956	0.912	91.64	0.956	0.913	96.96
	IV	82.21	0.903	0.834	83.34	0.909	0.844	92.55	0.96	0.93	100	1	1	94.84	0.974	0.947	91.83	0.958	0.915	95.37
	V	73.83	0.852	0.75	68.54	0.818	0.701	73.96	0.851	0.753	84.38	0.907	0.85	100	1	1	94.72	0.973	0.946	91.26
	VI	73.93	0.851	0.752	64.98	0.786	0.666	65.35	0.796	0.67	70.35	0.826	0.714	89.99	0.947	0.905	100	1	1	88.69

(Continued)

TABLE 7.8 (*Continued*)

Mean of R2, rx, and bx Values of Various Categories for the Transformation of RL Systems (V1–V6: I–VI) to S12 Systems [4]

		V1			V2			V3			V4			V5			V6			
		R^2	r_x	b_x	R^2	r_x	b_x	R^2	r_x	b_x	R^2	r_x	b_x	R^2	r_x	b_x	R^2	r_x	b_x	Average
VA	I	100	1	1	90.63	0.95	0.901	86.57	0.929	0.866	85.54	0.924	0.863	91.6	0.957	0.917	93.74	0.968	0.934	95.63
	II	93.73	0.967	0.942	100	1	1	94.25	0.97	0.945	89.97	0.948	0.906	91.89	0.959	0.921	93.78	0.968	0.934	96.93
	III	88.68	0.935	0.888	92.86	0.961	0.927	100	1	1	95.84	0.979	0.96	93.79	0.968	0.968	94.27	0.97	0.937	97.08
	IV	81.68	0.895	0.817	80.38	0.889	0.801	91.69	0.957	0.913	100	1	1	93.96	0.963	0.941	94.89	0.974	0.945	95.18
	V	74.11	0.834	0.748	68.59	0.816	0.691	77.72	0.877	0.779	90.66	0.95	0.908	100	1	1	97.14	0.985	1	92.31
	VI	78.02	0.874	0.782	69.18	0.819	0.692	72.02	0.841	0.723	80.46	0.888	0.816	96.02	0.979	0.967	100	1	1	91.27
ND	I	100	1	1	90.42	0.951	0.916	81.28	0.891	0.825	78.06	0.876	0.787	82.08	0.902	0.823	85.84	0.91	0.86	92.39
	II	90.92	0.949	0.913	100	1	1	93.54	0.967	0.931	83.72	0.91	0.838	84.61	0.916	0.847	87.28	0.92	0.876	94.44
	III	88.18	0.937	0.899	95.74	0.978	0.963	100	1	1	93.07	0.965	0.931	88.5	0.939	0.887	88.4	0.932	0.89	96.02
	IV	83.27	0.911	0.856	83.5	0.914	0.857	90.22	0.95	0.916	100	1	1	95.58	0.977	0.952	92.85	0.963	0.928	95.39
	V	76.77	0.874	0.796	68.32	0.825	0.723	67.11	0.802	0.704	90.12	0.948	0.909	100	1	1	96.89	0.984	0.969	91.52
	VI	78.51	0.883	0.816	72.83	0.853	0.766	69.69	0.831	0.728	83.32	0.911	0.843	94.97	0.974	0.951	100	1	1	91.51

Note: Average value depicted in this table has been taken from all the 12 leads.

TABLE 7.9

Visualizing Mean R^2, r_x, and b_x Values of Various Categories for the Transformation of FV System to S12 System [4]

	BB			HC			HY			MI			VA			ND		
	R^2	r_x	b_x	R^2	r_x	b_x	R^2	r_x	b_x	R^2	r_x	b_x	R^2	r_x	b_x	R^2	r_x	b_x
I	90.89	0.952	0.898	95.99	0.98	0.96	95.34	0.976	0.959	92.41	0.961	0.93	94.15	0.971	0.955	93.61	0.968	0.927
II	94.29	0.97	0.944	99.04	0.995	0.99	98.33	0.992	0.983	96.54	0.982	0.967	97.41	0.987	0.982	93.55	0.966	0.947
III	88.29	0.916	0.88	93.18	0.965	0.929	95.41	0.976	0.952	93.55	0.967	0.94	91.99	0.958	0.918	91.84	0.952	0.92
aVR	91.06	0.953	0.914	98.49	0.993	0.985	97.71	0.989	0.979	93.86	0.969	0.945	95.59	0.978	0.966	91.6	0.955	0.918
aVL	91.05	0.953	0.904	91.75	0.957	0.918	93.63	0.967	0.937	92.18	0.96	0.926	90.28	0.945	0.914	94.24	0.971	0.963
aVF	96.01	0.98	0.961	98.09	0.99	0.98	97.34	0.987	0.974	95.99	0.979	0.96	96.41	0.982	0.966	95.6	0.977	0.959
V1	91.12	0.954	0.897	96.41	0.982	0.966	97.29	0.986	0.977	94.15	0.97	0.949	94.83	0.974	0.953	92.07	0.959	0.928
V2	89.31	0.943	0.89	94.03	0.97	0.942	96.39	0.982	0.961	92.29	0.96	0.929	91.19	0.951	0.916	91.28	0.955	0.925
V3	94.44	0.973	0.953	96.9	0.984	0.968	96.91	0.985	0.971	94.56	0.972	0.95	95.2	0.976	0.959	94.72	0.973	0.953
V4	93.75	0.97	0.961	98.06	0.99	0.982	95.82	0.979	0.974	95.66	0.977	0.959	97.36	0.987	0.984	97.04	0.985	0.967
V5	97.79	0.989	0.976	99.27	0.996	0.993	96.19	0.98	0.964	97.53	0.988	0.975	98.8	0.994	0.991	96.64	0.983	0.964
V6	98.7	0.994	0.991	99.51	0.998	0.996	99.24	0.996	0.992	97.54	0.988	0.974	98.74	0.994	0.988	98.12	0.991	0.983
Average	93.06	0.9623	0.931	96.73	0.983	0.967	96.63	0.983	0.968	94.69	0.973	0.950	95.16	0.975	0.958	94.11	0.969	0.9436

TABLE 7.10

Mean RMSE Values of the ECG Features Extracted Using TDMG Algorithm [9] for All the RL Systems [4]

Sr.	Feature (in Unit)	I	II	III	IV	V	VI	FV
1	P duration (ms)	6.700	5.536	5.423	5.644	5.888	6.357	9.70
2	P height (mV)	55.47	40.04	37.30	51.00	82.70	86.57	60.24
3	PR interval (ms)	9.058	8.639	9.281	10.96	11.80	12.27	17.20
4	PR segment (ms)	8.982	8.567	8.595	9.36	10.68	11.45	15.81
5	Q peak (mV)	63.60	51.33	44.34	61.49	116.4	150.6	97.75
6	QRS length (ms)	5.997	4.710	4.355	4.966	7.407	8.527	9.40
7	QT interval (ms)	11.97	11.26	10.64	12.41	14.38	15.98	21.41
8	R height (mV)	79.96	65.48	54.85	65.47	102.3	133.1	92.78
9	S peak (mV)	14.25	12.87	16.67	20.79	42.10	59.59	29.73
10	ST interval (ms)	10.94	9.889	9.629	10.94	12.69	14.72	19.32
11	ST segment (ms)	10.45	9.542	9.316	10.96	13.36	15.00	18.49
12	T duration (ms)	13.33	11.85	11.03	12.04	12.21	14.88	21.16
13	T height (mV)	52.42	38.26	33.79	54.17	95.76	98.51	53.61

peak was measured in comparison to the features measuring the horizontal intervals. Table 7.11 presents a comparison between FRM and SRM in the context of remote healthcare monitoring applications. The aforementioned mean values and the values used in Figure 7.11 were taken over the complete database (all 275 patients).

Figure 7.12 shows a comparative study between the reconstructed (red) and the measured (blue) ECG signal; Figure 7.12a–c shows the reconstruction of lead set 1 using SRM, i.e. FV to S12 for the worst, 80% and the best case mean R^2 values. Figure 7.12d–i shows the reconstruction of lead set 2 (V1–V6) for all the RL systems with the basis leads starting from V1 (top) to V6 and FV system (bottom) [4]. (Image appears in color in eBook version of this book.)

Figure 7.13 provides the box plot of lead-wise mean RMSE for 13 different features extracted using TDMG over the complete database. Figure 7.13a–h presents thirteen boxes numbered 1–13 each corresponding to a particular feature. The correspondence between the labels in the horizontal axis and features has been described in Table 7.7. Figure 7.13a–f follows from the reconstruction results of RL systems for lead set 2 with basis lead following the order: a–V1, b–V2, c–V3,

TABLE 7.11

Comparison of FRM and SRM in Context of Remote Healthcare Applications [4]

	R3L System	FV System
1	Five electrodes system	Eight electrodes system
2	Three leads	Three leads
3	Inconsistent reconstruction of precordial leads	Consistent reconstruction of precordial leads
4	Comparatively bad reconstruction of V5 and V6	Comparatively better reconstruction of V5 and V6
5	Leads I, II, III, aVR, aVL, and aVF are obtained with approximately no information loss	Comparatively less accurate and information is lost in the reconstructed signal
6	Not much change in already existing acquisition system is required	A different system is required which can acquire both ECG and VCG
7	Online and offline registration possible	Online registration difficult

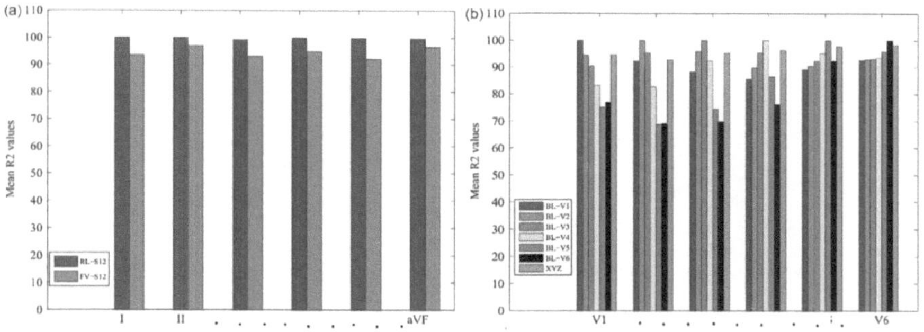

FIGURE 7.11

The bar plot of lead-wise mean R^2 values. (a) The RL to S12 (blue) versus FV to S12 for lead set 1 i.e. I, II, III, aVR, aVL, and aVF. (b) A comparative study between all the seven reconstruction methodologies for reconstruction of precordial leads (lead set 2): leftmost to sixth bar correspond to RL systems with basis lead in the order V1 (blue), V2 (green), V3 (magenta), V4 (yellow), V5 (red), V6 (black), and the rightmost corresponds to FV (cyan) system, respectively. (Image appears in color in eBook version of this book.)

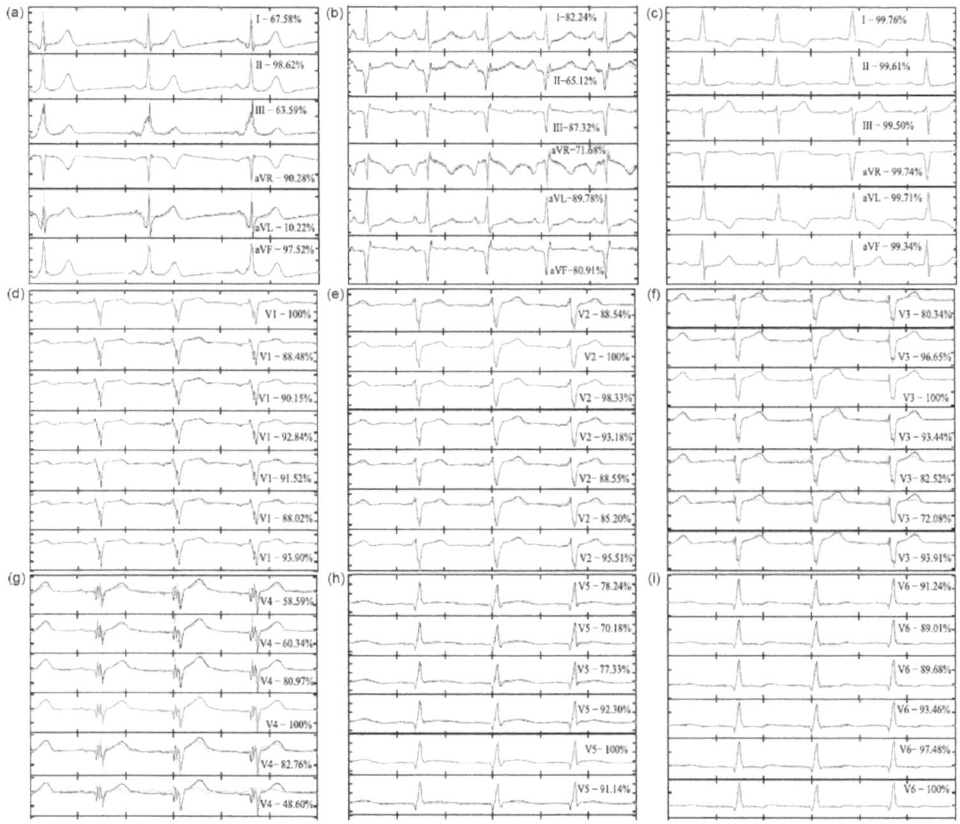

FIGURE 7.12

(a–c) Reconstructed signal (red) overlapping the original signal (blue), obtained using SRM, of three different patients with worst (71.3%), 80% and best case (99.61%) mean R^2 values for lead set 1. (d–i) A similar comparison for lead set 2 for both FRM and SRM, where the first six boxes in all the subfigures correspond to FRM in the order of R3L systems with leads V1–V6 as the precordial lead in the basis lead set and the last box corresponds to SRM. (Image appears in color in eBook version of this book.)

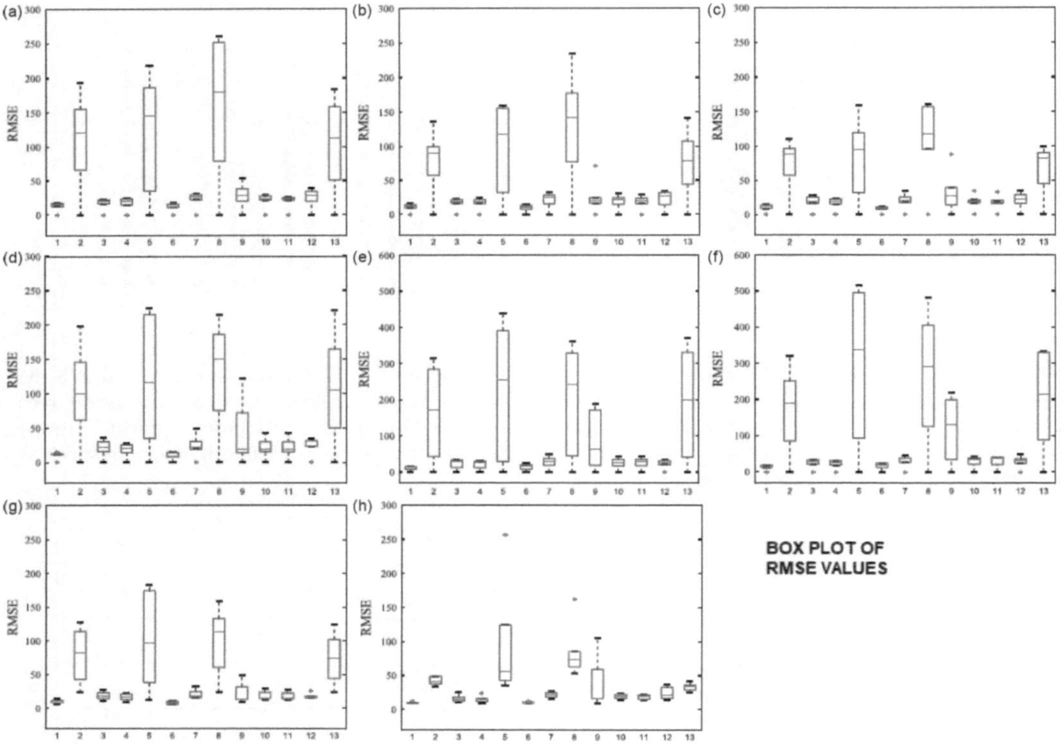

FIGURE 7.13
(a–h) A box plot of RMSE values [4]. Starting from left to right with R3L systems I–VI for lead set 2. The left subfigure last row corresponds to lead set 2 and right subfigure corresponds to lead set 1 when S12 was reconstructed from FV. The labels 1–13 on the horizontal axis correspond to the respective features extracted from TDMG as mentioned in the text. For details about denotations, please refer Table 7.7 [4].

d–V4, e–V5, and f–V6 (all of these basis lead sets essentially contain leads I and II). Figure 7.13g,h follows from the results of SRM for lead set 2 and lead set 1, respectively [4].

E. Figure 7.14 shows the summary of the methodology for this investigation [10]. First, the databases are denoised using the preprocessing module followed by PCA module for the construction of FV from S12 system. IDT and KT were also used to reconstruct Frank's leads. All the three sets of derived Frank leads were then compared with actually recorded Frank leads. Finally, the S12 leads were reconstructed from the derived FV leads, obtained using the proposed PCA-based methodology, and were compared with the originally measured S12 leads. The personalized reconstruction of S12 system from derived FV system further includes a module for generation of transformation coefficient [10] (Tables 7.12 and 7.13).

Figure 7.15 shows the working principle of PCA. For a 2D case, PCA finds a lower 1D linear vector which when the dataset is projected produces lowest mean square of the perpendicular distances from the points to the vector. This linear vector is known as a principal component. For higher dimensions, a lower dimensional plane or hyperplane represented by spanning vectors (principal components)

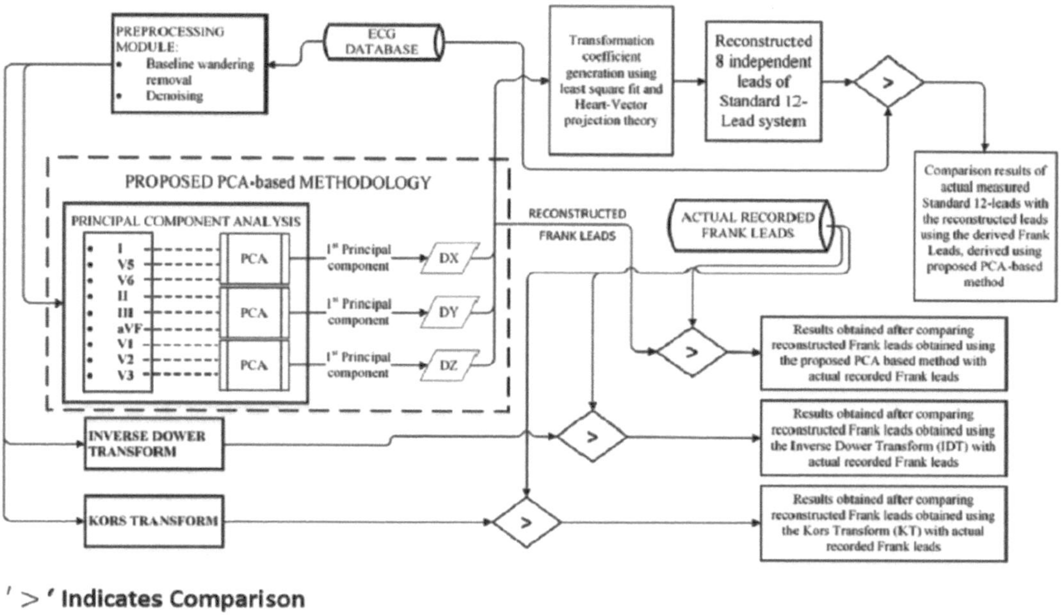

FIGURE 7.14
A complete summary of the methodology followed in this study for reconstruction of Frank system from standard 12-lead system and then using the derived Frank leads (DX, DY, and DZ) to reconstruct standard 12-leads using PT employing least-square fit method and heart-vector projection theory [10].

TABLE 7.12
Fractional Content of Heart Dipole Components in S12 Leads for PTB Database (PTBDB) [10]

12 Lead	*x*/*y*	*x*/*z*	*y*/*x*	*y*/*z*	*z*/*x*	*z*/*y*
I	7.338	10.06	0.136	1.370	0.099	0.73
II	0.164	8.155	6.112	49.85	0.123	0.02
III	0.589	1.296	1.699	2.202	0.771	0.45
AVR	0.129	0.245	7.767	1.904	4.079	0.52
AVL	0.797	3.560	1.255	4.468	0.281	0.22
AVF	0.355	1.583	2.821	4.465	0.632	0.22
V1	0.677	2.079	1.477	3.071	0.481	0.32
V2	2.773	3.352	0.361	1.209	0.298	0.82
V3	1.239	8.521	0.807	6.876	0.117	0.14
V4	0.834	17.23	1.199	20.67	0.058	0.04
V5	0.445	5.094	2.254	11.48	0.196	0.08
V6	0.996	4.884	1.004	4.905	0.205	0.20

is searched. When PCA was applied on a subset (I, V5, V6) of S12 system, the first principal component obtained was found to have 98.69% resemblance with the originally measured X lead of FV system, as shown on the right hand side of Figure 7.15.

The derived Frank leads, obtained using three different methodologies, namely, the proposed PCA-based method, IDT and KT have been compared independently with the originally recorded Frank lead using the evaluation metrics in Table 7.14.

TABLE 7.13

Fractional Content of Heart Dipole Components in S12 Leads for CSE Database (CSEDB) [10]

	x/y	x/z	y/x	y/z	z/x	z/y
I	7.207	13.79	0.139	1.914	0.072	0.523
II	0.315	2.829	3.178	8.990	0.354	0.111
III	0.540	2.927	1.850	5.417	0.342	0.185
aVR	1.141	126.5	0.877	110.9	0.008	0.009
aVL	1.082	7.522	0.924	6.953	0.133	0.14
aVF	0.149	0.833	6.722	5.594	1.202	0.19
V1	2.637	0.592	0.379	0.224	1.691	4.45
V2	0.295	0.086	3.385	0.290	11.65	3.43
V3	3.301	0.574	0.303	0.174	1.744	5.75
V4	81.77	1.401	0.012	0.017	0.714	58.36
V5	6.204	3.472	0.161	0.560	0.288	1.78
V6	3.155	14.89	0.317	4.718	0.067	0.22

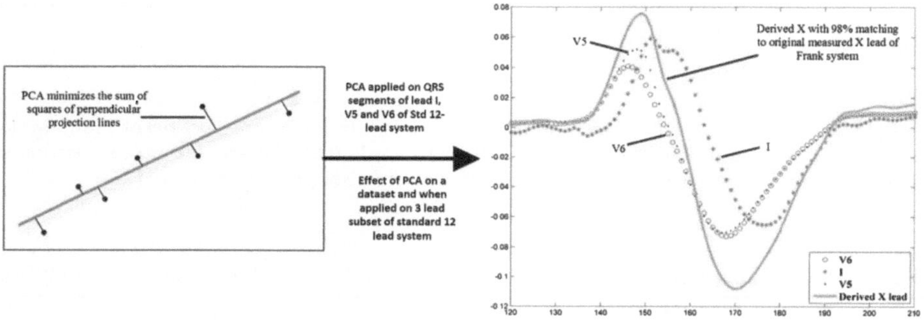

FIGURE 7.15
Effect of principle component analysis on a data set and when applied on 3-lead subset of standard 12-lead system. The subset in the figure includes I, V5, and V6. The resulting first principal component is shown as a continuous plot in green whose resemblance to originally measured Frank's X lead is 98.69%.

TABLE 7.14

Number of Subjects (in %) with Various Values of Reconstruction Accuracy of Frank System from Standard 12-Lead System for Both Physikalisch-Technische Bundesanstalt database and CSE database Using Our Proposed Methodology (PCA-Based), Inverse Dower Transform (IDT), and Kors Transform (KT) [10]

	No. of Patients in CSEDB (in %)			No. of Patients in PTBDB (in %)		
Mean R^2 Values	**PCA-Based**	**IDT**	**KT**	**PCA-Based**	**IDT**	**KT**
>0%	100%	98.8%	99.6%	97.63%	73.04%	56.47%
>50%	90.4%	94%	97.2%	73.95%	47.91%	47.18%
>80%	68%	66.8%	75.2%	40.98%	30.05%	34.97%
>90%	48.8%	40%	55.6%	19.85%	11.66%	19.31%
Overall mean R^2 value	81.6%	80.93%	85.52%	65.77%	34.15%	26.89%
Overall mean correlation coefficient	0.8289	0.6708	0.6344	0.9080	0.9046	0.9276

Note: CSEDB, CSE database; PTBDB, Physikalisch-Technische Bundesanstalt database.

Table 7.15 presents a comparison in performance of proposed PCA-based method on the first recordings of patients in PTBDB, which excluded patients with pace makers and other therapies, and the remaining recordings.

Table 7.16 presents the reconstruction results of S12 system from derived FV system. The Frank leads used for reconstruction of S12 system have been synthesized using our proposed PCA-based method.

Figure 7.16 shows the comparison between derived (red) and originally measured (blue) FV leads for three different patients in PTBDB. The mean R^2 values of the three patients lie close to mean, median, and maximum R^2 values, respectively, obtained from PCA-based methodology. Corresponding reconstruction performances for IDT and KT have also been shown. Figure 7.16a–c presents mean case patient's FV leads (X, Y, and Z) reconstruction using PCA-based method, IDT and KT, respectively. Similarly, Figure 7.16d–f shows the median case and Figure 7.16g–i shows the maximum case for PCA-based method, IDT, and KT, respectively. (Image appears in color in eBook version of this book.)

TABLE 7.15

Fraction (in %) of Subjects in Physikalisch-Technische Bundesanstalt database with Various Reconstruction Accuracy Values for Reconstruction of Frank System from Standard 12-Lead System for Healthy Control (HC), Unhealthy (UH) for 290 Records, and Remaining Records [10]

Mean R^2 Values	Healthy Control: 52	Unhealthy: 238	Remaining Records
>0%	100%	99.2%	96.14%
>50%	92.31%	92.44%	58.69%
>80%	82.69%	67.65%	17.76%
>90%	63.46%	47.06%	2.703%
Overall mean R^2 value	86.63%	80.90%	52.84%
Overall mean correlation	0.933	0.904	0.764

TABLE 7.16

Mean R^2 and Correlation Coefficient Values for the Reconstruction of Standard 12-Lead System from Derived Frank Leads, Derived Using PCA-Based Methodology and Its Comparison with the Reconstruction Result Using Originally Measured Frank Leads [10]

	PTB				CSEDB			
	Derive	Frank Leads	Original	Frank Leads	Derive	Frank Leads	Original	Frank Leads
Leads	R^2 (%)	CC (r_x)	R^2 (%)	CC (r_x)	R^2 (%)	CC (r_x)	R^2 (%)	CC (r_x)
I	48.85	0.527	46.39	0.494	92.06	0.9646	89.91	0.950
II	97.57	0.987	92.21	0.959	98.36	0.9928	95.29	0.977
V1	87.54	0.930	92.91	0.961	92.45	0.9609	91.61	0.961
V2	97.32	0.986	83.87	0.912	97.67	0.9884	87.41	0.954
V3	96.56	0.983	85.73	0.920	96.28	0.9840	92.86	0.967
V4	94.82	0.972	88.34	0.935	93.49	0.9676	95.41	0.977
V5	97.25	0.986	90.42	0.947	97.15	0.9871	97.46	0.988
V6	97.31	0.987	95.39	0.975	97.54	0.9912	97.46	0.988
Mean	89.65	0.9198	84.41	0.8878	95.62	0.9795	93.43	0.970

CC, correlation coefficient.

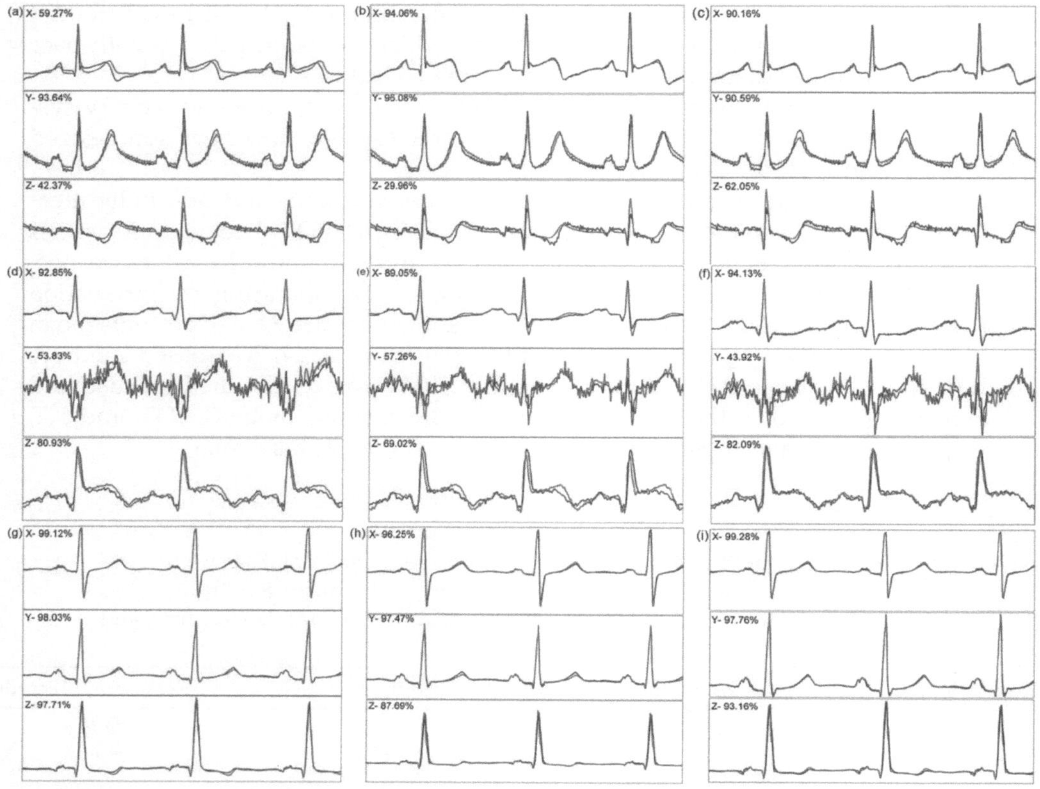

N.B. Comparison between original (in blue) and derived (in red) signals of Frank system when constructed from Standard 12-lead system.

FIGURE 7.16
A comparative study between original (in blue) and derived (in red) signals of Frank system when constructed from Standard 12-lead system. (a, d, and g) Construction using our proposed PCA-based method for the subjects, which had mean, median, and maximum R^2 value in PTBDB. (b, e, and h) Reconstruction for inverse Dower transform. (c, f, and i) Reconstruction for Kors transform for the same subjects. (Image appears in color in eBook version of this book.)

Vectorcardiography in general and Frank system, in particular, were postulated and realized six decades ago in preinformatics period devoid of modern computing capabilities and low-power low-cost electronic devices. The authors, hence, strongly believe that possible advantages of orthogonal leads were not fully explored. Unavailability of hassle-free Frank VCG was among the primary reasons behind dropping it from routine tests. In this chapter, authors have attempted to address this issue [10].

F. Since the ECG signal is a 1D data vector and the reconstruction process is time-independent, the first problem faced is turning it into a 2D matrix of data points [5]. Considering 1D-ECG to 2D-ECG conversion, references were made to the gradient-based method on image/shape reconstruction [25] and the traditional matrix reconstruction method based on slipping insertion [26]. In this study, the gradient of the input signal has been computed, then repeated and arranged them as Figure 7.17 (1 presents no. 1 original signal and 1′ presents the gradient of no. 1 signal). Overall, a total of 150,000 images are obtained. Designers have used

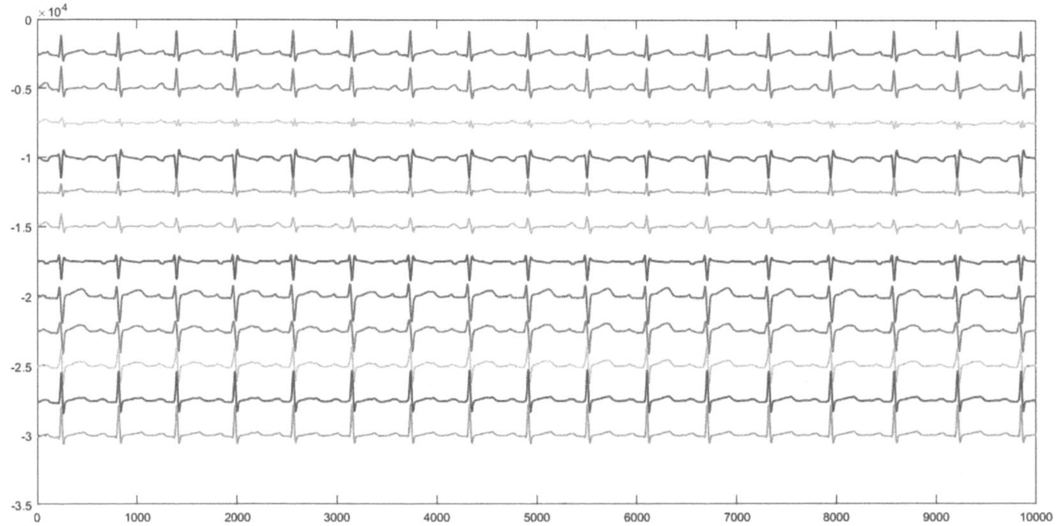

FIGURE 7.17
A set of 10s ECG signal of record patient 001.

120,000 images as a training set to train the CNNs and for validation and 30,000 images as a testing set to evaluate the proposed method. The aim of this transformation is not only to satisfy the input requirements but also to obtain more details of the input signal.

Authors computed the gradient of the input signal, then repeat and arrange them as Figure 7.18 (1 presents no. 1 original signal and 1′ presents the gradient of no. 1 signal).

Figure 7.19 is the structure schematic of the proposed method with CNN.

Table 7.17 shows the diagnostic classes of the remaining 268 subjects.

As shown in Table 7.18, the CNN-based method shows a better reconstruction result, compared to the ANN-committees-based method and the multiple-regression-based method.

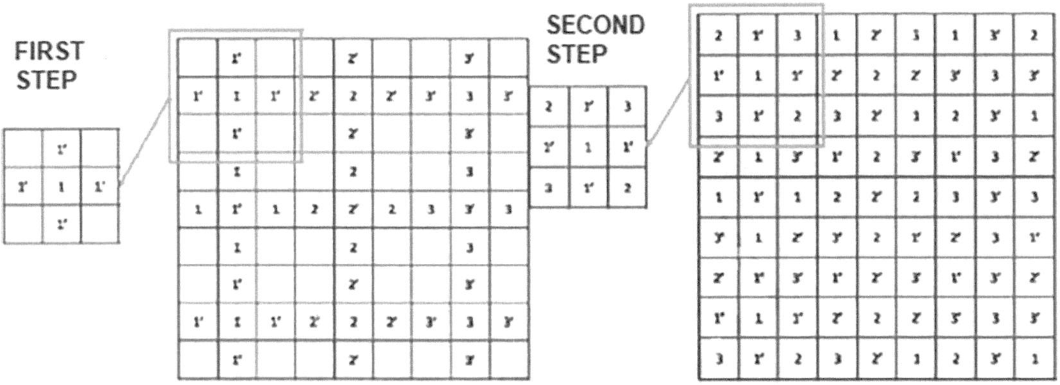

FIGURE 7.18
The morphology of the input signal.

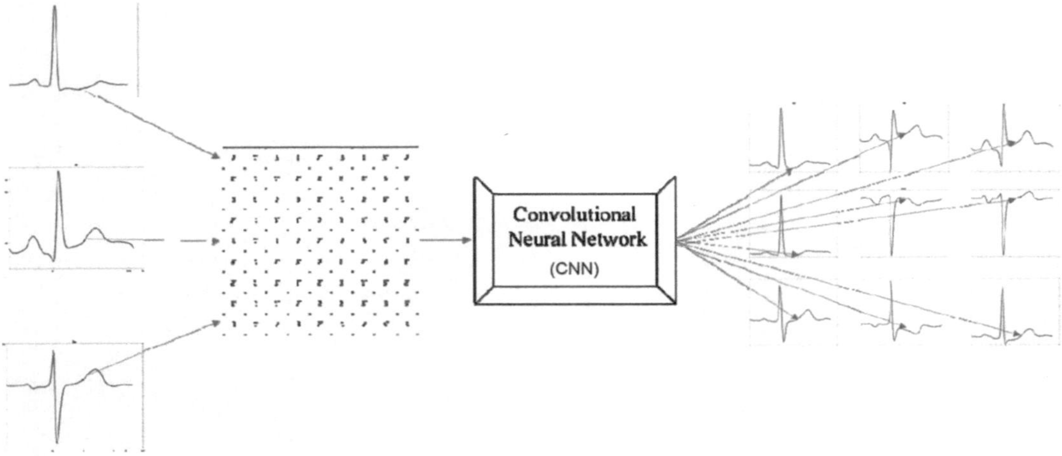

FIGURE 7.19
The schematic of the structure of the proposed method.

TABLE 7.17
The Diagnostic Classes of PTB Diagnostic ECG Database [5]

Class to Diagnose	Total Number of Subjects
Myocardial infarction	148
Cardiomyopathy/Heart failure	18
Bundle branch block	15
Dysrhythmia	14
Myocardial hypertrophy	7
Valvular heart disease	6
Myocarditis	4
Miscellaneous	4
Healthy controls	52

Figure 7.20 is the line chart of the results. The results show that the proposed method performs better in the reconstruction of lead V1, V4, V5, and V6.

A 12-lead ECG is one of the most important cardiac diagnostic tools. The need for synthesis of 12-lead ECG from fewer leads increases dramatically as portable devices provide often limited electrodes. The reduction of running time during synthesis is important, considering the large amount of signal and the real-time request. In this chapter, authors investigated the CNN method in detail and apply it to the reconstruction of an ECG signal. Compared to linear regression and ANN, this methodology shows not only better similarity between the reconstructed and the original ECGs but also less time consumption. Authors also believe that the generic synthesis method can be further modified by using a higher volume training database. Authors vowed to continue to improve the effectiveness of the generic approach and provide better support for more patient types [5].

G. To synthesize the missing (V1, V3, V4, V5, and V6) ECG signals from the recorded (I, II, and V2) 12-lead ECG subset, we use an ensemble of N multilayer feed-forward ANNs trained by means of a supervised back-propagation algorithm.

TABLE 7.18

Correlation Coefficients (r_x) between the Original ECG Signal and the Reconstructed ECG Signal Obtained by the below Three Methods [5]

Methods	Lead	*R* (%)
CNN-based methodology	III, aVR ,aVL, aVF	100
	V1	94.40
	V3	94.48
	V4	93.80
	V5	94.65
	V6	97.44
ANN-committees-based methodology	III, aVR ,aVL, aVF	100
	V1	91.40
	V3	95.66
	V4	89.74
	V5	94.54
	V6	97.15
Multiple-regression-based methodology	III, aVR ,aVL, aVF	100
	V1	91.42
	V3	96.51
	V4	89.20
	V5	93.01
	V6	96.58

ANN, artificial neural networks.

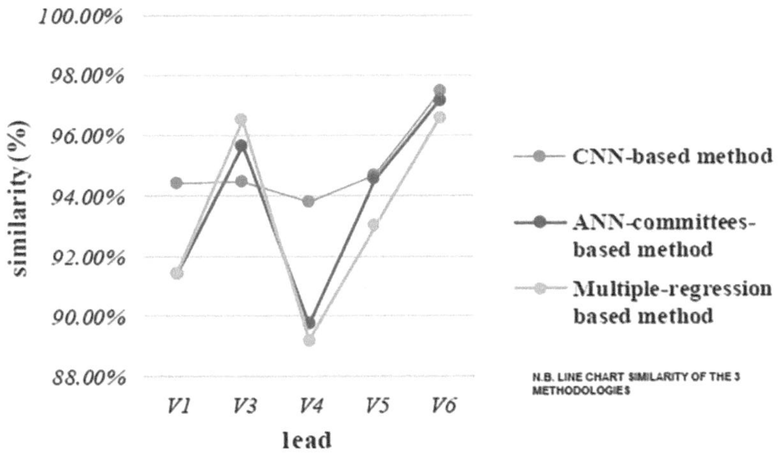

FIGURE 7.20
The line chart of similarity obtained by three methods.

Each individual ANN consists of one input layer with three input neurons (one for each recorded signal), one output layer with five output neurons (one for each derived signal), $h = 1$ hidden layer and $n = 15$ neurons per hidden layer, as shown in Figure 7.21.

Table 7.19 displays the rms values and the correlation coefficients between the first and second Cardiette ECGs of DS2, after removal of the 5% extreme values ($N = 35$) [6].

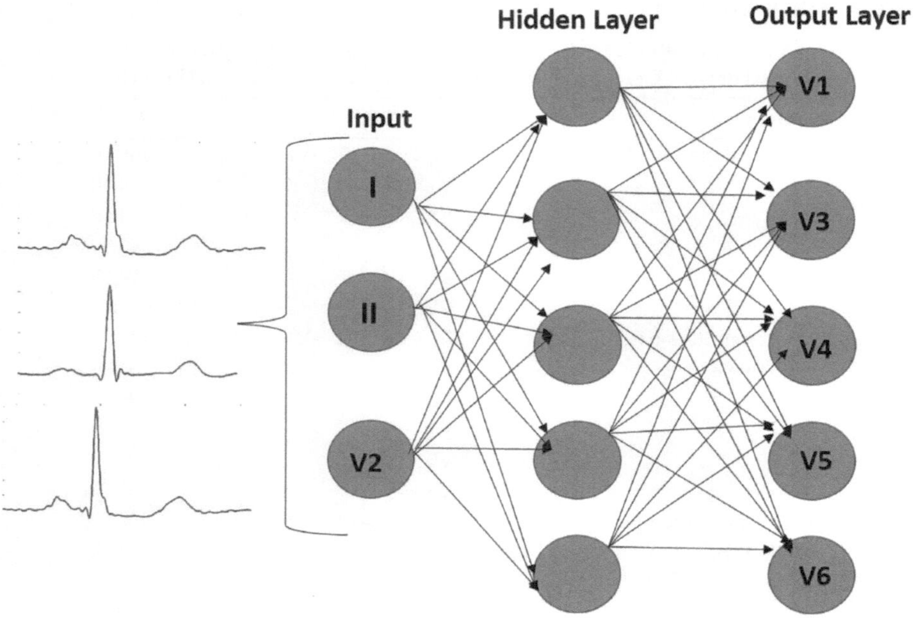

FIGURE 7.21
The architecture of the ANN used to synthesize the five missing V1, V3, V4, V5, and V6 leads (output layer) of the 12-lead ECG using a 3-lead ECG (I, II, and V2) as input layer, *h* (typically 1 or 2) hidden layers and *n* neurons per hidden layer [6].

TABLE 7.19
RMS (in μV) and Correlation Coefficients (r_x) Among the Original I, II, and V2 Leads of the First and the Second Standard 12-Lead ECGs Recordings for Every Patient of Dataset DS2 ($N = 35$) [6]

Lead	RMS (±SD)	*r* (±SD)
I	60 (±30)	0.93 (±0.09)
II	73 (±32)	0.85 (±0.18)
V2	125 (±58)	0.95 (±0.07)
V1	82 (±44)	0.95 (±0.08)
V3	135 (±57)	0.95 (±0.04)
V4	136 (±56)	0.92 (±0.08)
V5	139 (±59)	0.90 (±0.13)
V6	100 (±41)	0.91 (±0.10)

This study [6] investigates the accuracy of the reconstruction of the 12-lead ECG from a RL set using generic and patient-specific ANNs. The results suggest that ANN represents a rather interesting and very promising approach to improve the current 12-lead ECG synthesis method based on linear transforms [27,28], especially for the patient-specific approach. The differences between the original and the reconstructed ECGs were mainly ascribable to electrode displacements or to hour-to-hour changes in the ECG and not on the synthesis methodology (Figure 7.22).

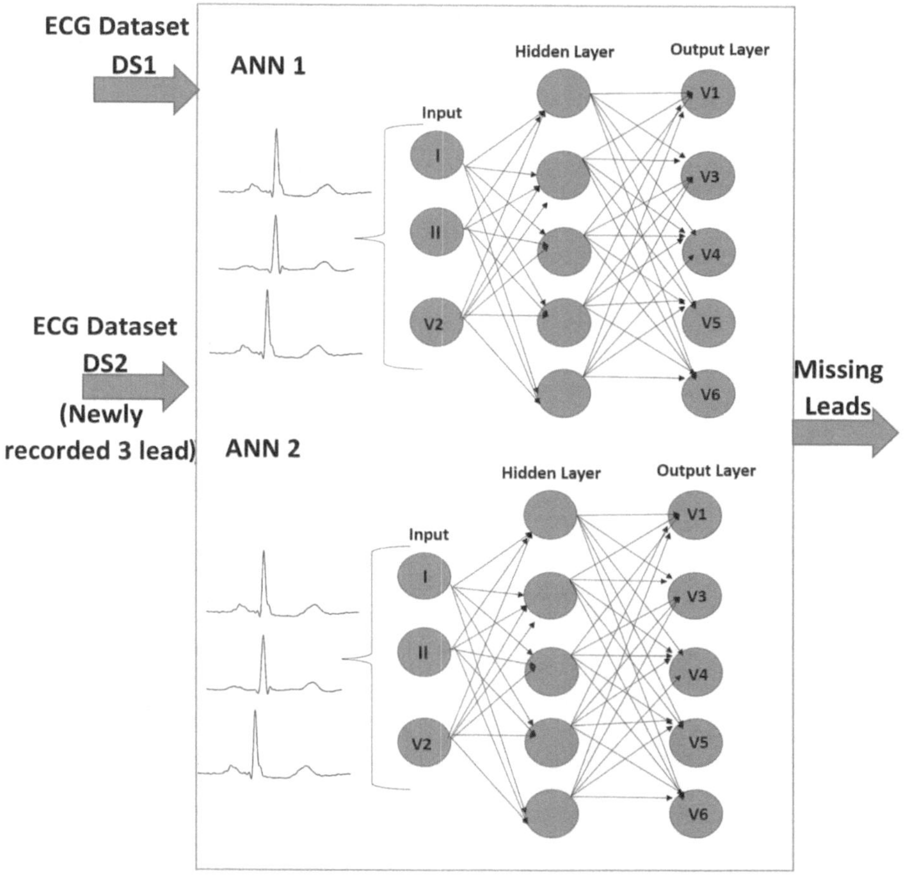

FIGURE 7.22
The reconstruction of a standard 12-lead ECG out of the three-lead PEM ECG. The synthesis of the five missing V1, V3, V4, V5, and V6 leads is obtained by averaging the outputs of a committee of 50 neural networks [6].

7.4 Conclusion

A 12-lead ECG is one of the most important cardiac diagnostic tools. The need for synthesis of 12-lead ECG from fewer leads increases dramatically as portable devices provide often limited electrodes. In this chapter, we have explored and have reported various studies, which provided solutions to the aforementioned problems.

The first remedy proposed a personalized reconstruction methodology for reconstruction of missing precordial leads of S12 and ML12 systems. Second solution proposed a robust and accurate method for the reconstruction of S12-lead system from FV system using PT matrices targeting personalized remote health-monitoring applications. Third strategy proposed a personalized R3L system formation methodology, which employs PCA, thereby reducing redundancy and increasing SNR ratio, hence making it suitable for wireless transmission. Forth, a technical methodology to the medical practitioners has been proposed for selection of RL systems suitable for personalized remote health-monitoring applications. Subsequently, a novel S12-lead ECG reconstruction methodology is also proposed which is shown to be more reliable than the state-of-the-art lead reconstruction methodologies. Subsequently, VCG-based substitution to complement S12 system in diagnosis of CVDs has been proposed. This is followed by CNN-based method, which proved to be more accurate and time-saving for deployment in nonhospital situations to synthesize a S12-lead ECG from a RL-set ECG recording. All these solutions have been summarized in this study along with their promising results. Every methodology is still in process of further exploration for better results.

References

1. Maheshwari S, Acharyya A, Rajalakshmi P, Puddu PE, Schiariti M. Accurate and reliable 3-lead to 12-lead ECG reconstruction methodology for remote health monitoring applications. *IEEE Healthcom* 2013;15:208–212.
2. Maheshwari S, Acharyya A, Puddu PE, Mazomenos EB, Schiar-itiM, Maharatna K. Robust and accurate personalised reconstruction of standard 12-lead system from Frank vectorcardiographic system. *Comput Methods Biomech Biomed Eng Imaging Vis* 2014. doi: 10.1080/21681163.2014.931029.
3. Maheshwari S, Acharyya A, Schiariti M, Puddu PE. Personalized reduced 3-lead system formation methodology for remote health monitoring applications and reconstruction of standard 12-lead system. *Int Arch Med* 2015;8. doi: 10.3823/1661.
4. Maheshwari S, Acharyya A, Schiariti M, Puddu PE. Reduced lead system selection methodology for reliable standard 12-lead reconstruction targeting personalized remote health monitoring applications. *Comput Methods Biomech Biomed Eng Imaging Vis* 2014;2(2):107–120.
5. Wang Lu-di, Zhou W, Xing Y, Liu N, Movahedipour M, Zhou XG. A novel method based on convolutional neural networks for deriving standard 12-lead ECG from serial 3-lead ECG, www.jzus.zju.edu.cn; engineering.cae.cn; www.springerlink.com. ISSN 2095-9184 (print); ISSN 2095-9230 (online).
6. Atoui H, Fayn J, Rubel P. A novel neural-network model for deriving standard 12-lead ECGs from serial three-lead ECGs: Application to self-care. *IEEE Trans Inf Technol Biomed* 2010;14(3):883–890.
7. Goldberger AL, Amaral LA, Glass L, Hausdorff JM, Ivanov PC, Mark RG, Mietus JE, Moody GB, Peng CK, Stanley HE. PhysioBank, PhysioToolkit, and PhysioNet: Components of a new research resource for complex physiologic signals. *Circulation* 2000;101(23):e215–e220.
8. Bousseljot R, Kreiseler D, Schnabel A. Nutzung der EKG-Signaldatenbank CARDIODAT der PTB überdas Internet. *Biomed Tech* 1995;40(1):S317.
9. Mazomenos EB, Chen T, Acharyya A, Bhattacharya A, Rosengarten J, Maharatna K. A time-domain morphology and gradient based algorithm for ECG feature extraction. *ICIT 2012: IEEE International Conference on Industrial Technology*, Athens, Greece, March 2012:pp. 117–122. doi: 10.1109/ICIT.2012.6209924,
10. Maheshwari S, Acharyya A, Schiariti M, Puddu PE. Frank vectorcardiographic system from standard 12 lead ecg: An effort to enhance cardiovascular diagnosis. *J Electrocardiol* 2016;49(2):231–242.
11. Dower GE, Machado HB, Osborne JA. On deriving electrocardiogram from vectorcardiographic leads. *Clin Cardiol* 1980;3(2):87–95.
12. Edenbrandt L, Pahlm O. Vectorcardiograms synthesized from 12-lead ECGs: Superiority of inverse Dower matrix. *J Electrocardiol* 1988;21(4):361–367. doi: 10.1016/0022-0736(88)90113-6.

13. Kors JA, Van Harpen G, Sittig AC, Van Bemmel JH. Reconstruction of the Frank vectorcardiogram from standard electrocardiographic leads: Diagnostic comparison of different methods. *Eur Heart J* 1990;11(12):1083–1092.
14. Man SC, Maan AC, Kim E, Draisma HHM, Schalji MJ, van der Wall EE, Swenne CA. Reconstruction of standard 12-lead electrocardiograms from 12-lead electrocardiograms recorded with the Mason–Likar electrode configuration. *J Electrocardiol* 2008;41(3):211–219.
15. Frank E. General theory of heart-vector projection. *Circulation* 1954;2:258–270.
16. Levkov CL. Orthogonal electrocardiogram derived from the limb and chest electrodes of the conventional 12-lead system. *Med Biol Eng Comput* 1987;25(2):155–164.
17. Dawson D, Yang H, Malshe M, Bukkapatnam STS, Benjamin B, Koman-duri R. Linear affine transformations between 3-lead (Frank XYZ leads) vectorcardiogram and 12-lead electrocardiogram signals. *J Electrocardiol* 2009;42(6):622–630.
18. Hernandez O, Olvera E. Noise cancellation on ECG and heart rate signals using the undecimated wavelet transform. *International Conference on eHealth, Telemedicine, and Social Medicine,* eTELEMED'09, Cancun, Mexico, 2009:pp. 145–150.
19. Zhang D. Wavelet approach for ECG baseline wander correction and noise reduction. *27th Annual International Conference of the Engineering in Medicine and Biology Society, IEEE-EMBS,* Shanghai, China, 2005;2:pp. 1212–1215. doi: 10.1109/IEMBS.2005.1616642.
20. Dower GE. A lead synthesizer for the Frank system to simulate the standard 12-lead electrocardiogram. *J Electrocardiol* 1968;1(1):101–116.
21. Nelwan SP, Kors JA, Meij SH. Minimal lead sets for reconstruction of 12-lead electrocardiograms. *J Electrocardiol* 2000;33:163–166.
22. Nelwan SP, Crater SW, Green CL, Johanson P, Dam TB, Meij SH, Simoons ML, Krucoff MW. Assessment of derived 12-lead electrocardiograms using general and patient-specific reconstruction strategies at rest and during transient myocardial ischemia. *Am J Cardiol* 2004;31:1529.
23. Nelwan SP, Kors JA, Meij SH, Bemmel JH, Simoons ML. Reconstruction of the 12-lead electrocardiogram from reduced lead sets. *J Electrocardiol* 2004;37:11–19.
24. Gregg RE, Zhou SH, Lindauer JM, Feild DQ, Helfenbein ED. Where do derived precordial leads fail. *J Electrocardiol* 2008;41:546–552.
25. Ettl S, Kaminski J, Knauer MC, Häusler G. Shape reconstruction from gradient data. *Appl Opt* 2008;47(12):2091. doi: 10.1364/AO.47.002091.
26. Lu C, Wang Z, Zhou B. Intelligent fault diagnosis of rolling bearing using hierarchical convolutional network based health state classification. *Adv Eng Inf* 2017;32:139–151. doi: 10.1016/j.aei.2017.02.005.
27. Modre R, Seger M, Fischer G, Hintermuller C, Hayn D, Pfeifer B, Hanser F, Schreier G, Tilg B. Cardiac anisotropy: Is it negligible regarding noninvasive activation time imaging? *IEEE Trans Biomed Eng* 2006;53(4):569–580.
28. Gulrajani R. The forward and inverse problems of electrocardiography. *IEEE Eng Med Biol Mag* 1998;17(5):84–101.

8

Trusted Digital Solutions and Cybersecurity in Healthcare

I. E. Lamprinos
Intracom S.A. Telecom Solutions

CONTENTS

8.1 Introduction: Background and Driving Forces

8.1.1 The Notion of Digitization in Healthcare

Digitization is considered as the major driving force in healthcare. Rather than simply indicating the process of converting information into a digital (i.e. computer-readable) format, digitization signifies something significantly broader, an enormous value creation as a result of a transformation process that affects the fundamentals of service delivery. Healthcare industry is an exemplary domain disrupted due to the ever-increasing digitization of its fundamental elements.

In this context, it does not come as a surprise that in Europe, for example, healthcare market comprises one of the core sectors that are set to realize the Digital Single Market notion [1]. There are several good reasons for this: the incorporation of Information and Communications Technologies (ICT) in almost every aspect of healthcare delivery, the digitization and automation of data exchange, the breakthrough of artificial intelligence and big data analytics, the enormous penetration of consumer-oriented wearable medical-grade and lifestyle-related sensing and control devices, all these are profoundly changing the way (and the dynamics) healthcare services are provided.

8.1.2 The Notion of Trust

Digitization and digital life are interwoven with trust. Indeed, trust is one of the hot topics in the implementation and adoption of digital solutions in healthcare. A recent study [2] revealed that two out of the three most important factors that increase the use of health apps is trustworthiness and guarantee of data security, with ease of use and simplicity being the third one. But what stands behind these types of trust-related requirements? The IoT-European Platforms Initiative published a guiding toolkit in the development of a broad range of IoT applications which highlights five different categories of requirements related to trust, namely *trustworthiness, transparency, privacy, compliance,* and *security* [3], Figure 8.1.

Trustworthiness pertains to issues regarding the right and accuracy of sharing personal and non-personal data with third parties. Transparency has similar goals to the trustworthiness; however, it focuses on gaining the trust of the end-user as opposed to the business actor which focuses on the trustworthiness of a digital solution. In that sense, it has to do less with technical aspects of data sharing and more with the approach of making certain processes transparent to the end-user. In its turn, the privacy of information is related to the protection of the information related to a subject's identity.

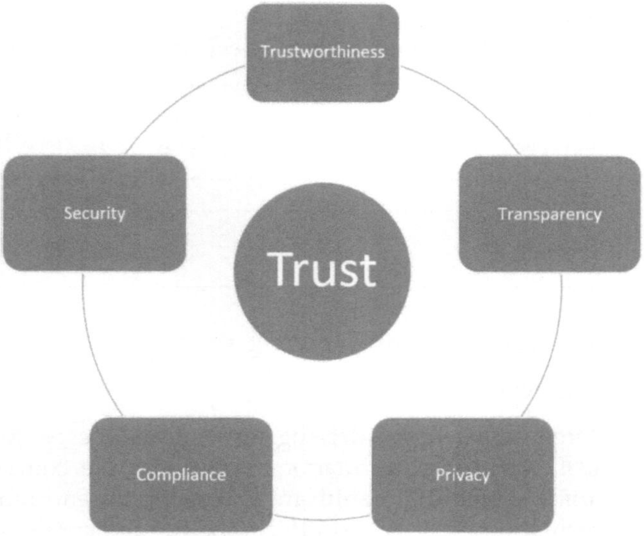

FIGURE 8.1

Injecting trust in digital solutions for healthcare.

Privacy has gained significant interest among end-users with the onset of sensors and devices in various walks of life. The issue of the General Data Protection Regulation [4] by the European Union (EU) is a characteristic action towards this direction (this topic is further analyzed in Section 8.1.4). While the issue of privacy is very broad and deep, key aspects of privacy that characterize a digital solution are the mechanisms put in place for blocking de-anonymization of data while at the same time guaranteeing data utility.

Compliance pertains to legal aspects related to digital solutions and particularly with data protection laws. To accomplish this need, digital solutions need to embed mechanisms for guaranteeing liability for data, system integrity, and security breaches, while being transparent as for the applicable data sovereignty laws.

Last but not least dimension of the trust notion is security. This dimension pertains mostly to the technical framework that ensures data protection against malicious attacks. This framework includes mechanisms such as end-to-end encryption of the communicated data and strict rules for control, audit, back up, update and upgrade of the software and hardware components of the digital solution. This particular issue is further analyzed from a technical perspective in Section 8.3.

8.1.3 Patient Role in Decision Making

The roles that patients' play in decision-making vary widely from not taking part in the decision process to seeking full control in the therapy selection. In Ref. [5], the authors present four models of patient-provider roles in healthcare decisions: the paternalistic model, the informed model, the collaborative model, and the deliberative model.

In the paternalistic model of clinical decision-making, a patient believes that the healthcare provider has sufficient knowledge of both the decision domain and the patient's values to make the appropriate decision. The patient vests the provider with complete authority for the decision. The decision domain consists of the possible options, outcomes, and associated risks and benefits of each. The provider functions as a guardian who is responsible for overseeing the wellbeing of the patient. In this model limited involvement of the patient is required, and it is assumed that the patient will be accepting and satisfied with whatever decision the provider chooses. It is up to the provider to know the patient's values, and it is assumed that both parties share common goals in the selection of treatment. It is estimated that only about 3%–8% of the patients seek a paternalistic role [6]. Older age, male gender, and a poor educational background generally characterize patients in this role.

In the informed model of clinical decision-making, the provider functions as a domain expert from whom relevant information about the patient's conditions and therapy options can be obtained. The patient seeks to be informed of her current disease state, the possible therapeutic options, and the risks and benefits of each therapy. The provider plays a little role in helping patients understanding their values. The provider ensures that the patient has an adequate educational base. It is assumed that the patient knows her own values and is skilled at the integration of risk and benefit. Once the patient has been provided the necessary facts by the provider, she will be able to make the appropriate decision on her own exercising full autonomy. The health provider's role is to support this decision. The proportion of patients seeking an informed role ranges from 20% to 30% [6].

In the collaborative model of clinical decision-making, the patient relies upon the healthcare provider to provide both facts and support in the decision-making process. The provider functions as an adviser, attempting to elicit the patient values for different health outcomes and assist the patient in selecting a therapy that is best matched to the patient's values. This role is similar to the informed model in that the provider is responsible for providing the necessary facts about the disease state and the available therapeutic options. Also, the provider has the responsibility to help the patients discover their own values and understand how these relate to the decision. It is different in that the provider must also assume the role often played by a decision analyst, such as in business consultations, where the provider's task is not only to educate the patients but to guide them through clarifications. The proportion of collaborative role is believed to be 50%–60% [6]. Like the informed group, these patients tend to be in a younger age group and have a strong educational background.

Finally, in the deliberative model of clinical decision-making, the healthcare provider abandons objectivity. The provider's goal is to influence the patient's beliefs about the likelihood of outcomes and values so that they reflect what the provider feels are the patient's best interests. The provider functions in the role of friend or a teacher in that the provider may suggest to the patient which decision they think is the best for the patient. Further, the provider may attempt to persuade the patient to change values the provider believes harmful or change mistaken beliefs about the likelihood of outcomes if the provider believes this to be the best in the interest of the patient. This differs from a collaborative model, where the provider primarily seeks to discover patients' values and from the paternalistic model, in that authority is not delegated to the physician. For an action to occur, both the patient and the provider must believe that the chosen path is in the patients' best interest. Studies show that the number of patients desiring a deliberative role ranges from 10% to 20% [6]. These patients tend to be in a younger age group, highly educated, and female gender.

8.1.4 Patient Data Protection Framework

Patients' health and genetic data are considered as a special category of data called *sensitive data*. This encompasses all personal data which are, by their nature, particularly sensitive concerning fundamental rights and freedoms. Patients' fundamental right to protection of their health data is an important issue in diverse healthcare application domains (e.g., hospital care, eHealth, cross-border healthcare, clinical trials, epidemiological research, etc.) [7].

As of May 2018, the General Data Protection Regulation (GDPR) [4], that is, Europe's new framework for data protection laws has entered into force, having a significant impact also on healthcare service providers. The regulation defines personal data as information that relates to an identified or identifiable living individual [4]. Different pieces of information, which collected together can lead to the identification of a particular person, also constitute personal data [8]. Personal data that have been de-identified, encrypted, or pseudonymized, but can be used to re-identify a person remains personal data and fall within the scope of the law. Personal data that have been rendered anonymous in such a way that the individual is not or no longer identifiable are no longer considered personal data. For data to be truly anonymized, the anonymization must be irreversible [4]. The law protects personal data regardless of the technology used for processing that data [8]. It also doesn't matter how the data are stored – in an IT system, through video surveillance, or on paper; in all cases, personal data are subjected to the protection requirements set out in the GDPR.

The GDPR framework sets clear principles that apply to all use of patients' data and to all data controllers. These principles relate to the aspects of lawfulness, fairness and transparency, purpose limitation, data minimization, accuracy, limited storage, integrity and confidentiality, and accountability.

When it comes to patient privacy and security in the USA, the healthcare industry is mainly governed by the framework set by Health Insurance Portability and Accountability Act (HIPAA). HIPAA is designed to prevent unauthorized data access within a healthcare ecosystem vs. addressing broader digital trust and data ownership issues. In that sense, the GDPR differs from HIPAA in its scope and intent, with GDPR designed to help individuals gain more control over how their complete digital data trails are used, by whom and for how long vs. regulating a single industry, as HIPAA does [9].

8.2 Trusted Digital Solutions in Healthcare

One of the most characteristic trends of modern society is the aging of the population. This is particularly intense in Asia and Europe where a significant number of countries face population aging in the near future. Within 20 years, many countries in these regions will face a situation where the largest population cohort will be those over 65 and the average age will be approximately 50 years old.

Aging of the population does not come without an impact on the needs for healthcare. Senior citizens are at greater risk of cognitive impairment, frailty, and multiple chronic health conditions with considerable negative consequences for their independence, quality of life, and for the sustainability of the healthcare system as a whole.

To address this challenge, several digital solutions are available or emerging. Indicative such solutions are presented in the following sub-sections.

8.2.1 Digital Solutions for Physical Activity Monitoring

Physical activity monitoring is an important aspect of modern, often out-of-hospital disease management. Physical activity monitoring is important also for healthy people, as it contributes to maintaining an adequate level of a healthy lifestyle and also enhances their interaction with trainees, teammates, and social contacts. Regular and moderate training provides many physiological benefits. A balanced daily physical activity is essential for an active retirement; it represents a fundamental indicator of good health and life quality, and it is an important factor in primary and secondary prevention of diseases, as well as in rehabilitation. However, daily physical activity is far below the recommended level, especially among the elderly population. Moreover, some people are not used to physical activity practice.

The nature of physical activity can be categorized as aerobic activity promoting cardiovascular fitness and strength exercises promoting musculoskeletal fitness. Particularly for the frail and elderly population, the balanced activity of both types is important for keeping or regaining functional independence. Also, it is essential to promote the practice of physical activity for patients who undergo a physical rehabilitation treatment.

In the latter case, physical activity has to be supervised in order to improve physical fitness while minimizing the risk of injury by incorrect execution or overuse. Moreover, this supervision has to be achieved while providing feedback to the user, since this is required for preserving or increasing motivation and program adherence. To address this challenge, several initiatives have delivered innovative solutions.

Lamprinos et al. [10] present an ICT system for accurately monitoring and promoting the physical activity of elderly in both structured contexts and in daily life at home for both personal (primary prevention) and professional use (secondary prevention and rehabilitation) (Figure 8.2). The system consists of five major components: body-worn sensors (IMUs, HR monitor) and a mobile unit are used to acquire information; innovative information processing technology is used to extract the relevant parameters of physical activity; motivational user interfaces provide instant feedback and guidance during exercising and encourage elderly to improve their level of physical activity and learn associated good practices; and finally, a personal health record (PHR) with web and smart-TV interfaces enable management, sharing and reviewing of collected activity data and facilitate healthcare professionals in the follow-up of personalized training plans.

In the commercial sector, an increasing number of providers of physical activity monitoring hardware and software solutions provide open APIs for the integration of their artifacts with third-party service providers. For example, the health and sports tracker, Runkeeper, provides an API to save, store, read, and modify personal health information of a user in a cloud-based service environment. FitBit is a provider of tracking sensors and apps for activity, health, nutrition, and sleep monitoring. MapMyRun also provides a health service API called MapMyHealth and collaborates with various providers of sports equipment such as Nike, Polar, Garmin, CycleOps, or Wahoo.

8.2.2 Digital Solutions for Personalized Medicine

8.2.2.1 Person-Centric Decision Support Systems

Personalized medicine, also referred as precision medicine, is a way of practicing medicine that separates patients into different groups, with medical decisions, practices, interventions, and/or products being tailored to the individual patient, based on their predicted response or the risk of disease [11].

A critical step towards personalized medicine is the implementation of digital solutions for decision support. This goes hand-in-hand with the evolution in the ability to collect, filter, process, and analyze data associated with numerous different aspects of everyday

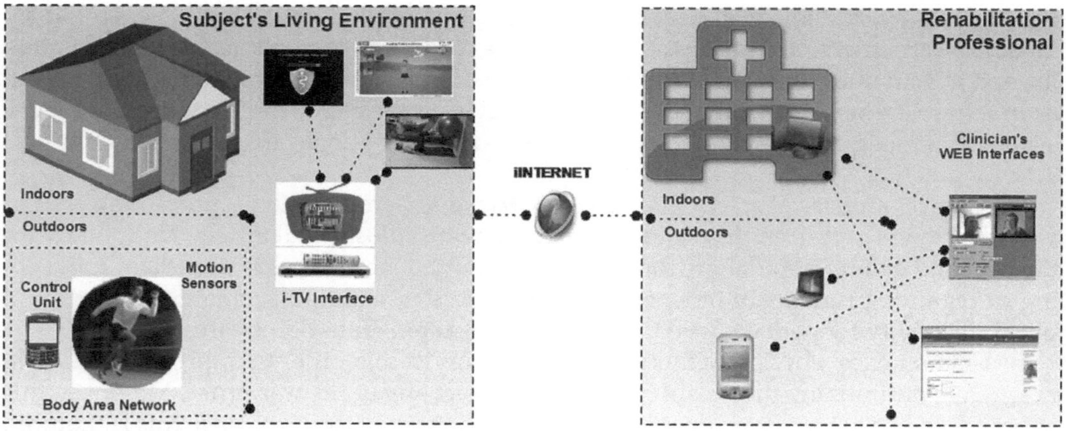

FIGURE 8.2
An indicative platform for out-of-hospital rehabilitation and physical activity monitoring.

life. These data, often rich with vital information, are scattered all over the place and often remain underutilized.

By integrating carefully selected data sources into a single platform, these data, and eventually the information derived from it can be used more effectively and in conjunction with established medical knowledge to help patients better self-manage their disease. In fact, the combination of dynamically collected data with established knowledge, properly co-processed by an intelligent decision support system (DSS), forms an ideal framework for wellness and disease management (Figure 8.3).

Health promotion and disease self-management can be served and promoted by leveraging several different sources of data (patient specific, disease specific, ambient environmental data, etc.) originating from personal health systems (i.e. observations of daily living, lifestyle, and behavioral data), knowledge sources (e.g., clinical guidelines, predictive models), and other sources (biological, therapeutic, environmental, or occupational exposure, etc.).

All these data and knowledge sources feed DSSs that are used either directly by the patients or indirectly via their supervising medical team, to raise awareness and empower the patients to participate in various aspects of health management (wellbeing and prevention, disease monitoring and management).

Such DSSs provide timely information and recommendations for disease self-management, improve patients' interaction with their healthcare professionals, and even

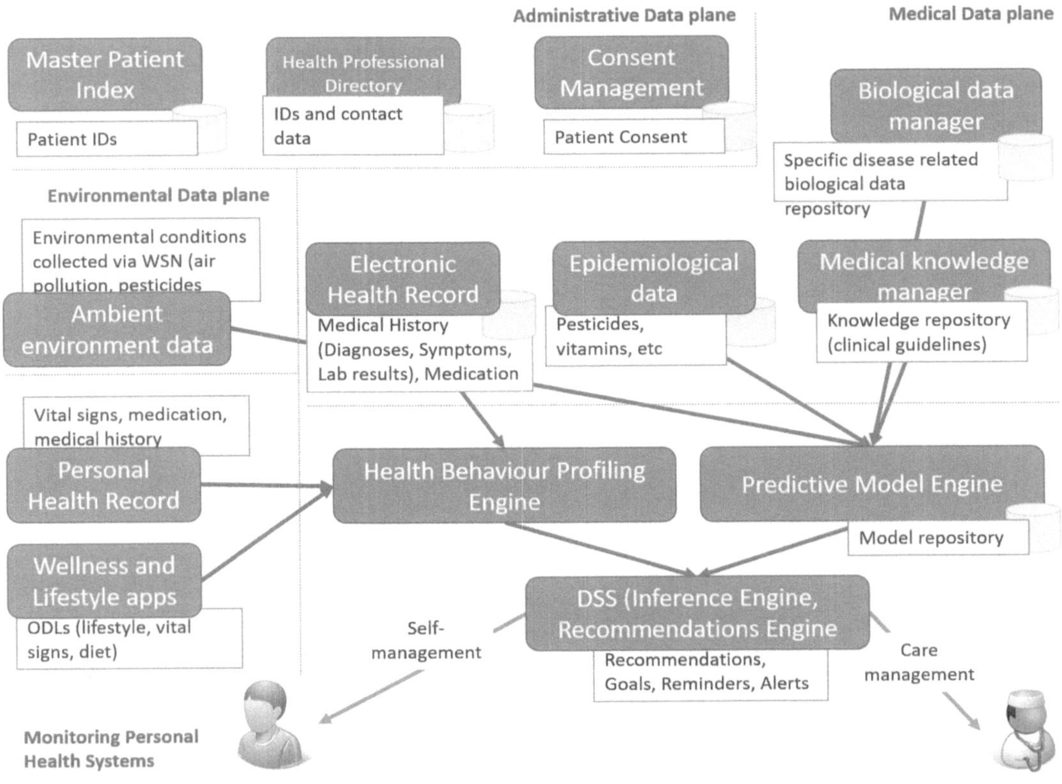

FIGURE 8.3
Leveraging health, environmental, and administrative data and medical knowledge in decision support systems.

foresee an adverse event for high-risk individuals. Often, the recommendations are related to medication and aim at establishing personal changes in lifestyle (e.g., increasing physical activity, lowering caloric intake).

8.2.2.2 The Role of Machine Learning and Artificial Intelligence

Although prediction research in medicine has traditionally used statistical techniques such as regression models, data mining techniques such as machine learning are a form of artificial intelligence that are used with increasing frequency [12]. The high-level process includes steps related to predictive model derivation, validation, and impact assessment (Figure 8.4).

Several different techniques to develop machine-learning algorithms exist (neural networks, support vector machines, decision trees), using a variety of prediction analytics tools/software. Machine learning algorithms have several advantages over traditional statistical modeling, such as lack of a predefined hypothesis, making it less likely to overlook unexpected predictor variables or potential interactions. Approaching a prediction problem without a specific causal hypothesis can be quite effective when many potential predictors are available (increasingly common due to the use of Electronic Health Records) and when there are interactions between predictors (rather common in biological processes) [12].

Predictive models using machine-learning algorithms may, therefore, facilitate the detection of clinically important risk factors in patients with several marginal risk factors that may otherwise not be identified. Predictive models are developed to provide estimates of outcome probabilities to complement the healthcare provider's clinical intuition [12]. They should ideally recommend decisions instead of simply providing risk estimates for an outcome. Predictive models that estimate risk without recommending particular

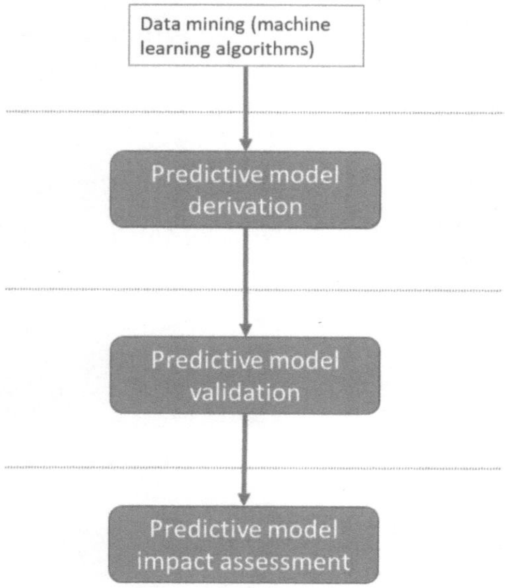

FIGURE 8.4
High-level process flow related to machine learning in predictive modeling.

decisions are less likely to change healthcare provider's behavior and disease management outcomes than those that translate risk into a decision recommendation [13].

Predictive models can be used to identify high-risk individual cases and transmit annotated data back to the healthcare provider, triggering updates to their clinical assessment [14]. If properly validated in several different populations, predictive models can also form the basis for patient-oriented decision support systems. Such models are particularly attractive, as they can optimize user-friendliness and may be introduced quickly and efficiently in disease self-management solutions.

8.2.2.3 Indicative Application Domains

Personalized medicine comprises a major trend in healthcare, already applied in various cases. An indicative set of such cases include the prediction and management of the treatment of Parkinson's disease and chronic obstructive pulmonary disease (COPD).

Parkinson's disease is a degenerative disease and medication must be adjusted from time to time. Treatment plans are currently based on subjective patient questionnaires. An alternative, digitally empowered approach combines wearable devices that collect relevant patient data in real time, feed them to cloud infrastructure, and enable clinicians to adapt their care plan, often with the intervention of a DSS. Brohanec et al. [15] have developed decision-support models that identify situations in which the disease has progressed to the point which requires a change of medical therapy. The input data include motor symptoms, non-motor symptoms, and epidemiologic data.

Due to its significant prevalence among the population and also the cost of delivering relevant treatment services, an increasing amount of research is also being directed towards predicting the Parkinson's disease. For example in the work of Challa et al. [16], various machine learning models are used towards the implementation of an automated diagnostic tool for early prediction of Parkinson's disease on the basis of the analysis of non-motor symptoms of the individual.

COPD is a chronic inflammatory lung disease that causes obstructed airflow from the lungs. COPD is treatable. With proper management, most people with COPD can achieve good symptom control and quality of life, as well as reduced risk of other associated conditions. Wearable devices are used to monitor cardio-respiratory data and associate it with patient's daily activity. Such systems may carry embedded intelligence based on innovative machine learning algorithms capable of predicting a COPD exacerbation well before the onset of the first symptoms.

8.2.3 Personal Health Records

A key issue for patient empowerment is to facilitate the process of collecting and managing their health-related data. A common application that allows individuals to manage their health information is a PHR. In principle, PHR refers to both paper-based and electronic records. However, the current usage of the term usually refers to an electronic resource. Switching from paper records to digital records has a lot of advantages. Patients may be able to share simply and securely their entire medical history with all the healthcare professionals they get in touch with. This allows a more personalized and thus more effective treatment. Moreover, the content of the PHR can be both sourced by the individuals or IT systems of the healthcare providers via appropriate interfaces.

The PHR is a lifelong record of health information, a tool for collecting, tracking, and sharing important, up-to-date information about an individual's health that is maintained by individuals themselves. The common goal of a PHR is to provide access to and management of personal health data and eventually support personal health management and enable better decision making. A PHR provides a single, detailed, and comprehensive track record of a person's health status and healthcare activities. It facilitates informed care decisions. Additionally, a PHR can help people to prepare for appointments, facilitates care in emergencies, and helps track health changes.

According to the American Health Information Management Association and the American Medical Informatics Association, a set of basic principles when selecting and using a PHR should be applied [17]. In detail, every person is ultimately responsible for making decisions about his or her health and should have access to his or her complete health information. This should ideally be consolidated in a comprehensive record. The information in the PHR should be understandable to the individual, accurate, reliable, and complete. The integration and secure communication of PHRs with EHRs of providers allows data to be shared between a consumer and his or her healthcare team. People should have control over how their PHR information is accessed, used, and disclosed to concerned parties. All secondary uses of PHR data must be disclosed to the consumer, with an option to opt-out, except as required by law. The operator of a PHR must be accountable to the individual for unauthorized use or disclosure of personal health information.

A PHR should contain any information relevant to an individual's health. For example, a PHR should contain information related to personal identification, including name and birth date, people to contact with in case of emergency, names, addresses, and phone numbers of physicians, dentists, and specialists, health insurance information, a list and dates of significant illnesses and surgical procedures, current medications and dosages, immunizations and their dates, test and physical examination results, allergies or sensitivities to drugs or materials, health-related information of the family members, nutrition and physical activity logs, and educational material.

8.3 Cybersecurity in Healthcare

8.3.1 The Needs

The incorporation of ICT in healthcare and the digitization of data collection and exchange transform the way healthcare services are provided. Health service providers collect, analyze, store, and exchange large amounts of data collected by wearable, portable, or stationary medical and wellness devices or imported by the patients and their caregivers into IT systems.

Being characterized by openness and broad accessibility, this IT infrastructure also inherits various cyber risks and vulnerabilities. For example, the universal proliferation of consumer electronic equipments such as smartphones and wearables in health services brings substantial risks due to the engagement of an audience with low digital literacy and significant ignorance to the dangers of the digitally connected world. It is not unusual that such portable devices, being used to manage a personal health record and collect and locally store vital signs measurements, are secured by a four numbers password that can be "cracked" in less than 20 s [18].

The situation is not significantly more promising when it comes to corporate systems of hospital and care centers. These entities offer services the availability of which is of paramount importance. Such services can be divided into two major types: healthcare services and administrative services [19]. The first type ensures continuity of care for the patients, the disruption of which may have a significant impact on their health status. The administrative services are dedicated to the smooth hospital workflow. Systems handling work orders, medicine inventories, prescriptions, bills, or appointments are part of these services. Their unavailability is, however, less critical as long as their downtime remains of short duration. In this context, healthcare service providers are primary targets of cyber-attacks, mainly related to data theft, denial-of-service, and ransomware. A retrospective observational study of all available reported data breaches in the United States from 2013 to 2017 revealed 1,512 data breaches that affected circa 154 million patient records [20].

Following the above analysis, it is obvious that there is a tremendous need for novel solutions that can bring trust and security in the IT systems used in healthcare services. Indicative such solutions are being described in the subsequent sections.

8.3.2 Technology Trends Related to Cybersecurity in Healthcare

8.3.2.1 Homomorphic Encryption

The need for privacy protection of health and organizational data collected and managed by healthcare service providers and consumers triggered research on the use of cryptographic techniques in the processing of these data. Homomorphic encryption is a form of encryption that allows computation on ciphertexts, generating an encrypted result which, when decrypted, matches the result of the operations as if they had been performed on the plaintext. The purpose of homomorphic encryption is to allow computation on encrypted data. Cloud computing platforms can perform difficult computations on homomorphically encrypted data without ever having access to the unencrypted data. A list of potential applications for homomorphic encryption in healthcare includes applications in genomics, for example, when sharing genomics data to understand the clinical significance of genetic variants or when executing combined analysis of genotype and phenotype data in powerful cloud infrastructures [21].

8.3.2.2 Blockchain Applied in Cyber-Security Solutions for Healthcare Applications

A blockchain is a growing list of records, called blocks, which are linked using cryptography [22]. Though blockchain records are not unalterable, blockchains may be considered secure by designing and exemplifying distributed computing systems [23]. Blockchains which are readable by the public are widely used by cryptocurrencies while private blockchains have been proposed for use in various business domains, one of which is the healthcare services' domain.

The concept of blockchain addresses a common but complex issue, that of enabling different entities trusting each other for having the same copy of relevant data. In the healthcare services' domain, relevant parties are, for example, healthcare service providers and consumers, national and regional authorities, etc. Such entities may be legally obliged to keep a log of all privacy-critical operations performed while being engaged in healthcare services. These data must be protected against tampering and must be available to each participating entity; also, there should be a mechanism to reach consensus between all the stakeholders regarding the operations that have been performed by the users on the system.

In this context, a single centralized log management and auditing system is not feasible, both for technical reasons (as it introduces a single point of failure) and for organizational and possibly legal reasons (each entity must have a copy of the logs under its direct control and each entity should not be forced to trust others' management procedures).

A blockchain-based approach can successfully overcome such limitations. Each entity has its own log system that concentrates and stores all the relevant descriptive information of the operations performed by users that are relevant to a healthcare service. Each operation that is performed by users of the system and results in an exchange of data is logged by all participating stakeholders into a timestamped and digitally signed block.

Many of the immediate challenges present in healthcare cybersecurity will not be met by blockchain technology [24]. For instance, blockchain can do little against attackers attempting to steal patient data, phishers looking for credentials or authorization to transfer money or steal financial documents, insiders looking at the records of patients outside of their care, as all of these data must somehow be accessible and readable within the service provision physical environment.

However, blockchain may address a threat that is rapidly approaching: integrity-based attacks. In these attacks, malicious insiders or external actors modify data, in such a way that is not trackable, leading to major patient safety and institutional trust concerns. Such an intervention may be related, for example, to the integrity of information related to a patient's history of drug allergy. Blockchain technology is an excellent tool to address such a challenge, enabling an immutable record of changes that can be retroactively examined to see precisely what was changed, when it was changed and who changed it.

8.3.2.3 Sandboxing

The concept of sandboxing pertains to the case where a predefined and tightly controlled environment is being set up in which software applications run. The operating system gives the application just enough access to perform activities that it deems "safe" and nothing more. Also, the application is only given access to a predefined area of computer and memory access that is segregated and unique to its owner [25].

Sandboxing is a highly effective means of detecting previously unknown threats. Operating at key locations in the security fabric, a sandbox provides an isolated, secure environment to examine unrecognized code. If it determines it's a threat, it automatically propagates the threat information throughout the fabric, immunizing the entire network against further damage [26].

Sandboxing is a promising technique to be applied against medical devices hijack. This is particularly true for legacy medical devices running on outdated operating systems for which the security is no longer supported, which renders them good candidates as entry points for cyber-attacks.

Despite the main advantages of sandboxing, this technique suffers from two important drawbacks [27]: first, sandboxing is quite time- and resource-intensive, and, second, sandboxing can be evaded. For example, a threat might be programmed to remain dormant until a future date, so that during sandboxing it appears benign. Another effective evasive technique is to make the malware able to detect whether it is in a virtual environment and to remain dormant until it finds itself in a real desktop or other device.

8.3.3 Service Provision Trends Related to Cybersecurity in Healthcare

One of the main trends in modern health IT is the exploitation of cloud infrastructure and services. Cloud services enable healthcare service providers to move data processing and storage onto the internet, provided that security and data ownership concerns are addressed. Towards this direction, the emerging security as a service (SecaaS) model offers a novel security solution for cloud services' users. This model encompasses a variety of advantages such as authentication as a service (AaaS), encryption as a service (ENCaaS), as well as dedicated virtual firewalls and web application firewalls to the cloud infrastructure costumers. It also assists healthcare organizations in strengthening their virtual private clouds with controls applicable to their business domain.

8.4 Conclusion

In this chapter, we presented major driving forces behind digitization in healthcare. We also explored the state of play regarding trust, analyzing its key dimensions as well as the patients' data protection framework. Cyber-security is a must in all digital solutions. With this in mind, we presented multidisciplinary technologies and solutions (e.g., homomorphic encryption, blockchain-based cyber-security solutions, and sandboxing) that are capable to assure data privacy, security, and protection of healthcare-related applications and services, such as physical activity monitoring and personal health records. All these aspects are relevant and timely taking into account the transformation of our world into a digitally driven environment of products, processes, and services.

References

1. European Commission (2017), Commission staff working document Accompanying the document, Communication from the Commission to the European Parliament, the Council, the European Economic and Social Committee and the Committee of the Regions on the Mid-Term Review on the implementation of the Digital Single Market Strategy, A Connected Digital Single Market for All, Brussels. Available at http://eur-lex.europa.eu/legal-content/EN/TXT/?uri=SWD:2017:0155:FIN. [Last accessed 09/09/2019].
2. Deloitte Centre for Health Solutions. Connected Health: How digital technology is transforming health and social care. 2015. Online report available at https://www2.deloitte.com/content/dam/Deloitte/uk/Documents/life-sciences-health-care/deloitte-uk-connected-health.pdf. [Last accessed 09/09/2019].
3. Online report available at: https://iot-epi.eu/ 2017/04/12/iot-and-trust/. [Last accessed 09/09/2019].
4. WP 136 Opinion 4/2007 on the concept of personal data Article 29 Working Party Opinion 05/2014 on Anonymisation Techniques. Available at: http://eur-lex.europa.eu/legalcontent/EN/TXT/?uri=OJ%3AL%3A2016%3A119%3ATOCWP01245/07/EN. [Last accessed 09/09/2019].
5. Scott, G. C. and Lenert, L. A. What is the next step in patient decision support? In *Proceedings of the AMIA Symposium*, Los Angeles, CA, USA, November 4-8, 2000, p. 784. American Medical Informatics Association.
6. Deber, R. What role do patients wish to play in treatment decision making? *Archives of Internal Medicine*, vol. 156, no. 13, 1414, 1996. doi: 10.1001/archinte.1996.00440120070006 [Last accessed 09/09/2019].
7. European Patient's Forum. The new EU Regulation on the protection of personal data: What does it mean for patients? A guide for patients and patients' organisations. Online report available at www.eu-patient.eu/globalassets/policy/data-protection/data-protection-guide-for-patients-organisations.pdf. [Last accessed 09/09/2019].

8. https://ec.europa.eu/info/law/law-topic/data-protection/reform/what-personal-data_en. [Last accessed 09/09/2019].
9. Cognizant 20–20 Insights. The U.S. healthcare implications of Europe's stricter data privacy regulations. 2018.
10. Lamprinos, I. & Steffen, D. Physical activity monitoring for the aging population. In *IADIS International Conference e-Health, located at MCCSIS 2012 IADIS Multi Conference on Computer Science and Information Systems*, July 17–19, 2012, Lisbon, Portugal.
11. Academy of Medical Sciences. Stratified, personalised or P4 medicine: A new direction for placing the patient at the centre of healthcare and health education. 2015.
12. Waljee, A., Higgins, P., and Singal, A. A primer on predictive models, *Clinical and Translational Gastroenterology*, vol. 5, no. 1, e44, 2014. doi: 10.1038/ctg.2013.19 [Last accessed 09/09/2019].
13. Moons, K., Altman, D., Vergouwe, Y., and Royston, P. Prognosis and prognostic research: Application and impact of prognostic models in clinical practice, *BMJ*, vol. 338, no. 042, b606–b606, 2009. doi: 10.1136/bmj.b606 [Last accessed 09/09/2019].
14. Singal, A., et al., Machine learning algorithms outperform conventional regression models in predicting development of hepatocellular carcinoma, *The American Journal of Gastroenterology*, vol. 108, no. 11, 1723–1730, 2013. doi: 10.1038/ajg.2013.332 [Last accessed 09/09/2019].
15. Bohanec, M., et al., A decision support system for Parkinson disease management: Expert models for suggesting medication change, *Journal of Decision Systems*, vol. 27, no. 1, 164–172, 2018. doi: 10.1080/12460125.2018.1469320 [Last accessed 09/09/2019].
16. Challa, K., Pagolu, V., Panda, G., and Majhi, B. An improved approach for prediction of Parkinson's disease using machine learning techniques. In *International Conference on Signal Processing, Communication, Power and Embedded System* (*SCOPES*), India, 2016, pp. 1446–1451. doi: 10.1109/scopes.2016.7955679 [Last accessed 09/09/2019].
17. American Health Information Management Association, and American Medical Informatics Association. The value of personal health records. A joint position statement for consumers of health care, *Studies in Health Technology and Informatics*, vol. 137, 402, 2008.
18. www.popsci.com/technology/article/2012-09/infographic-day-fastest-way-crack-4-digit-pin-number. [Last accessed 09/09/2019].
19. State of cybersecurity & cyber threats in healthcare organizations. Applied cybersecurity strategy for managers. http://blogs.harvard.edu/cybersecurity/files/2017/01/risks-and-threats-healthcare-strategic-report.pdf. [Last accessed 09/09/2019].
20. Ronquillo, J., Erik Winterholler, J., Cwikla, K., Szymanski, R., and Levy, C. Health IT, hacking, and cybersecurity: National trends in data breaches of protected health information, *JAMIA Open*, vol. 1, no. 1, 15–19, 2018.doi: 10.1093/jamiaopen/ooy019 [Last accessed 09/09/2019].
21. Archer, D., et al., Applications of homomorphic encryption. Technical Report, HomomorphicEncryption.org, Redmond WA, 2017.
22. Narayanan, A., et al., *Bitcoin and Cryptocurrency Technologies: A Comprehensive Introduction*. Princeton, NJ: Princeton University Press, 2016.
23. Kim, S. The trailer of blockchain governance game, arXiv.org, 2019. [Online]. Available at: https://arxiv.org/abs/1807.05581 [Last accessed 09/09/2019].
24. Lord, R. Blockchain in health care: The good, the bad and the ugly. Available at: www.forbes.com/sites/forbestechcouncil/2018/04/13/blockchain-in-health-care-the-good-the-bad-and-the-ugly/#4035b14a6278.
25. Improving health data security with a 'sandbox' approach. https://healthitsecurity.com/news/improving-health-data-security-with-a-sandbox-approach. [Last accessed 09/09/2019].
26. Advanced threat protection for the healthcare industry. Advancing medicine needs advanced security. www.fortinet.com/content/dam/fortinet/assets/white-papers/WP-ATP-for-Health care.pdf. [Last accessed 09/09/2019].
27. What is sandboxing? Technical Report. Available at: www.barracuda.com/glossary/sandboxing. [Last accessed 09/09/2019].

9

Novel e-Health and m-Health Services

I. E. Lamprinos
Intracom S.A. Telecom Solutions

CONTENTS

9.1 Introduction: Background and Driving Forces

Information and communication technologies (ICTs) and their applications contribute to the revolution of the practice of healthcare at all levels. ICT is a major driver for change in an environment of eventually ageing population and dramatic increment in chronic diseases along with a compelling demand for healthcare services' cost containment.

As life expectancy increased and birth rates decreased over the last decade the proportion of elderly people in the population has also increased. The fastest growing segment of the European, American and Asian population is over 65 years, and within that segment, the most rapidly growing group is over 86 years. By 2050, a quarter of all Americans and Europeans will be older than 65 years, an increase of 14% from 1995. The result of this change in demographics is that the population in more need of healthcare services is from the elderly segment. Given that people older than 65 years utilize 3–5 times more healthcare services than their younger counterparts the impact of ageing population is more than obvious.

With increasing life expectancy and changing lifestyles, the disease patterns are also changing; for instance, chronic diseases become much more common. Almost 86% of US healthcare spending in 2010 was for patients with at least one chronic condition [1]; whereas in Europe, recent figures raise it to 70%–80% [2]. These numbers signify a great need to reduce costs, for example, by seeking for solutions that reduce the number of hospital visits, that comprise more cost-efficient care alternatives, and that reduce the number of acute hospitalizations due to complications.

Apart from the cost, the quality of healthcare services is an equally crucial factor. The accumulated experience in large and complex systems such as that of healthcare delivery indicates that in many cases the interests of individuals may be secondary to inherent issues of the system (e.g., administrative issues) or to the system inertia (e.g., inadequate services that have always been delivered that way). One of the ways to overcome this issue is by organizing healthcare services around the needs of patients rather than the needs of the healthcare delivery system.

In the context described above, e-Health can play an important role in rendering the healthcare delivery system more citizen-centered; it can improve the access to clinical and specialized services, bringing these services to citizens rather than moving people to them. Additionally, it can shift part of the responsibility of care to the citizens' hands, particularly for wellness and disease prevention.

In line with the above-described socio-economic context, there are various different entry points for the e-Health business domain, giving emphasis on either cost containment or health promotion or both. A typical one when focusing on the cost aspect is the provision of health services by decentralized service providers. The information technology (IT) systems used in such services support a tiered multilayer architecture (aligned with the primary, secondary and tertiary care notions). While supporting timely access to the specialists' medical advice, such systems enhance rural healthcare delivery by integrating low cost, scalable technological components into existing infrastructures, while favoring the continuity of healthcare provision by supporting the integration of patients' medical data to globally and instantly accessed Electronic Health Records.

Offering citizen/patient-centered health services is another entry point which supports remote health status monitoring of those with limited access, notably people living in rural regions, citizen empowerment to manage their own health status and conduct independent living, personalized care and treatment plans, preventative lifestyle, and continuity of care (from prevention to rehabilitation).

Other attractive use cases include home-based or mobile patient monitoring with ICT systems that offer various functions. Such systems may:

- Support remote monitoring of vital signs (e.g., electrocardiograph, blood pressure, temperature, pulse and blood glucose levels);
- Allow patients to communicate clinical information to their care team;
- Provide care teams with diagnosis assisting tools and communications systems;
- Enable users to share information and carry out video-based face-to-face communication and observation;
- Enable care teams to get access to clinical information from any setting, even outside hospitals;
- Provide care teams with evidence-based clinical protocols or care plans; and
- Provide consumers with education and decision support tools.

Finally, another significant group of e-Health services relates to personal health management and wellness services that focus on preventive healthcare. Typical services of this family include optimal nutrition and weight management, physical activity and exercise monitoring, stress management, personalized medical care, and pregnancy and maternity calendar.

In line with the above, in the subsequent sections, we present the most promising e-Health and m-Health applications (apps) and services, after first having presented an overview of the most significant advances in technology that favors the setup and delivery of those services.

9.2 Technology Advances behind Novel e-Health and m-Health Services

9.2.1 The Role of Big Data and Cloud Computing

In most of the novel e-Health services nowadays, it is important to collect and analyze information not only related to the patient health condition, but also related to ambient environment conditions, clinical knowledge and administrative data, which may have impact on the onset and evolution of a disease. This need has brought us to the era of big data.

Health-related big data refer to large, routinely or automatically collected datasets, which are electronically captured and stored. These datasets are reusable as containing multipurpose data. Big data analytics also pertains to functions related to fusion and connection of existing databases. It does not refer to data collected for a specific study [3]. The collected datasets may pertain to inpatient and outpatient hospital data, prescription data, claims information and patient-reported data, socio-demographic data, clinical trial data, "omics data", wellness and lifestyle tracking data, remote monitoring data, etc.

Linking the above-listed data can lead to many powerful insights. Applications span from disease prevention to risk assessment related to the onset of a disease and eventually to the management of the disease after its onset.

A conceptual framework for big data analytics in healthcare is depicted in Figure 9.1 indicating components such as data sources' adapters, data analyzers and visualization, and events' processing.

Big data analytics may contribute to widening the possibility for prevention of diseases by detecting several marginal risk factors at population, cohorts and individual subjects as well as by improving the effectiveness of interventions that aim to help people achieve healthier behaviors [3]. Moreover, it can contribute in increasing the effectiveness and quality of the treatments.

Big data analytics could be a breakthrough also for the pharma domain. The McKinsey Global Institute estimates that applying big data strategies to better-informed decision making could generate up to $100 billion in value annually across the US healthcare system, by optimizing innovation, by improving the efficiency of research and clinical trials, and by building new tools for physicians, consumers, insurers, and regulators to meet the need for more individualized approaches [4]. It can also improve the pharmacovigilance and patient safety through the ability to make more informed medical decisions based on directly delivered information to the patients.

Big data analytics goes hand in hand with cloud computing. In fact it was the emergence of cloud computing which made it easier to provide the best of technology in a most

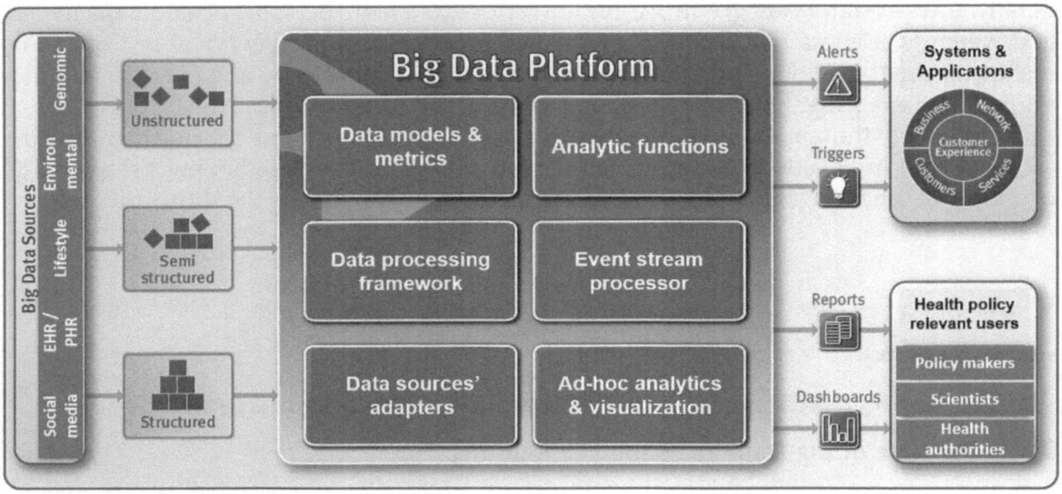

FIGURE 9.1
Framework for big data analytics in healthcare.

cost-effective manner. Cloud computing not only reduces costs, but also makes a wide array of applications available to companies and researchers as it closes the gap between generation, management and processing of data. Here is an example: patient genomic data analysis allows one-step association of biomarkers with therapies and enables the detection of new actionable biomarkers, or clinical trials compatible with patients, thereof saving time and money and increasing treatment success [5]. Classic research laboratories do not possess sufficient storage and computational resources for processing omics data. Laboratory-hosted servers require investments in informatics support for configuring and using software. Such servers are not only expensive to setup and maintain, but do not meet the dynamic requirements of different workflows for processing omics data, leading to either extravagant or sub-optimal servers [6].

The above trend for novel cloud computing-based big data analytics services is favored by several initiatives that support the public availability of data through open datasets. Indicatively, we refer to the European Union (EU) Open Data portal [7] or the European Data portal [8]. This is a major step forward towards enabling the development of data-driven innovative personalized health solutions.

9.2.2 The Role of Internet of Things in Healthcare

The Internet of Things (IoT) concept injects internet connectivity beyond standard devices, such as desktops, laptops, smartphones and tablets, to any range of traditionally dumb or non–internet-enabled physical devices. This paradigm shift means that numerous devices surrounding us are now connected to the internet and able to collect, process, and exchange data.

The exploitation of the IoT concept and relevant technologies has the potential to transform the practice of delivering healthcare services in terms of novelty, efficiency, and quality. The transformation of the significantly increased data volumes to valuable information and subsequently to knowledge enables the transition to personalized and customizable, person-centric healthcare services. It also favors the shift of the services' provision environment closer to the service receiver and enables a more efficient management of the financial cost of those services.

The application of IoT in the healthcare domain is boundless. Indicative applications are the following:

- Smart indoor infrastructures enable citizens with special health profiles (e.g., elderly, chronic patients) to live in their familiar environment in an assisted living manner, prolonging their independence and enhancing their quality of life. Wearable or implantable sensors with wireless transmitters ranging from blood pressure and heart rate monitors to advanced devices capable of monitoring specialized implants, such as pacemakers, electronic wristbands, or advanced hearing aids, facilitate the monitoring of crucial vital signs and favor ubiquitous patient monitoring.
- A broad range of IoT applications can be found also within the hospital environment. For example, IoT is a promising technology for medical asset tracking and management. In combination with real time geolocation systems it facilitates the management of medical devices, supplies and biological material therefore conducting on preventing medical errors and improving patient care and safety inside hospitals. IoT is also a valuable enabling technology for rendering hospital beds into "speaking beds" that detect and communicate the presence of a patient or her attempt to get up. Such beds can also adapt to ensure appropriate pressure and support applied to the patient without the manual intervention of nurses.
- Several use cases are based on the convergence of IoT with robotics. Among the many robots that are already in use today there are well-known examples of medical robots in surgery (precision surgery or tele-surgery), robots that are used for rehabilitation, robots which carry out tasks such as medication delivery, food delivery, and delivery of supplies in general. All these robots comprise internet-connected devices that are able to collect, process, and exchange data.

IoT-based applications make use of an underlying infrastructure that includes components and services at the devices' plane, as well as various cloud and edge-positioned items, as indicated in Figure 9.2.

9.2.3 The Role of Blockchain in Health Information Sharing

Blockchain is a particular type of distributed ledger technologies (DLTs), enabling parties to exchange any type of digital data on a peer-to-peer basis with fewer to non-existent intermediating third parties. DLTs are particular types of databases in which data are recorded, shared and synchronized across a distributed network of computers or participants. The specificity of blockchain is that it employs cryptographic techniques to record and synchronize data in "chains of blocks." Blockchain relies on public-private keys to ensure the authenticity and integrity of data exchanges between different parties, all recorded in an irrevocable and time-stamped record or ledger [9].

Applications of blockchain and other DLTs in the healthcare domain relate among others to electronic health records, identity management, data authentication and sharing, pre-authorization of payments, claims management, clinical trials, and counterfeit drug prevention and detection.

Data confidentiality and security is a major concern in the healthcare sector, so any blockchain solutions need to put in place strong privacy mechanisms and ensure compliance

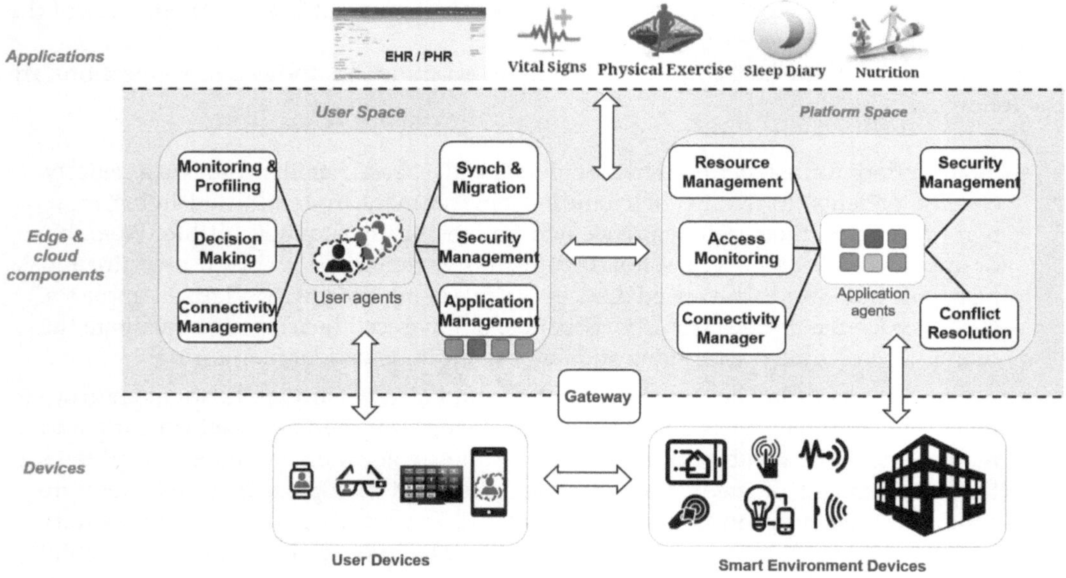

FIGURE 9.2
Leveraging IoT platforms and connected devices in e-Health and m-Health applications.

with data protection regulation. Particularly for the latter, an extensive discussion over the impact of the EU General Data Protection Regulation (GDPR) [10] on the adoption of blockchain technologies in healthcare applications is ongoing.

GDPR governs the use and protection of personal data collected from or about any EU citizen. In GDPR, personal data are defined very broadly and include any information relating to an identified or identifiable natural person. Under current EU legal interpretations, this includes encrypted or hashed personal data as well as public cryptographic keys that can be tied to an individual. One of the foundational aspects of GDPR is the fact that citizens have more control over their personal information and are given the right to erase, that is, the right to have all their personal data permanently deleted. This comes in contrast with blockchain's structural element of immutability. In other words, since data that is stored on the blockchain, including personal data, can't be deleted, there is no way to exercise the right to erasure that people are granted under GDPR [11].

Blockchain is not designed to be GDPR-compatible. Or rather, GDPR is not blockchain-compatible the way it is written today [11]. Several ideas are proposed, for example, rendering the solution of making personal data permanently inaccessible equivalent to the right to erasure; or moving sensitive personal data "off chain." The latter would mean that personal data are moved outside of the blockchain and only a reference to the data and a hash of the data are stored on the blockchain. This would allow the removal (erasure) of the personal data without breaking the blockchain. However, this approach defeats many of the benefits of DLT such as security and resilience through redundancy [5].

Anyway, e-Health includes several application domains other than the Electronic Health Records (EHRs) and similar applications where blockchain technology could play a significant role. An indicative such domain is the one that intersects healthcare with logistics: provenance-based blockchains can track the manufacturing and distribution of medication across the supply chain, ensuring medications are properly stored and handled, and that counterfeit drugs do not enter the market [12].

9.3 Novel e-Health and m-Health Services

9.3.1 The Trajectory of a Disease: From Risk Assessment to Early Intervention and Patient Empowerment

The application of ICT offers new opportunities to empower healthy individuals or patients and their caregivers in relation to health promotion and disease management, to facilitate wellbeing and delay the onset of chronic diseases, and to significantly enhance the quality of life. Successful ICT tools about citizen engagement in health, wellbeing and disease prevention are not just about exploiting novel, and at some level "exotic" access means (smartphones, tablets, smart watches, and smart glasses), but also about how technology enables them to have access to educational information, and how they can interact with health promotion services and healthcare providers.

One particularly challenging aspect of disease prevention and wellbeing where ICT tools may especially prove successful is enhancing health literacy skills and patient's decision making capacity. Citizens and citizens-at-risk need to be engaged in disease prevention tasks and educated as to why this is important. In order to end up with a ground-breaking ICT toolbox one has to mix a proper set of ingredients such as user engagement, context awareness, persuasiveness, attractiveness, and simplicity.

On the other hand, there is a need for robust tools that facilitate healthcare providers in assessment of risk for the onset of a disease, followed by early intervention in the management of the disease through solutions that empower patients in this process. The following subsections analyze this chain.

9.3.1.1 Risk Assessment Phase

With rising healthcare costs, all healthcare stakeholders (payers, physicians, and patients shift the onus from a "pay for intervention" to a "pay for performance" model [13]. This change in focus towards overall outcomes drives a paradigm shift towards disease prevention and early-diagnosis, that is, move from a "diagnose and treat" to a "predict and pre-empt" approach.

Several research and innovation activities address the issue of risk assessment for the onset of a disease. Particular emphasis is given on the assessment of risks associated with non-communicable diseases (e.g., cardiovascular diseases such as heart attacks and stroke, cancer; chronic respiratory diseases such as chronic obstructed pulmonary disease and asthma, and diabetes) which are characterized by long duration and slow progression.

The evolution of ICT enables the implementation of innovative disease onset risk assessment tools and interventions which can be based on either centralized or decentralized technology infrastructure. For example, Abbas et al. [14] propose a cloud-based framework that manages health-related big data and benefits from the ubiquity of the Internet and social media. The framework provides a disease risk assessment service based on the comparison between the profiles of enquiring users and the profiles of existing patients. The typical profile attributes that are provided as input to the framework include age, gender, ethnicity, weight, height, family disease history, and other commonly observed symptoms for a disease. Based on the attributes specified in the users' query, the enquiring users' profiles are compared with the existing users and the users are returned a risk assessment score for that disease.

9.3.1.2 Early Intervention

Early intervention in the disease onset and progression process to prevent serious effects is a critical aspect of healthcare service delivery. Additionally, as the trajectory of chronic diseases is often cyclical, such an approach offers multiple interception opportunities to prevent serious decline. The role of the citizen in this model of disease prevention and management is pivotal. In fact, several clinical situations can be prevented or better monitored and managed with active participation of the patient. This in turn highlights the importance of delivering e-Health and m-Health solutions that raise individual's awareness and empower the patient to participate in the management of his or her health.

Towards this direction the use of ICT tools and e-Health services is of paramount importance. Such solutions are typically based on a multi-layered architecture including components for data collection, analysis, visualization, and action. The data collection layer is based on a sensors' data fusion approach towards the collection of physiological, physical, cognitive, environmental, and social data related to the monitored individual. Measuring physiological and activity-based parameters remotely and continuously via wearable or ambient sensors has the potential to revolutionize our ability to predict and pre-empt harmful changes in a disease trajectory.

The collected data are fed to a decision support engine that analyses it and produces new knowledge. An important aspect of this part of the chain is the development of methods for real-time identification of behavioral and physiological patterns since early detection and communication of alerts to both patients and providers can prompt help-seeking behavior and deployment of just-in-time interventions that may prevent relapse episodes and effectively alter the disease progression [15].

The above architecture pertains to a closed loop built around the citizen. Different forms of this loop exist, mostly depending on the action step: some interventions render the knowledge extraction into recommendations that may be prompted either to the monitored individuals or their health caregivers; others use this knowledge to support highly personalized tools for the individuals.

9.3.1.3 Patient Empowerment

Patient empowerment interventions aim at involving patients to a greater extent in their own healthcare and disease management cycle. A concept often applied in patient empowerment is that of ICT-enabled self-management pathways. This comprises a cyclical process proposed by Lorig et al. [16] that includes medical consultations followed by self-management goal setting, then self-management actions, then feedback collection, and finally evaluation and readjustment of the self-management process. Self-management can be further decomposed into information collection, decision making, and action-taking steps (Figure 9.3).

In line with this concept, Lamprinos et al. [17] propose a modular ICT framework for the establishment of a personalized yet evidence-based patient guidance strategy, leveraging various sources of personal health data and clinical knowledge. This approach was applied in diabetes self-management, using an ICT-based self-management framework, as a means to seek lifestyle and behavioral changes that facilitate disease management. This framework is composed of four phases:

a. the phase of recommendations' specification where treatment goals are specified taking into account the output of a decision support tool that blends commonly applied, evidence-based clinical guidelines, the physician's experience, and the actual patient's profile;

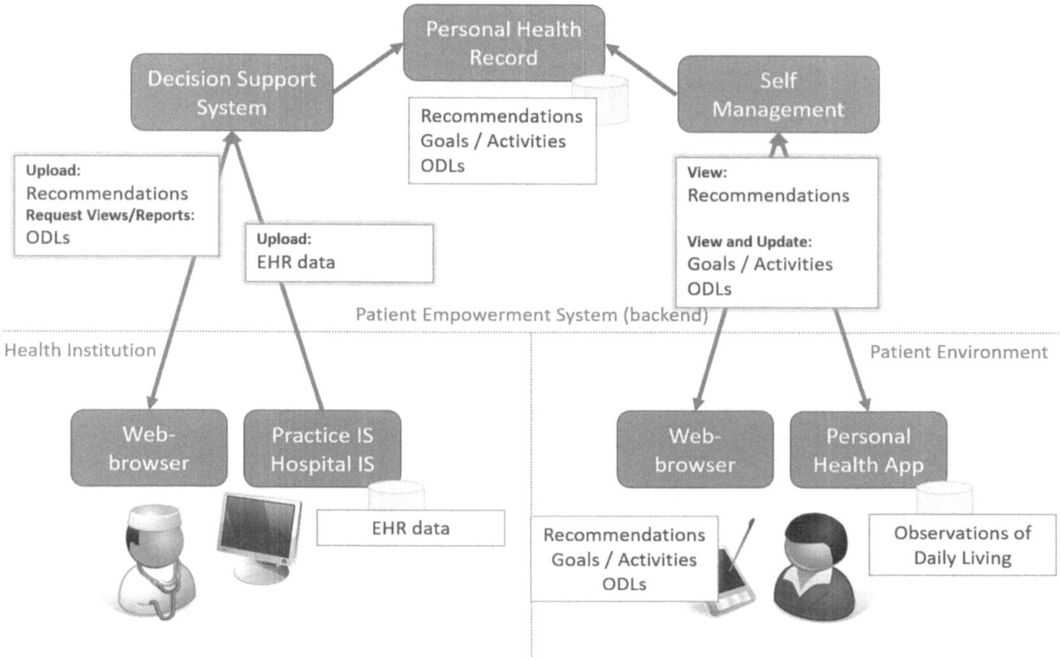

FIGURE 9.3
Framework for self-management applications.

b. the phase of goals' definition or modification, where the patient breaks down the physician's recommendations into short-term self-management goals, stored in her/his individual Personal Health Record (PHR);
c. the phase of actions' specification, where the patient breaks down the defined goals into small, achievable portions; and
d. the phase of observations of daily living (ODLs) recording, where the patient collects and records ODLs in alignment with the planned activities using a web or mobile application (Figure 9.4).

The loop closes with an evaluation and feedback step where the patient is given an indication on how realistic the planning was as well as hints for planning adaptation and rationalization.

9.3.2 Trends in m-Health Applications and Services

m-Health pertains to the delivery of medical and public health services supported by mobile devices. The term is most commonly used with regards to using mobile devices, such as smart phones and tablets, wearable and implantable devices. The m-Health field has emerged as a sub-segment of e-Health, leveraging the tremendous evolution of the mobile devices regarding connectivity and processing power. This evolution has rendered the mobile devices into open platforms, initially for delivering innovative health-related consumer applications, but eventually also as certified medical applications [18].

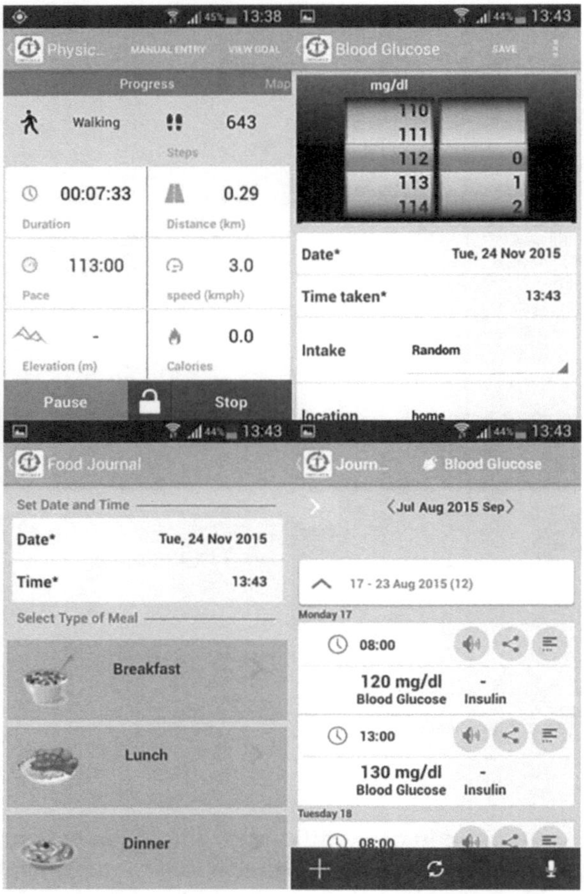

FIGURE 9.4
Mobile applications for collection of ODLs by diabetic patients.

Departing from the mobile devices' evolution, m-Health nowadays encompasses a much broader concept, capitalizing the tremendous evolution in key enabling technologies. This evolution is based for example on the increased accuracy of sensors and measurement methods, on the injection of internet connectivity to miniature-measuring devices and sensors, on the ability to interface various different devices and fuse data from different sources, and to the increased user friendliness due to wearing comfort and invisibility of the devices (being either wearables, such as smart watches and textiles, or implantable or ingestible).

The above evolution at the technical foundations' plane enabled to flourish a marketplace with numerous applications and services, related to personal health monitoring and healthcare delivery via certified medical apps. In the subsequent sections we analyze these two dimensions.

9.3.2.1 Consumer Mobile Applications for Personal Health Monitoring

Personal health monitoring is usually based on solutions that consist of hardware-supported software applications, seamlessly integrated into a person's everyday life. Based on this architecture, several applications are available to monitor an individual's vital signs (e.g., blood pressure, heart rate, and body temperature), and behavior (e.g., medication adherence and physical activity). According to recent market studies, at least 325,000 mobile health apps were available in 2017 [19].

This evolution goes hand in hand with the ever increasing penetration of smart phones and wearable devices such as wristbands and smartwatches. Goode Intelligence forecasts that by 2019 there will be 5.5 billion users of mobile and wearable biometric technology around the globe [20].

The growth of wearable devices that can be coupled to a smartphone influences various market segments. The "killer" application is a heart rate-based performance measurement for fitness and cardio training [21]. The latest trend is to remove the chest strap, using heart rate sensors in earphones, a headband, or a second arm-band. Other applications include activity monitoring and management, weight-diet management, with extensions to sleep. Commercially, the wearing position is a differentiator, which affects the accuracy of the calculation of the energy spent during activities. Other parameters that affect this dimension are the quality of sensors, and the sampling methods [21].

9.3.2.2 Consumer Mobile Applications for Telemonitoring

m-Health apps exist not only for wellness and fitness' monitoring but also for monitoring of chronic diseases. Chronically ill people eventually become the focus of m-Health, as traditional health industry players move into the m-Health app market to serve a potential customer base that exceeds 1 billion subjects worldwide (based on projection of WHO estimates for patients with non-communicable diseases prevalence [22]).

The aim of these applications is to improve the healthcare delivery to chronically ill patients by supporting sophisticated monitoring of treatment compliance, time-management and an abundance of other functionalities. In addition, m-Health apps can be used to support communications with and consultation from healthcare service providers as well as health education, eliminating the need for physical presence at the clinic. All the relevant information can be transmitted and stored in the cloud being incessantly accessible by the patients and their caregivers.

Among the various different chronic disease management and telemonitoring we indicatively refer here to diabetes for which hundreds of apps exist in the various app stores for patient-centric management. Diabetes-specific apps enable patients to monitor their health condition, track blood sugar, body weight, food consumption, plan meals, and find diabetes-friendly recipes. Typical functionality of such applications include symptoms' registration, ability to locate a physician or hospital in an event of emergency, food database for patients to track water and nutritional ingredients' intake, access to recipes and nutritional information, barcode scanners and meal planners, blood sugar, and weight tracker. Other features include data visualization and statistics reporting, potentially with medication correlations, integration with personal and electronic health records, access to online health communities, and various gamification elements, particularly interesting for young patients.

Besides chronic disease monitoring, another remarkable trend nowadays is to further extend the usage of smart phones, as platforms for diagnostics and disease screening functions, by carrying out measurements of biomarkers and rendering them to point-of-care devices. Lan et al. [23] present an approach for transforming the blood glucose meters including those already integrated with smartphones, into a general healthcare meter for in-vitro diagnostics in mobile health.

Overall, one of the major drivers of the increasing impact that consumer mobile applications for telemonitoring can have is that of the cost-saving potential they create. Taking as an example diabetes mellitus, the direct costs per patient in countries like Germany, not including the cost of long-term care (nursing insurance), are between EUR 4,000 and EUR 5,000 per year [24]. In the same research, the indirect costs, which include loss of productivity and early retirement, are calculated to more than EUR 5,000 per patient per year. In the US, the total cost of diagnosed diabetes is expected to rise to $336 billion by 2034 [25].

9.3.2.3 Certified Mobile Medical Apps

The emergence of the m-Health marketplace also brought the need for certification. In the US, the Food and Drug Administration (FDA) issued the Mobile Medical Applications Guidance [26] according to which mobile medical apps are medical devices in the form of mobile apps, and as such they meet the definition of medical device, either being an accessory to a regulated medical device or transforming a mobile platform into a regulated medical device. The guidance focuses on apps that present a greater risk to patients if they do not work as intended and on apps that cause smartphones or other mobile platforms to impact the functionality or performance of traditional medical devices. Such elements should be properly certified.

An indicative use case is that of a system used for transcutaneous electrical nerve stimulation (TENS). TENS devices have been prescribed by doctors as an effective way to reduce or even eliminate pain, (e.g. back pain). The TENS devices work by sending tiny electrical signals through the skin to intercept pain signals from reaching the brain and help to release endorphins, the body's own natural pain fighting chemical. For example, iTENS [27] harnesses the power of a smart phone to wirelessly sync with a Bluetooth-based electrotherapy device, instead of using a separate control unit and wires. This device works via an iPhone- or Android-based app. In that sense, iTENS comprises a mobile medical device accompanied by a mobile medical app, and indeed it is certified by FDA as such.

9.3.2.4 Best Practices for Building an m-Health App

Given the fact that thousands of mobile apps claim to be the best tool either for wellness and fitness tracking or for chronic disease monitoring and management, there is an unavoidable need for injecting proper attraction hooks to their potential users (Figure 9.5).

From a functionality perspective, it is important to provide proper motivation for installing and start using the app. A key aspect has to do with the ability to share collected information with the healthcare provider and get feedback and guidance. In fact, this comprises a key incentive for keep using an app, after an initial phase of enthusiastically keeping logs of data. Not only it motivates the patient on keep using the app but it also contributes in better compliance with the prescribed therapy, and in dynamic guidance from the clinical supervisors.

Another key aspect of a successful m-Health app has to do with minimizing the effort for data collection. Modern sensors and wearables are great tools for efficient data collection. Seamless self-logging can be powerful as long as it is easy and fast to use. Actually, quite some of these apps are bundled with specific medical peripherals, for example, glucose monitors or scales. Nevertheless, manual logging should also be an option supported by the application. Eventually, this ends up to be not just a matter of functionality, but also a matter of the applicable business model and the key players participating in the applicable value chain (device manufacturers, integrators, solution providers, etc.).

Another interesting aspect is the use of gamification to motivate patients to monitor and potentially improve their own health condition. While no long-term studies indicate benefits

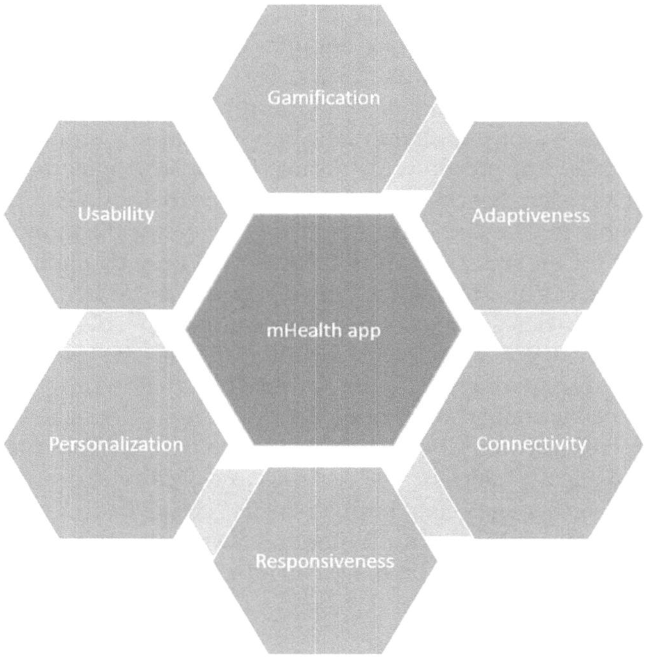

FIGURE 9.5
Rendering an m-Health app attractive to potential users.

of gamification for patients using health applications, there is a growing interest in going beyond simply improving usability of an application, by motivating the user to incorporate the application into her everyday routine and reap the intrinsic benefits of using it. Gamification seeks to improve user engagement, help solve problems, learn, and help complete tasks. From the health perspective, gamification techniques might help the patient manage their general health, their chronic disease(s) and perhaps improve their health condition.

Integration with social networks provides patients with significant benefits. Patients with similar disease can interact with each other and share experiences. The idea is that the mobile app serves as a bridge to social networks to communicate news and relevant information related to the monitored disease. However, the use of social networking involves privacy and security implications and may have controversial impact on the subjects. So, while the feature of bridging disease management mobile app with social networks is an appealing feature for many users, prior consent of them should be mandatory, as well as the option to disconnect the link between the apps whenever desired.

Personalization is a very important feature when it comes to digital artifacts that support our life nowadays. The one-size-fits-all model is a guarantee for failure. Chronic patients who decide to use a mobile app to support their disease management are actually users with very diverse characteristics (digital and health literacy, usage preferences, experiences, and competences in disease self-management). A dynamic combination of offered functionalities as well as a variation in the timing and ordering of how they are applied (i.e. user driven customization) is a key factor for success. Personalization is a key dimension also when considering access to health-related information material. Holtz and Lauckner [28] indicate that although Internet health information is growing rapidly, the average person lacks the skills for finding and using the health information strategically for his/her benefit. This indicates the need for personalized information, leveraging user's profile, desires, and actual health status.

Finally, the design of the presentation layer of the applications is a key success factor. Older people comprise a segment of population who can benefit only from accessible interfaces. They have some overlapping needs with disabled people in general, due to age-related impairments affecting vision, hearing, physical, and cognitive abilities. Therefore, user interfaces that are accessible to disabled users are also more accessible to older users. However, an effective interface for the older people is achieved not only by applying the generic web accessibility guidelines, but also by implementing accessibility aids specifically addressed to their special needs. When designing user interfaces for older people, usability is a major concern. Apart from making them accessible and multimodal, for the certain benefit of enabling access to a broader public, perhaps more important than that is simply make them usable. Achieving usability needs some special design considerations when the target audience are older people. The decrease in cognitive capabilities, especially the short-term memory, is the biggest obstacle for using an interface that requires exploratory learning and building conceptual models. While other age-related impairments (vision, hearing, and physical ability) can be addressed by accessibility or multimodality features, the difficulties derived from intellectual decline can only be overcome with an adaptive design of the user interface.

9.4 Conclusion

In this chapter, we presented a compact yet inclusive set of novel e-Health and m-Health apps and services. We briefly showed that the evolution is healthcare delivery is propelled by the enormous advances in key enabling technologies and infrastructure, such as big data and cloud computing, IoT, and blockchain. This evolution will contribute to the efficient management of health and wellbeing, while empowering the participation of citizens, and facilitate the transformation of health and care services to more digitized, person-centered and community-based care models, thereby enabling better access to healthcare and the sustainability of health and care systems.

References

1. Gerteis, J., Izrael, D., Deitz, D., LeRoy, L., Ricciardi, R., Miller, T., and J. Basu. *Multiple Chronic Conditions Chartbook.* Rockville, MD: Agency for Healthcare Research and Quality, 2014.
2. Ceretti, E. et al. 356 European Journal of Public Health, Vol. 26, Supplement 1, 2016, *European Journal of Public Health*, 26, 1, 2016. [Online] Available at: https://eupha.org/repository/sections/HIA/2016_HIA_HSR_ECO_PHPP_10K.pdf [Last accessed 09/09/2019].
3. Österreich, G. Study on big data in public health, Telemedine and healthcare. *European Commission, Directorate-General for Health and Food Safety. Brussels*, 2016. [Online] Available at: https://eupha.org/repository/sections/HIA/2016_HIA_HSR_ECO_PHPP_10K.pdf [Last accessed 09/09/2019].
4. Dhankhar, A., Matthias, E., and M. Møller. Escaping the Sword of Damocles: Toward a New Future for Pharmaceutical R&D, McKinsey Perspect Drug Device R&D, 2012.

5. Grant, A. Blockchain and GDPR: Between a Block and a Hard Place. Online report available at: www.tripwire.com/state-of-security/security-data-protection/blockchain-gdpr. [Last accessed 09/09/2019].
6. Alyass, A., Turcotte, M., and D. Meyre. From big data analysis to personalized medicine for all: Challenges and opportunities, *BMC Medical Genomics*, 8, no. 1, 33, 2015.
7. https://data.europa.eu/euodp/en/home.
8. www.europeandataportal.eu.
9. Nascimento, S., Pólvora, A., and J. Sousa Lourenço. 2018. Blockchain4EU: Blockchain for Industrial Transformations, EUR 29215 EN, Publications Office of the European Union, Luxembourg, 2018.
10. Working Party 136 set up under Article 29 of Directive 95/46/EC as a European advisory body on data protection and privacy. Opinion 4/2007 on the concept of personal data. Available at: http://eur-lex.europa.eu/legalcontent/EN/TXT/?uri=OJ%3AL%3A2016%3A119%3ATOC WP 01245/07/EN. [Last accessed 09/09/2019].
11. Toth, A. Will GDPR block blockchain, *World Economic Forum*, 24, 2018. Available at: www.weforum.org/agenda/ 2018/05/will-gdpr-block-blockchain. [Last accessed 09/09/2019].
12. IBM Institute for Business Value. Healthcare Rallies for Blockchains: Keeping Patients at the Center. [Online]. Available at: www.ibm.biz/blockchainhealth. [Last accessed 09/09/2019].
13. Introduction to the Remote Assessment of Disease and Relapse (RADAR) Programme. Innovative Medicines Initiative. Report available at: https://intranet.imim.es/files/news/Topic_Feb2016_RADAR_AD.pdf. [Last accessed 09/09/2019].
14. Abbas, A., Ali, M., Khan, M. U., and S. U. Khan. Personalized healthcare cloud services for disease risk assessment and wellness management using social media, *Pervasive and Mobile Computing*, 28, 81–99, 2016.
15. Innovative Medicines Initiative 3rd Call for proposals. Report available at: https://www.imi.europa.eu/sites/default/files/uploads/documents/apply-for-funding/call-documents/imi2/IMI2Call3_TopicTextWebFINAL.pdf. [Last accessed 09/09/2019].
16. Lorig, K., Holman, H., Sobel, D., Laurent, D., Minor, M., Sobel, D., Holman, H., and V. M. Gonzalez. *Living a Healthy Life with Chronic Conditions: Self-Management of Heart Disease, Arthritis, Diabetes, Asthma, Bronchitis, Emphysema & Others*. Boulder, CO: Bull Publishing Company, 2000.
17. Lamprinos, I., Demski, H., Mantwill, S., Kabak, Y., Hildebrand, C., and M. Ploessnig. Modular ICT-based patient empowerment framework for self-management of diabetes: Design perspectives and validation results, *International Journal of Medical Informatics*, 91, 31–43, 2016. doi: 10.1016/j.ijmedinf.2016.04.006 [Last accessed 09/09/2019].
18. Sousa P.S., Sabugueiro D., Felizardo V., Couto R., Pires I., and Garcia N.M. mHealth Sensors and Applications for Personal Aid. In: Adibi S. (Ed.) *Mobile Health*. Springer Series in Bio-/Neuroinformatics, vol 5. Springer, Cham, 2015, 265–281.
19. Research2Guidance. Mhealth App Economics 2017: Current Status and Future Trends in Mobile Health, 2017. Report summary available at: https://research2guidance.com/325000-mobile-Health-apps-available-in-2017/. [Last accessed 09/09/2019].
20. Goode, A. Biometric tech to drive take-up of wearables by 5.5 bn people by 2019, *Biometric Technology Today*, 7, 1, 2014.
21. Magnus, B. Innovation Radar – Active Healthy Ageing. Health and Wellbeing White Paper. EIT Digital, 2013. Online report available at https://pdfs.semanticscholar.org/a76f/2785d9320516f9ab1d7f1d1e6503bcb11c58.pdf. [Last accessed 09/09/2019].
22. *Noncommunicable Diseases Country Profiles 2018*. Geneva: World Health Organization, 2018. License: CC BY-NC-SA 3.0 IGO.
23. Lan, T., Zhang, J., and Y. Lu. Transforming the blood glucose meter into a general healthcare meter for in vitro diagnostics in mobile health, *Biotechnology Advances*, 34, no. 3, 331–341, 2016. doi: 10.1016/j.biotechadv.2016.03.002 [Last accessed 09/09/2019].
24. Köster, I., Huppertz, E., Hauner, H., and I. Schubert. Direct costs of diabetes mellitus in Germany – CoDiM 2000–2007, *Experimental and Clinical Endocrinology & Diabetes*, 119, no. 6, 377–385, 2011. doi: 10.1055/s-0030-1269847 [Last accessed 09/09/2019].

25. Huang, E.S. et al. Projecting the future diabetes population size and related costs for the US, *Diabetes Care*, 32, no. 12, 2225–2229, 2009.
26. Mobile Medical Applications, Guidance for Industry and Food and Drug Administration Staff. Onlineavailableat:www.fda.gov/downloads/MedicalDevices/DeviceRegulationandGuidance/GuidanceDocuments/UCM263366.pdf. [Last accessed 09/09/2019].
27. iTENS Wearable Electrotherapy Review, http://itenseurope.com/itens-wearable-electrotherapy-review/.
28. Holtz, B. and C. Lauckner, Diabetes management via mobile phones: A systematic review, *Telemedicine and e-Health*, 18, no. 3, 175–184, 2012. doi: 10.1089/tmj.2011.0119 [Last accessed 09/09/2019].

10

Activity Monitoring of Elderly Patients

Dwaipayan Biswas
IMEC

CONTENTS

10.1 Introduction

Human activity detection and classification is a key component for remote health monitoring applications. Remote health monitoring systems aided by technological advancements have enabled continuous monitoring of inhabitants within the home environment, in effect cutting down hospitalization and care expenses [1]. The developments in microelectromechanical systems (MEMS) technology along with wireless sensor networks (WSN) and Internet of Things (IoT) have proliferated the miniaturization of wearable sensors capable of pervasive long-term monitoring [2]. Wearable sensors have become popular means of providing accurate and reliable information on people's activities and behaviors. They also measure physiological parameters (e.g., heart rate, blood pressure, etc.) which are combined with mobility information to infer the state of critical patients during home rehabilitation [3].

Human movement typically entails a variety of periodic activities including walking, running, stair ascent, and stair descent. Monitoring of movements performed in an ambulant environment, termed as 'activities of daily living (ADL)', for example, brushing, drinking, cooking, bathing, etc., has also gained prominence [4,5]. ADL monitoring helps determine the activity levels of elderly inhabitants and ascertain motor activities especially for subjects with physical impairments. Typical modalities for activity monitoring can be categorized into – vision [6] and wearable sensor-based monitoring [7,8]. Although more accurate, camera-based sensing systems are intrusive and involve the processing/storage of large amounts of data [9]. In comparison with the former, time-series data generated by low-cost wearable sensors provide an opportunity for reliable and pervasive monitoring. Smart homes that provide intelligent systems for assistance in the subject's living environment, also referred as ambient assisted living (AAL), are developed by combining wearable sensors with ambient sensors – such as motion detectors on doors, radio-frequency identification tags on objects of daily use, sensitized cutlery and acoustic sensors to monitor activities pertaining to food intake, pressure sensors proximal to the bed to monitor sleep durations, etc. [10–12]. Information collected by wearable sensors is augmented by the ambient sensors and distributed throughout the home to provide holistic feedback of the activities performed and living behavior of the subject aimed toward improved health management [13].

10.2 Activity and Sensing

Human activity recognition (HAR) is a well-researched topic across AAL [14–16], context-aware computing [17], pervasive computing, and mobile computing [18,19]. The varying approaches to HAR are representative of the underlying sensing techniques, the signal processing methodologies, and the environment from where the activity data is acquired. The key research steps involved in a successful HAR system are – (a) activities to be monitored, (b) types of sensors, (c) methodology to determine/classify the activities performed, (d) embedded implementation of low-complexity algorithms for real-time information extraction, and lastly, (e) reconfigurability of the system for new users [20]. In this chapter, we will focus on HAR using wearable sensors, primarily in the home environment.

10.2.1 Activities to Be Monitored

It is imperative to comprehend the granularity inherent in human behavior; correspondingly, movements can be categorized as (but not limited to) – (a) gestures, (b) actions, or (c) interactions. Gestures represent elementary components comprising a holistic movement, for example, raising an arm, moving a finger, or nodding. Actions, on the other hand, comprise multiple gestures leading to a meaningful activity, for example, drinking a glass of water (involves grabbing/holding an object, raising the arm, and using the mouth) [8,21]. Finally, human–object/human–human interactions take place to complete a dedicated set of tasks, for example, making coffee or preparing breakfast [22,23].

10.2.2 Sensing Techniques

Vision-based activity recognition uses video sequences or digitized visual data to monitor a subject's movement in a designated area [9,24]. Computer vision techniques are used to analyze visual observations. Popular camera-enabled gaming consoles and Microsoft

Kinect systems have been used in the field of rehabilitation. However, their surveillance is intrusive on user privacy and restricted within a specific zone [25], and the processing of the data involves complex image processing algorithms [26]. Moreover, vision sensors are affected by the variability of light leading to a decrease in performance due to visual obstructions [27]. Monitoring ADL in ambulant environment has necessitated a shift in paradigm toward mobile and wearable sensor-based monitoring. They provide a comparative advantage over vision sensors in terms of unobtrusiveness, low cost, and ease of deployment, enabling pervasive monitoring [23,28,29]. Moreover, mobile phones have become part of daily life and hence can double up for monitoring applications. Alternatively, the use of social networks [30] that exploit appropriate users' information from multiple sources to understand user behavior and interest has also been proposed recently. An illustration of a mobile phone attached to the user's arm is shown in Figure 10.1.

Sensors, the key elements of wearable and mobile monitoring systems, are used to measure the physiological parameters of interest accurately and reliably over long durations. Sensors used for HAR generate time-series data reflecting physiological parameters or state changes which are further statistically analyzed for recognizing/inferring the underlying activity. Inertial measurement units (IMUs) comprising accelerometers and gyroscopes are the most commonly used sensors for HAR [32]. IMUs are generally augmented by magnetometers to derive orientation information with respect to earth's magnetic field. However, magnetometers are susceptible to electromagnetic interference from ferromagnetic materials, which could be present in ambulant home environment [33]. IMUs help measure acceleration, distance, rate of rotation, time, etc. which act as health indicators for gait, posture, tremor, and spasticity, aiding clinical assessment [34]. In addition to IMUs, smartphones provide access to a wide range of built-in sensors such as Bluetooth, Wi-Fi, microphones, pedometer [36], and proximity and light sensors [35]. These sensors can be exploited to infer coarse-grained and context-aware activity recognition, user location, and social interaction between users. For example, the pedometer in Samsung Galaxy smartphones allows step counts and, correspondingly, heart rate monitoring [37–39]. Other sensors such as barometers, thermometers, air humidity, etc. have also been applied in assisted living and wellness monitoring of elderly citizens [36].

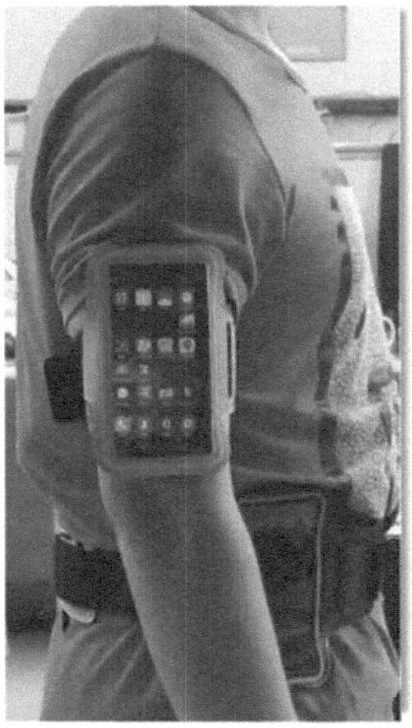

FIGURE 10.1
Smartphone attached to the elbow joint for detecting activities [31].

We take an in-depth look at a MEMS accelerometer and gyroscope given their popularity in HAR. In general, MEMS-based transducers generate resistive/capacitive variations when excited by external forces, for example, compression, pressure, acceleration, etc. The change in geometry of the piezoresistive material or the change in distance between two capacitor plates as an effect of external force is transduced to an electrical signal obtained at the front end of the sensor device. An accelerometer is the most popularly used sensor in HAR, which measures force rather than acceleration. This is achieved by measuring the displacement of an internal proof mass and using Newton's second law of motion ($F = ma$). It is not necessary to know the value of the proof mass since this is accounted for during the calibration process (and it is assumed that it remains constant); and thus a calibrated MEMS accelerometer produces an output that is directly proportional to the acceleration experienced by it. Since an accelerometer responds to force, it is always subjected to gravity which implies that gravitational acceleration (g) can be used as the reference value during the calibration of accelerometers. During the absence of external forces acting on the accelerometer (e.g., when stationary), analyzing the recorded value of gravitational acceleration as experienced by the accelerometer helps determine orientation information for postural tracking [40]. Calibrating the accelerometer with the use of g as a reference standard is a very simple process. The accelerometer is placed on a flat surface with its sensing axis either aligned, opposed, or orthogonal to the direction of g, causing it to experience accelerations equal to $+1g$, $-1g$, or $0g$, respectively. Hence, a simple three-point calibration procedure implemented without additional or specialist equipment is adequate if the accelerometer exhibits a linear response to acceleration. However, the three calibration points that define the upper, lower, and midrange values of a measurand can be deceptive since both a linear or sine function can be fitted to them with high correlation. It is therefore highly recommended that, where practical, calibrations are performed with more than three reference values [40].

A MEMS gyroscope is used to measure the rate of rotation (°/s) without a fixed point of reference. It works on the principle of a tuning fork and comprises a pair of identical masses (m) that are driven to oscillate with equal amplitude but in opposite directions, as shown in Figure 10.2. The positive X, Y, and Z directions are shown, and the negative direction for each axis represents the opposite direction. If one of the masses moves in

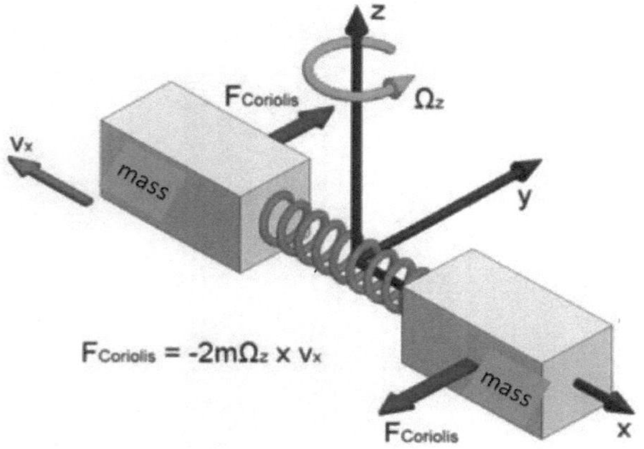

FIGURE 10.2
Principle of operation for MEMS vibrating gyroscope [32].

the positive *X*-axis direction with velocity V_x and an angular rotation Ω_z applied about the *Z*-axis, then the mass will experience an 'F_{Coriolis}' force in the direction of the arrow due to the Coriolis effect. Correspondingly, the other mass moving in the negative *X*-axis direction with the same velocity (V_x) will also experience a Coriolis force of the same magnitude but in the opposite direction [32]. These two forces can be measured by sensing mechanisms built into the MEMS structure (e.g., strain change measured with piezoresistor or deflection measured by change in capacitance). Since the gyroscope measures the angular rate of rotation, it requires a source of rotation to excite the device. In relevant literature, turntables (rotating at predefined speeds) are generally used for calibrating a gyroscope [41]. Each axis of the gyroscope can be calibrated by rotating the sensor through 360° about that axis, integrating the total response obtained over the time taken (angle × rate) and dividing the result by 360 (rate). Rotations can be performed both clockwise and counterclockwise, and corresponding measurements are also averaged when the gyroscope is stationary. The calibration coefficients thus determined are used to calibrate the raw sensor data before any further processing.

10.2.3 Challenges of HAR

The challenges related to HAR – experimental setup for data collection, choice of sensor, sensor placement, and data analysis techniques depending on the application scenario – have been discussed in detail in this section [20]. Sensor data representing physical activities demonstrate a wide range of variability across individuals. Variability is inherent within the same activity morphology performed by different individuals. Furthermore, variability could also be exhibited when the same individual repeats the same activity over time due to fatigue and environmental factors. Hence, an activity recognition system needs to be robust enough to incorporate these variabilities. There can be two approaches to building a successful HAR model. First, a generalized/subject-independent methodology built using data from more than one subject; however, it could be susceptible to inter-subject variability. This can be mitigated by using a large number of data samples pertaining to each subject, formulating a second subject-dependent/personalized model approach. This model is based on the data from each individual subject. A personalized model necessitates collecting a large dataset from an individual, incorporating maximal variability [42]. Generally, for health monitoring applications, formulating a personalized model is beneficial, demonstrating the difference in levels of impairment depending on the stage of rehabilitation [43].

Data segmentation is a further challenge in HAR systems for real-time monitoring applications based on continuously streaming data; this data needs to be segmented depending on the relevance of activities to be monitored. Data pertaining to activities out of interest are referred as null class and difficult to model since it is representative of a wide range of activities. It can be identified if the corresponding signal morphology is uncorrelated/different from the activity being monitored. Class imbalance is another problem especially for data collected over long durations where the monitored activities have dissimilar occurrences, which can be mitigated by generating augmented data or adopting techniques such as oversampling/interpolation to balance out the different class (activities) representations. Activities performed in ambulant environment also suffer from the unavailability of ground truth annotation required for verifying the activity recognition/detection systems [29]. This is particularly relevant to wearable sensor data as opposed to vision data.

Activities are generally performed in the following situations: (a) controlled environment – for example, in the laboratory, where the annotations of the activities with minimal variability are known, (b) semi-naturalistic environment – the subject performs activities as they would normally do in daily life while another person annotates the activities through visual inspection, and lastly, (c) naturalistic environment – the subject performs activities voluntarily, and no annotations for actions performed are available [20]. Corresponding to the activities, designing an experimental protocol for data collection presents another challenge in HAR. Although there are publicly available datasets for a varied range of activities [44–49], for example, walking, running, different gestures, etc., it is important to acquire custom data for specific-use cases.

Sensor characteristics present a significant challenge for long-term monitoring, for example, hardware failures, sensor drifts, and errors in the software aimed at streaming or capturing the data, leading to erroneous situations. External factors such as temperature, pressure, and change in position due to loose straps (used to attach sensors) can require frequent calibration, thereby affecting recorded data. Energy efficiency is one of the last notable challenges. Remote monitoring systems often transmit the data acquired from wearable sensors to a remote station (server/back-office service platform) wirelessly, where the signals are analyzed. Besides incurring energy loss, continuous data transmission along with key processing steps – such as analog to digital conversion, quantization, filtering, and microcontroller operation – would affect the longevity of battery-operated sensors. Hence, for long-term continuous monitoring using wireless body area networks comprising heterogeneous sensors, it is important to select low-complexity data analysis techniques that can be carried out on the sensor node itself, thereby negating continuous data transmission. Therefore, it is imperative to perform low-power, on-node processing for applications requiring real-time operation [50] while for applications supporting long-term behavioral analysis offline data processing may be sufficient [51].

10.3 HAR Processing Overview

The basic processing steps in HAR are shown in Figure 10.3. The sensor data is acquired, pre-processed, segmented/partitioned, and then processed to infer the performed movement. Since human activities are hierarchical and inherently translational, there is an ambiguity in the temporal segmentation of the sub-activities constituting the main activity. Hence, for capturing the spatial and temporal variation in the activities, it is important to use adequate processing techniques. Feature engineering and classification is a popular technique for inferring performed activities from sensor data [1,51]. However, with

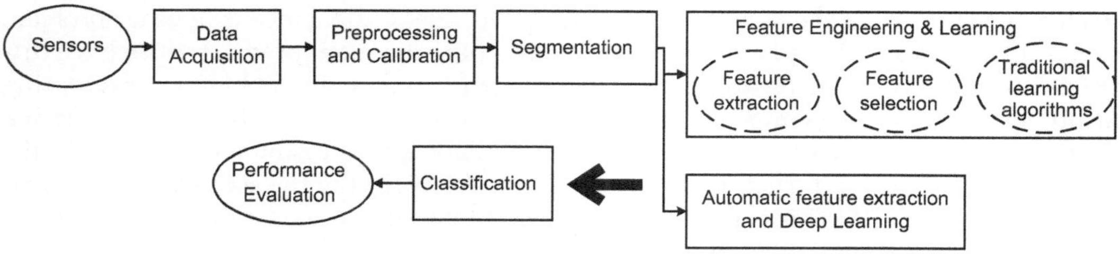

FIGURE 10.3
Overview of data processing for HAR.

advancement in deep learning techniques, traditional feature-engineering based classification methods are fast being replaced by automatic feature extraction and neural networks [19]. We present a brief overview of both approaches adopted for HAR in relevant literature.

10.3.1 Data Acquisition and Pre-Processing

Raw data from multiple sensors yield a multivariate time series, which might need to be synchronized for sensors with differing sampling rates. IMU data is generally sampled at low frequencies, that is, 20–50 Hz, depending on the monitored activities. The raw data contains noise and artifacts caused generally from sensor malfunctioning (e.g., drifts) or induced by motion of the body parts they are attached to. These low-frequency artifacts and high-frequency noise components are attenuated using band-pass filters. IMU data is generally calibrated, their units converted, normalized, resampled, synchronized, and then filtered or fused in the pre-processing stage [32]. For example, specific axes of a tri-axial accelerometer or gyroscope can be fused for detecting a particular motion [52].

10.3.2 Segmentation

The pre-processed signal is segmented or windowed to contain information on the activity of interest. It is challenging to determine the boundaries indicating the start and stop time of the activities to be monitored during ADL. Sliding windows are popularly used, where a fixed window size representing a time duration is used to extract segments of a signal [53]. A trade-off is required on the window size since a smaller window causes missing out on relevant activity while a longer window could cause an overlap between successive activities influencing classification. Hence, dynamic windows based on a data-driven or probabilistic approach could lead to optimal solutions, however, incur high computational complexity [8].

10.3.3 Feature Engineering and Classification

Features characterize the data qualitatively as well as quantitatively by transforming the data into a feature space, resulting in separable classes (activities). Typically, time- and/or frequency-domain features and heuristic features are used for HAR. Some of the commonly used time-domain features are – mean, standard deviation, skew, kurtosis, interquartile range, root mean square, and the correlation between sensor axes [22]. Similarly, frequency-domain features can be extracted from various transforms – Short-time Fourier transform, fast Fourier transform (FFT), or discrete wavelet transform. Activities of different intensities can be discriminated by analyzing the signal energy at different frequency bands obtained through one of the transforms. Some common features include spectral spread, frequency peak estimation, and frequency peak's power estimation, and information entropy of the discrete FFT component magnitudes [54].

A high-dimensional feature space makes classification a difficult task because it requires a large number of training samples for parameter extraction of the model, which increases computational complexity. The performance of the classification algorithm is directly proportional to the feature space dimension, and hence it is important to reduce the dimension. Particularly for online HAR system, it is important to use minimal number of features, resulting in reduced complexity and memory requirements. Ideally, selected features should exhibit a large inter-class and small intra-class distance in the respective feature space for successful classification. The extracted features are typically pre-processed to remove outliers and normalized (e.g., zero mean and unit variance also known as standardization). An outlier, as the name suggests, refers to a point/sample in the feature vector that lies away from the mean. For normally distributed data, three standard deviations from the mean is usually considered as the threshold. Normalization of the features is further essential to avoid the effect of feature values lying in different numeric ranges on the cost function of the classifier. The normalized features are ranked, where a higher-ranked feature has more discriminatory ability between competing classes as compared to a lower-ranked feature in the respective feature space [20]. Some common techniques like Fisher discriminant ratio and Bhattacharya distance are used to quantify the discriminatory ability of features between two equi-probable classes [55]. Similarly, scatter matrices [56], ReliefF algorithm, and clamping technique are some of the popular alternatives for multiple-class problems. Principal component analysis [57] or Karhunen–Loeve transform is also a popular dimensionality reduction technique used in HAR. Principal component analysis can be used directly on the raw data or extracted features to transform the data into a small number of uncorrelated variables referred as principal components. The extraction, ranking, and selection of features, commonly referred as feature engineering, is entirely dependent on the type of activities, number of sensors, and application scenarios.

The classification phase helps map extracted/selected features into sets of activities (classes) using machine learning or pattern recognition methods [29]. A wide range of classifiers are used; the primary selection factor being accuracy, complexity, and the feasibility of real-time execution [19,23]. There can be two broad approaches undertaken for the classification task: (a) supervised learning – the training dataset comprises extracted/selected feature vectors with prior information on the class label associated with them; (b) unsupervised learning – the number of classes could either be known or unknown, and a class label is assigned to each training set. Activities performed in a controlled or semi-naturalistic environment with annotations available for each activity are best suited for supervised learning techniques [20]. Whereas, activities performed in a completely naturalistic/ambulant environment with no annotations available are best suited for unsupervised training methodology. There could or could not be information available regarding the type/number of activities performed. In such scenarios, clustering-based unsupervised learning has been effectively used to determine the classes in HAR [58].

The classification mechanisms can be broadly categorized as:

a) Probabilistic – effective for modeling temporal dependencies and variations in the data. Gaussian mixture models are used for offline learning on large training sets. Hidden Markov models (HMMs) are also popular means to model activity sequences using first-order Markov assumption, where the current state depends only on the previous state. They have been extended to semi- and coupled-HMMs where the state at time t for a given HMM is dependent on all the states of other HMMs in the collection at time $t - 1$ [59].

b) Discriminative – a decision boundary in the feature space segregating regions for each class is constructed through geometric considerations. Support vector machines (SVMs) and *k*-Nearest Neighbor (*k*-NN) are popularly used techniques producing high accuracy with moderate computational complexity (depending on the number of classes). SVMs work by constructing boundaries that maximize margins between the nearest features relative to two distinct classes. Decision tree is a popular alternative, where at each step a decision is made based on a criterion (e.g., threshold or a template) [42].

c) Template based – exploits the similarity between the test dataset and pre-computed class (activity) templates. Dynamic time warping is a popularly used method for assessing the similarity between two time series with varying window sizes [60].

The classification process is commonly involved with cross-validation during the estimation of model parameters (or training) to judge its robustness. The available data is split into training and testing datasets (e.g., in a ratio of 80:20 or 70:30). The training dataset is used to train the classifier or formulate the model. The model parameters thus estimated are used to validate the effectiveness of the model on unseen test dataset. Cross-validation or resampling is used to split the training dataset into k partitions/folds, where $k - 1$ folds are used to formulate the model or train the classifier, and the remaining one fold is used to test the model. Hence, all segments are iteratively trained and tested upon, thereby formulating a robust model. The k-fold partitioning could be further extended to leave one data sample out of testing, and the rest of the data for training, resulting in an exhaustive 'leave-one-sample-out' cross-validation technique. The error computed over each iteration is averaged to estimate the overall error. The cross-validation technique (partitioning mechanism) adopted depends on the size of the available dataset. A large partition helps achieving near accurate estimation of the classifier's performance, however, is limited by a longer computation time; whereas smaller number of folds is a sacrifice on the performance of the classifier. Common practices involve using ten partitions of the data which are iterated over ten times, yielding ten runs of tenfold cross-validation to estimate a robust trained model [20,55].

10.3.4 Deep Learning

The translation invariance of human activity data involves different people performing the same kind of activity in a different manner or even the same person performing the same activity in different ways over multiple repetitions that causes a fragment of activity to manifest at different points in time [61]. The use of hand-crafted features has a high impact on the highly temporally correlated data structure; hence, a method that addresses this strong bias in feature engineering is needed, especially on the backdrop of high volume of data generated by smart wearable platforms (e.g., mobile phones). Expert-driven, hand-engineered features, although representative of the activity and the application scenario, are limited in their ability to faithfully represent the salient characteristics of complex activities and involve time-consuming feature selection procedures. In effect, the whole process of feature engineering involves extensive heuristic knowledge to extract and select appropriate features for HAR. Deep learning, alternatively known as deep neural network, is a new branch in machine learning that allows automatic feature representation from raw data, capturing local dependencies and scale invariance. With the current influx of unlabeled sensor data from IoT, crowdsourcing, and cyber-physical systems, implementing an HAR system with manual feature extraction and classification could be extremely challenging and incur a high design time and effort [19].

Deep learning methods are particularly useful where a large amount of data needs to be processed and can be broadly classified into restricted Boltzmann machine [62], autoencoder [63], convolutional neural network (CNN), and recurrent neural networks (RNN) [64], as shown in Figure 10.4. Nweke et al. provide a detailed description of these methods [19]. The taxonomy of deep learning popularly used for HAR includes CNN and RNN. CNNs are characterized by an initial layer of convolutional filters (a set of weights which slides over the input), followed by non-linearity (activation function – rectified linear units), sub-sampling (pooling), and a fully connected layer which realizes the classification.

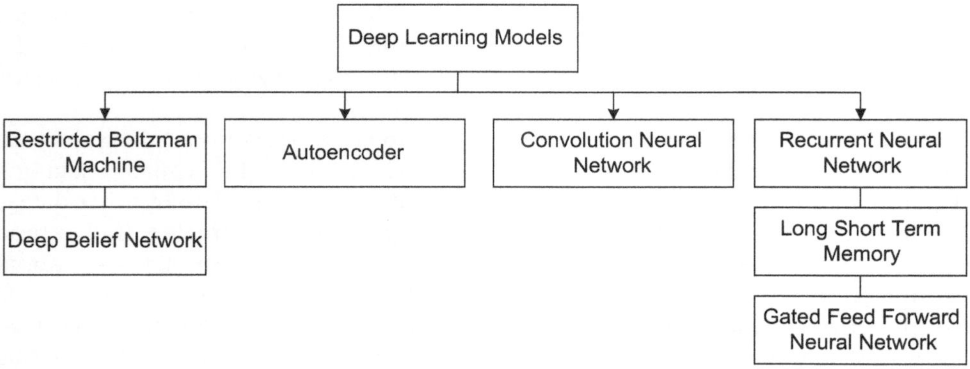

FIGURE 10.4
Different algorithms of deep neural networks [19].

The stacking of multiple convolutional layers helps achieve automatic feature extraction, where downstream layers capture more complex or differentiating features [65,66]. This integrates information from different filters and levels of abstraction. RNNs are an effective choice to analyze time-series data for inferring sequential/time-variant information since they incorporate contextual information from past inputs and are robust to localized distortions in the input sequence along time. A bottleneck for deep CNN structures (with large network size) is the vanishing gradient problem, wherein information from previous computations is rapidly attenuated with progression through the data flow. RNNs applied to long sequential data suffer similarly since every time step has the same weight, and consequently, the contribution of an input in the hidden state is subjected to exponential decay. Long short-term memory (LSTM), a variant of RNN, uses a memory block inspired by a computer memory cell, where context-dependent input, output, and forget gates control what is written, read, and kept in the cell in each time step [67]. Hence, it becomes convenient for the network to store a given input over many time steps, helping LSTM layers capture temporal correlations, a key feature of inertial-sensor-data-representing activities. Hence, CNN and LSTM present a data-driven approach to learning discriminative features from raw sensor data to infer complex, sequential, and contextual information in a hierarchical manner.

Deep neural network performance is highly influenced by the hyper-parameters selected for modeling, which involve architectural parameters and training parameters. Architectural parameters include the number of layers stacked together (e.g., convolution or LSTM layers), number of filters/kernels in each layer, size of each filter, the stride rate (during convolution of the input data and the filter), and type of pooling used (e.g., max pooling, average pooling, etc.), all of which are key toward model formulation. The training parameters include the learning rate, type of backpropagation algorithm, activation function, loss function, dropout, and number of epochs. It is cumbersome to manually iterate through all these parameters to determine the optimal selection; hence, a grid search cross-validation technique [68] is generally used that works through multiple combinations of parameters and helps arrive at the optimal selection.

Recent studies have indicated superior results of deep learning over conventional handcrafted features for HAR [69,70]. CNN outperformed other state-of-the-art data mining techniques in HAR for the benchmark dataset collected from 30 volunteer subjects [71], achieving an overall performance of 94.79% on the test set with raw sensor data and 95.75%

	Predict 'A' $j = 1$	Predict 'B' $j = 2$	Predict 'C' $j = 3$
True 'A', $i = 1$	0.9	0.1	0
True 'B', $i = 2$	0	0.9	0
True 'C', $i = 3$	0.02	0	0.98

FIGURE 10.5
An illustrative confusion matrix with three classes.

with additional information on temporal FFT of the HAR data set. Lastly, LSTM was successful used to mode the sequence of time-correlated samples of activity pertaining to inertial sensor data of ADL tasks [69,72].

10.3.5 Performance Evaluation

The performance of an HAR system can be evaluated in terms of correct classification through true positive (TP) and true negative (TN); and false detection of activities that did not occur, that is, false negative (FN) and false positive (FP). These measures are used to estimate a few well-known matrices, which are explained as follows:

a) Confusion matrix – helps summarize the misclassification of different activity classes, where the matrix rows represent the actual number of instances in each class and the columns represent the predicted instances for each class.
b) Accuracy – the ratio of correctly predicted observations to total observations, that is, [(TP + TN)/(TP + TN + FP + FN)].
c) Precision – the ratio of correctly predicted positive observations to the total predicted positive observations, that is, [TP/(TP + FP)].
d) Recall – the ratio of correctly predicted positive observations to all observations in the actual class, that is, [TP/(TP + FN)].
e) F1 Score – the weighted average of precision and recall.
f) Receiver operating characteristic curve – illustrates the TP rate (sensitivity) against the FP rate for the different threshold points.

Further, normalized confusion matrices could be used for unbalanced datasets having a significant amount of difference in the number of ground truth annotations of the activity classes [73]. An illustration of a confusion matrix for multiple classes is shown in Figure 10.5. Here, only the accuracy might not be a true evaluation of the classifier due to possible dissimilar classification rates of different classes affecting the overall performance measure. The sensitivity S_i of the i-th class is a measure of the number of correct predictions with respect to the total number of observations in that class [74], as shown in Equation 10.1:

$$S_i = \frac{N_{ii}}{f_i} \times 100 \tag{10.1}$$

$$f_i = \sum_{j=1}^{c} C_{ij} \tag{10.2}$$

where $i = 1\ldots c$, and c is the total number of classes. The diagonal and off-diagonal elements of the confusion matrix represent correctly classified and misclassified patterns, respectively. C_{ij} represents the number of times the patterns are predicted to be in class j when they really belong to class i.

Figure 10.5 demonstrates a near-accurate classification since all left-to-right diagonal elements approach unity and all off-diagonal elements approach zero. The sensitivity of class A can be computed as $S_A = \frac{0.9}{(0.9 + 1.0)} = 90\%$ and the overall accuracy as $\text{Accuracy} = \frac{(0.9 + 0.9 + 0.98)}{3} = 92.67\%$ [20].

10.4 Case Study

In this section, we present a short use case scenario on recognizing arm movements performed by stroke survivors during ADL. Cerebrovascular accident, more popularly referred to as stroke, ranks second to heart diseases [75,76]. Stroke is a medical emergency since it causes some degree of physical disability and potentially leads to death. The effect of stroke is different for each individual based on the degree and region of the brain that suffers damage and can result in partial paralysis, impaired vision, memory loss, and speech problems [77,78]. Such aftereffects pose a socio-economic challenge in terms of loss of lives, disability among the survivors, and expenses incurred toward their rehabilitation. Physiotherapy is a popular treatment to restore the motor functionality of stroke survivors. Rehabilitation therapies in clinical settings have also relied on functional electrical stimulation, virtual reality [79], and constraint-induced movement therapy [80,81]. Although these approaches are successful in achieving motor recovery in patients, their use is restricted to clinical settings, requiring unnecessary patient transfer to the clinic. Physical activity has traditionally been monitored by questionnaires, citing their usefulness in covering large subject groups and cost-effectiveness. These questionnaires can be completed by the patients themselves on a daily basis at home, thereby helping to formulate a patient activity log which is monitored periodically by respective clinicians. In addition, therapists perform a wide range of tests to assess the patient's motor ability within a clinical environment – the Wolf motor function test [82], box and block test [83], and the nine-hole peg test [84]. These tests involve a subjective scoring system to quantify the performance of the subjects. However, subjective measures of physical activity may overestimate activity levels compared to assessments made by objective measures. Hence, sensor-based telemedicine modalities provide an alternative toward objective measurement for patient monitoring. A specialized telemedicine application for stroke rehabilitation called 'Telestroke' has been in practice with major success in delivering 'around-the-clock' specialist clinical evaluation of stroke survivors in home settings, thereby also reducing costly human intervention [1].

10.4.1 Experimental Protocol and Data Acquisition

In this case study, the classification of three elementary arm movements, generally associated with ADL, is discussed, using wrist-worn wireless tri-axial accelerometers. The objective was to detect the occurrence of these specific arm movements in an ambulant environment using minimal number of sensors. Enumerating occurrences of particular arm movements (e.g., prescribed exercises) during daily activities over time could be particularly useful in monitoring rehabilitation progress in neurodegenerative diseases such as stroke or cerebral palsy. The reported exploration focuses on three elementary arm movements: (a) Task A – reach and retrieve (extension and flexion of the forearm), (b) Task B – lift arm (rotation of the forearm about the elbow), and (c) Task C – rotate arm (rotation of the wrist about the long axis of the forearm). In principle, the three chosen movements constitute a significant proportion of the complex movements performed with the upper limb during ADL, besides resembling tasks 8, 9, and 15 of the streamlined Wolf motor function test. Here, a summarized view has been presented with respect to the contextual details presented in Section 10.3. For a detailed description of the methodology refer to [22].

Data were collected in two settings from four stroke survivors (age range 45–73, both sexes, both left and right arm dominant) within a treatment center at the Brandenburg Klinik, Germany. Firstly, in a controlled laboratory environment where, for patients 1 and 4, 80 trials of Task A and 40 trials each of Tasks B and C; and for patients 2 and 3, 40 trials of Task A and 20 trials each of Tasks B and C are considered. It is interesting to note that there is a higher number of trials for Task A in comparison with Tasks B and C; this is in line with the fact that Task A is performed more frequently in daily lives. Secondly, in a semi-naturalistic environment, that is, in the kitchen proximal to the laboratory. A customized activity list (Table 10.1), emulating the process of 'making tea', a common ADL, with repeated occurrences of Tasks A, B, and C, was performed by the subjects [8]. The activity list in the experimental protocol comprises 20 individual actions including 10 occurrences of Task A, and 5 each of Task B and C. Data was collected for 5 days from

TABLE 10.1

Use Case Activity List – 'Making Tea'

Activity Sequence		Task
1.	Fetch cup from desk	A
2.	Place cup on kitchen surface	A
3.	Fetch kettle	A
4.	Pour out extra water from kettle	C
5.	Put kettle onto charging point	A
6.	Reach out for the power switch on the wall	A
7.	Drink a glass of water while waiting for kettle to boil	B
8.	Reach out to switch off the kettle	A
9.	Pour hot water from the kettle in to cup	C
10.	Fetch milk from the shelf	A
11.	Pour milk into cup	D
12.	Put the bottle of milk back on shelf	A
13.	Fetch cup from kitchen surface	A
14.	Have a sip and taste the drink	B
15.	Have another sip while walking back to the desk	B
16.	Unlock drawer	C
17.	Retrieve biscuits from drawer	A
18.	Eat a biscuit	B
19.	Lock drawer	C
20.	Have a drink	B

each of the four stroke survivors, where two repetitions of 'making tea' were performed by the subjects each day. Therefore, the semi-naturalistic dataset comprises 10 repetitions (2 repetitions/day), adding up to 200 arm movements (100 Task A, 50 each of Task B and C). Shimmer 9DoF wireless kinematic sensor module containing mutually orthogonal tri-axial accelerometer, magnetometer, and gyroscope was used for sensing. We use a tri-axial accelerometer (range $\pm$ 1.5g) on the impaired arm, proximal to the wrist, chosen in view of its sensitivity to arm movements compared to other potential positions like the elbow or sternum [42]. Data is sampled at 50 Hz, deemed sufficient for human upper limb movement, and streamed along with timestamp to a host computer using the Bluetooth wireless standard. The data pertaining to each task was segmented in accordance with annotations from an accompanying researcher during the experiment. The participants gave their kind consent for the experiments and were encouraged to perform the movements naturally without restricting physical factors such as height/distance/position of tables/chairs/working surface with respect to the subject's position and pace of performing the designated task. This was done to ensure a wide range of variability within the kinematic data that aided the development of a robust activity classification mechanism.

10.4.2 Arm Movement Classification

Here, we demonstrate two methodologies for arm movement classification: (a) feature engineering and clustering, and (b) deep learning – CNN.

10.4.2.1 Feature Engineering and Clustering

For the first approach, controlled data was used for training and semi-naturalistic data was used for testing. The data collected in the laboratory represents the scenario where patients are instructed to follow a particular exercise regime involving the impaired arm in a controlled environment (clinic or home) and later monitored to track these specific movements while they perform daily activities (e.g., 'making tea') with their impaired arm which facilitates a measure of rehabilitation progress in natural settings. Here we adopt a personalized approach for arm movement detection as a rehabilitation indicator, thereby training and validating each subject's data. A subject-dependent/personalized-training testing methodology is important to cater to the varying levels of dexterity among patients. The fundamental concept of the first approach is to form three clusters in multi-dimensional feature space, where each cluster represents a particular movement using features generated from subject-specific data collected during the controlled (training) phase. Subsequent data collected during the 'making-tea' session (testing phase) is tested for its proximity to each of the clusters using a minimum distance classifier. The basic philosophy and the approach adopted are further discussed in detail in the following section. The *k*-means clustering algorithm is employed, owing to its computational simplicity, making it an attractive choice for a wide variety of applications.

Ten one-dimensional (1-D) features are computed on each individual accelerometer (AccX, AccY, AccZ) and gyroscope (GyroX, GyroY, GyroZ) data segment for each movement trial of each subject. The features are: (a) standard deviation, (b) root mean square, (c) information entropy, (d) jerk metric, (e) peak number, (f) maximum peak amplitude, (g) absolute difference, (h) index of dispersion, (i) kurtosis, and (j) skewness. The subsequent process of feature selection and cluster formation is performed on the sensor-specific feature space (comprising 30 features, i.e., 10 features computed on each axis). The extracted features are linearly normalized and the best features of each subject are selected using the low-complexity class-separability measure based on scatter matrices [55] which ranks the 30 features for each movement (class). The scatter matrices quantify the scatter of feature vectors in the feature space and the ranked features are sorted in descending manner. A sequential forward selection technique is employed, selecting the first i features of the ranked feature set in each iteration ($i = 2,\ldots,30$), and it is checked if the data from the training phase can be correctly clustered in a multi-dimensional feature space. Clustering forms groups of similar objects as a means of distinguishing them from each other and can be applied in any discipline involving multivariate data [55]. A regularized Mahalanobis distance [85] is used to form clusters on the training feature vectors. The cluster formation on the training dataset is associated with ten runs of tenfold cross-validation to determine the best feature combination.

The test data from two trials of 'making-tea' session (40 test vectors for each stroke survivor [(10A + 5B + 5C) × 2 trials]) is pre-processed for each type of sensor, and only those features are extracted from each test vector which resulted in the best accuracy in the cross-validation of the training data. A Euclidean distance-based classifier is used to compute the distance of each test vector from the centroid of each cluster in a multi-dimensional feature space. The test vector is assigned to a particular cluster depending on the minimum distance computed for each of the two measures. The predicted label is verified with respect to the annotations in the activity list of Table 10.1. The feature engineering and clustering methodology were implemented using MATLAB® 2016.

The overall accuracy for the stroke survivors is between 40% and –88% (average of 70%) using accelerometer data, as shown in Table 10.2. For Subject 1, the sensitivity for individual actions are above 80%, but for Subjects 2 and 3, although the overall accuracy is above 70%, the sensitivity for Action B and C are quite low (20%). For Subject 4, the overall results are not high because the subject was at an early stage of rehabilitation, and the impaired arm being tested was not the naturally dominant arm which resulted in poor repeatability. This further emphasizes how detection accuracies may be improved by considering more than one sensor type (e.g., the gyroscope) or multiple sensors at different locations for specific cases. It is evident from the results that there is a need for a completely personalized approach to detect elementary arm movements. Various factors like the stage of rehabilitation and the affected/natural arm play a significant role since they affect the level of repeatability of individual movements as well as introduce a high degree of temporal variation.

For patients with arm dexterity, the specific movements (e.g., exercises) that need to be tracked need to be performed multiple times, during an exercise regime, in a controlled environment (clinic or home). The sensor data collected during this phase can be analyzed through cross-validation to determine the best cluster forming features and obtain the centroids of each cluster corresponding to each movement. This helps perform clinical

TABLE 10.2

Sensitivity and Accuracy for Arm Movement Recognition Using Clustering for Stroke Survivors

		Sensitivities (%)			
Subjects	**Features**	**A**	**B**	**C**	**Overall Accuracy (%)**
S1	19	80	90	100	88
S2	19	90	20	100	75
S3	21	95	100	20	78
S4	8	10	80	60	40

profiling of the individual patient with respect to their movement quality. Movements performed in an ambulant environment (e.g., ADL) can be associated with the proximal cluster centroid using minimum distance classifier. It is important that the methodology adapts to changing movement patterns of the patients with time, reflective of their motor functionality. The patient's training data can be collected periodically, and the cluster centroids and the associated features (new selected feature set) can be recomputed to reflect the changing movement patterns. This information (new cluster centroid and feature set) will be subject-specific due to the inter-subject variability in movement profiles, variation in the rehabilitation profile, and the associated functional ability of each subject. This information can be further used by the minimum distance classifier to recognize movements performed in daily life. This methodology can be implemented for online detection of arm movements. For the training phase, the key steps of cluster formation and feature selection (being computationally intensive) can be performed in an offline mode when requested by the clinician, depending on the rehabilitation progress. Online detection can be used to compute only the required features and the distance to the pre-computed cluster centroids in near real-time, thereby providing an energy-efficient solution toward operation of wearable sensors for long durations [50].

10.4.2.2 Deep Learning – CNN

The second approach for the detection of the three arm movements revolves around developing a personalized, low-complex, and customized CNN model using data from a wrist-worn tri-axial accelerometer [86]. Deep learning algorithm requires a large amount of data to ensure better generalization as well as performance. Given the application area, data collection for a long duration from stroke survivors was challenging due to fatigue. Data augmentation helps build a powerful model with fewer data in deep learning [87]. It ensures that the variance applied to the labeled data only changes the way of representation while keeping the semantic meaning of the labels unaltered, ensuring a robust network on real test data. Gaussian noise [88] is added to the raw accelerometer data to create possible variations, yielding a 20-times-large dataset. The input data is segmented to a fixed window of 5.12s, deemed sufficient to capture the different arm movements of stroke survivors. Further, the data is downsampled by a factor of four, which helps compress the data size. The tri-axial data is also transformed into a 1-D signal by computing the modulus, followed by normalization (unit mean and zero variance).

The developed network topology comprises two convolution layers followed by ReLU activation function [89], two pooling layers, and one fully connected (FC) layer. The first convolutional layer has 20 unique 9×1 (stride 1) filters which are convolved with the input data, resulting in 20 feature maps. These feature maps are passed through ReLU, to introduce non-linearity, followed by the first pooling layer that downsamples by a factor of two. This progressively reduces the spatial size of each rectified feature map and allows a form of translation invariance, besides controlling overfitting. A similar process is repeated for the second convolution and pooling layer, the output of which is flattened and acts as an input to the FC Layer. Finally, the output of the FC layer is passed to the softmax activation function to classify the particular task.

The proposed model is trained in a subject-dependent/personalized manner with a dropout probability of 0.5, that is, to drop a random set of activations in a given layer with 50% probability, avoiding overfitting. Sparse categorical cross-entropy loss function and 'Adam optimizer' [56] is used to update the weights faster than the popularly used stochastic gradient descent. The training was done with a maximum of 100 epochs and

TABLE 10.3

Results for Arm Movement Recognition Using CNN for Stroke Survivors

Subjects	Accuracy	Precision	Recall	F1-Score
S1	95.42	95.23	94.92	95.18
S2	98.89	99.82	99.68	99.52
S3	98.34	97.52	98.75	98.24
S4	98.91	99.46	99.8	99.27
Average	97.89	98.01	98.29	98.05

mini-batch size of 15. After each training epoch, model performance is evaluated on the testing set and an early stopping criteria is set to halt the training process when there is no decrement in training error during the 100 epochs. The classification results for each subject are shown in Table 10.3.

In this study, ten trials of the semi-naturalistic 'making-tea' dataset is used both for training and testing. Hence, a direct comparison, although not fair, presents the advantage of deep learning over the feature-engineering-based approach. Table 10.3 demonstrates that the second approach outperforms the previous method based on *k*-means clustering (Table 10.2). While the former approach involves feature extraction and computation of the minimum distance, the second approach involves convolution operations. Hence, a trade-off between the complexity and accuracy needs to be analyzed before adopting an approach for a given application.

10.5 Conclusion

In this chapter, HAR using wearable inertial sensors has been discussed, addressing the key areas of processing sensor data and inferring activities performed in the ambulant environment. Activity monitoring has primarily gained importance with the advent of wearable sensors and development of telemedicine technologies for monitoring elderly patients. Feature-engineering-based classification techniques that have dominated the literature and the more recently used deep learning techniques have been presented. In addition, a brief case study on arm activity recognition of stroke survivors while they perform archetypal ADL has been included. The data processing details and results are demonstrated for both approaches on the same dataset which could help potential readers understand the context of the methodology with respect to the application.

References

1. D. Biswas, Recognition of elementary upper limb movements in nomadic environment, Doctoral Dissertation, University of Southampton, UK, 2015.
2. M. Konijnenburg, R. van Wegberg, S. Song, H. Ha, W. Sijbers, J. Xu, S. Stanzione, C. van Liempd, D. Biswas, A. Breeschoten, and P. Vis, 22.1 A 769uW battery-powered single-chip SoC with BLE for multi-modal vital sign health patches, in *2019 IEEE International Solid-State Circuits Conference-(ISSCC)*, 2019, pp. 360–362.

3. J. Li, H. Zhou, D. Zuo, K.-M. Hou, and C. De Vaulx, Ubiquitous health monitoring and real-time cardiac arrhythmias detection: A case study, *Bio-Medical Materials and Engineering*, vol. 24, no. 1, 1027–1033, 2014.
4. S. Katz, T. D. Downs, H. R. Cash, and R. C. Grotz, Progress in development of the index of ADL, *The Gerontologist*, vol. 10, no. 1 Part 1, 20–30, 1970.
5. L. Bao and S. S. Intille, Activity recognition from user-annotated acceleration data, in A. Ferscha and F. Mattern (eds.) *Pervasive Computing*, Berlin, Heidelberg: Springer, pp. 1–17, 2004.
6. S.-R. Ke, H. L. U. Thuc, Y.-J. Lee, J.-N. Hwang, J.-H. Yoo, and K.-H. Choi, A review on video-based human activity recognition, *Computers*, vol. 2, no. 2, 88–131, 2013.
7. Y. Liu, L. Nie, L. Liu, and D. S. Rosenblum, From action to activity: Sensor-based activity recognition, *Neurocomputing*, vol. 181, 108–115, 2016.
8. L. Chen, J. Hoey, C. D. Nugent, D. J. Cook, and Z. Yu, Sensor-based activity recognition, *IEEE Transactions on Systems, Man, and Cybernetics, Part C: Applications and Reviews*, vol. 42, no. 6, 790–808, 2012.
9. L. Onofri, P. Soda, M. Pechenizkiy, and G. Iannello, A survey on using domain and contextual knowledge for human activity recognition in video streams, *Expert Systems with Applications*, vol. 63, 97–111, 2016.
10. M. A. Islam, S. Halder, M. A. Uddin, and U. K. Acharjee, Periodic activity monitoring using radio frequency tag arrays, in *2018 International Conference on Innovation in Engineering and Technology (ICIET)*, 2018, pp. 1–5.
11. C. R. Plateau, S. Bone, E. Lanning, and C. Meyer, Monitoring eating and activity: Links with disordered eating, compulsive exercise, and general wellbeing among young adults, *International Journal of Eating Disorders*, vol. 51, no. 11, 1270–1276, 2018.
12. H. G. Kang, D. F. Mahoney, H. Hoenig, V. A. Hirth, P. Bonato, I. Hajjar, and L. A. Lipsitz, In situ monitoring of health in older adults: Technologies and issues, *Journal of the American Geriatrics Society*, vol. 58, no. 8, 1579–1586, 2010.
13. N. M. Garcia and J. J. P. Rodrigues, *Ambient Assisted Living*. CRC Press, Boca Raton, FL, 2015.
14. A. Rghioui, S. Sendra, J. Lloret, and A. Oumnad, Internet of Things for measuring human activities in ambient assisted living and e-health, *Network Protocols and Algorithms*, vol. 8, no. 3, 15–28, 2016.
15. G. Singla, D. J. Cook, and M. Schmitter-Edgecombe, Recognizing independent and joint activities among multiple residents in smart environments, *Journal of Ambient Intelligence and Humanized Computing*, vol. 1, no. 1, 57–63, 2010.
16. M. Philipose, K. P. Fishkin, M. Perkowitz, D. J. Patterson, D. Fox, H. Kautz, and D. Hahnel, Inferring activities from interactions with objects, *Pervasive Computing, IEEE*, vol. 3, no. 4, 50–57, 2004.
17. U. Alegre, J. C. Augusto, and T. Clark, Engineering context-aware systems and applications: A survey, *Journal of Systems and Software*, vol. 117, 55–83, 2016.
18. M. M. Hassan, M. Z. Uddin, A. Mohamed, and A. Almogren, A robust human activity recognition system using smartphone sensors and deep learning, *Future Generation Computer Systems*, vol. 81, 307–313, 2018.
19. H. F. Nweke, Y. W. Teh, M. A. Al-Garadi, and U. R. Alo, Deep learning algorithms for human activity recognition using mobile and wearable sensor networks: State of the art and research challenges, *Expert Systems with Applications*, vol. 105, 233–261, 2018.
20. D. Biswas, A. Cranny, and K. Maharatna, Body area sensing networks for remote health monitoring, in E. Vogiatzaki and A. Krukowski (eds.) *Modern Stroke Rehabilitation through e-Health-Based Entertainment*. Springer, Switzerland, 2016, pp. 85–136.
21. N. C. Krishnan and D. J. Cook, Activity recognition on streaming sensor data, *Pervasive and Mobile Computing*, vol. 10, 138–154, 2014.
22. D. Biswas, A. Cranny, N. Gupta, K. Maharatna, J. Achner, J. Klemke, M. Jöbges, and S. Ortmann, Recognizing upper limb movements with wrist worn inertial sensors using k-means clustering classification, *Human Movement Science*, vol. 40, 59–76, 2015.
23. M. Cornacchia, K. Ozcan, Y. Zheng, and S. Velipasalar, A survey on activity detection and classification using wearable sensors, *IEEE Sensors Journal*, vol. 17, no. 2, 386–403, 2017.

24. R. M. Cichy, A. Khosla, D. Pantazis, A. Torralba, and A. Oliva, Comparison of deep neural networks to spatio-temporal cortical dynamics of human visual object recognition reveals hierarchical correspondence, *Scientific Reports*, vol. 6, 27755, 2016.
25. P. P. San, P. Kakar, X.-L. Li, S. Krishnaswamy, J.-B. Yang, and M. N. Nguyen, Deep learning for human activity recognition, in H.-H. Hsu, C.-Y. Chang and C.-H. Hsu (eds.) *Big Data Analytics for Sensor-Network Collected Intelligence*, Amsterdam, Netherlands: Elsevier, 2017, pp. 186–204, https://plu.mx/plum/a/?doi=10.1016/B978-0-12-809393-1.00009-X&theme=plum-sciencedirect-theme&hideUsage=true.
26. C. Zhu and W. Sheng, Motion-and location-based online human daily activity recognition, *Pervasive and Mobile Computing*, vol. 7, no. 2, 256–269, 2011.
27. A. Shahroudy, J. Liu, T.-T. Ng, and G. Wang, NTU RGB+ D: A large scale dataset for 3D human activity analysis, in *Proceedings of the IEEE Conference on Computer Vision and Pattern Recognition*, 2016, pp. 1010–1019.
28. S. Bhattacharya and N. D. Lane, From smart to deep: Robust activity recognition on smartwatches using deep learning, in *2016 IEEE International Conference on Pervasive Computing and Communication Workshops (PerCom Workshops)*, 2016, pp. 1–6.
29. A. Bulling, U. Blanke, and B. Schiele, A tutorial on human activity recognition using body-worn inertial sensors, *ACM Computing Surveys (CSUR)*, vol. 46, no. 3, 33, 2014.
30. Y. Nie, Y. Jia, S. Li, X. Zhu, A. Li, and B. Zhou, Identifying users across social networks based on dynamic core interests, *Neurocomputing*, vol. 210, 107–115, 2016.
31. Y. Lu, Y. Wei, L. Liu, J. Zhong, L. Sun, and Y. Liu, Towards unsupervised physical activity recognition using smartphone accelerometers, *Multimedia Tools and Applications*, vol. 76, no. 8, 10701–10719, 2017.
32. A. Cranny, A. Beriain, H. Solar, G. Tartarisco, and G. Pioggia, Vital sign sensing technology, in K. Maharatna and S. Bonfiglio (eds.) *Systems Design for Remote Healthcare*, New York, NY: Springer, 2014, pp. 55–92.
33. C. L. Kendell and E. D. Lemaire, Effect of mobility devices on orientation sensors that contain magnetometers, in *CMBES Proceedings*, vol. 31, 2008.
34. B. Najafi, K. Aminian, A. Paraschiv-Ionescu, F. Loew, C. J. Bula, and P. Robert, Ambulatory system for human motion analysis using a kinematic sensor: Monitoring of daily physical activity in the elderly, *IEEE Transactions on Biomedical Engineering*, vol. 50, no. 6, 711–723, 2003.
35. P. Burns, C. Lueg, and S. Berkovsky, Activmon: Encouraging physical activity through ambient social awareness, in *2012 30th ACM Conference on Human Factors in Computing Systems (CHI)*, 2012, pp. 2363–2386.
36. J. Gong, L. Cui, K. Xiao, and R. Wang, MPD-model: A distributed multipreference-driven data fusion model and its application in a WSNs-based healthcare monitoring system, *International Journal of Distributed Sensor Networks*, vol. 8, no. 12, 602358, 2012.
37. L. Kanaris, A. Kokkinis, A. Liotta, and S. Stavrou, Fusing bluetooth beacon data with Wi-Fi radiomaps for improved indoor localization, *Sensors*, vol. 17, no. 4, 812, 2017.
38. D. Natarajasivan and M. Govindarajan, Filter based sensor fusion for activity recognition using smartphone, *International Journal of Computer Science and Telecommunications*, vol. 7, no. 5, 26–31, 2016.
39. N. Zouba, F. Bremond, and M. Thonnat, Multisensor fusion for monitoring elderly activities at home, in *2009 6th IEEE International Conference on Advanced Video and Signal Based Surveillance*, 2009, pp. 98–103.
40. D. Biswas, D. Corda, G. Baldus, A. Cranny, K. Maharatna, J. Achner, J. Klemke, M. Jöbges, and S. Ortmann, Recognition of elementary arm movements using orientation of a tri-axial accelerometer located near the wrist, *Physiological Measurement*, vol. 35, no. 9, 1751, 2014.
41. W. Fong, S. Ong, and A. Nee, Methods for in-field user calibration of an inertial measurement unit without external equipment, *Measurement Science and Technology*, vol. 19, no. 8, 085202, 2008.
42. D. Biswas, A. Cranny, A. F. Rahim, N. Gupta, K. Maharatna, N. R. Harris, and S. Ortmann, On the data analysis for classification of elementary upper limb movements, *Biomedical Engineering Letters*, vol. 4, no. 4, 403–413, 2014.

43. D. Biswas, K. Maharatna, G. Panic, E. B. Mazomenos, J. Achner, J. Klemke, M. Jöbges, and S. Ortmann, Low-complexity framework for movement classification using body-worn sensors, *IEEE Transactions on Very Large Scale Integration (VLSI) Systems*, vol. 25, no. 4, 1537–1548, 2017.
44. R. Chavarriaga, H. Sagha, A. Calatroni, S. T. Digumarti, G. Tröster, J. del R. Millán, and D. Roggen, The Opportunity challenge: A benchmark database for on-body sensor-based activity recognition, *Pattern Recognition Letters*, vol. 34, no. 15, 2033–2042, 2013.
45. H. Sagha, S. T. Digumarti, J. del R. Millán, R. Chavarriaga, A. Calatroni, D. Roggen, and G. Tröster, Benchmarking classification techniques using the opportunity human activity dataset, in *2011 IEEE International Conference on Systems, Man, and Cybernetics*, 2011, pp. 36–40.
46. P. Zappi, C. Lombriser, T. Stiefmeier, E. Farella, D. Roggen, L. Benini, and G. Tröster, Activity recognition from on-body sensors: Accuracy-power trade-off by dynamic sensor selection, in R. Verdone (ed.) *Wireless Sensor Networks*, Berlin, Heidelberg: Springer, 2008, pp. 17–33.
47. B. Barshan and M. C. Yüksek, Recognizing daily and sports activities in two open source machine learning environments using body-worn sensor units, *The Computer Journal*, vol. 57, no. 11, 1649–1667, 2014.
48. J. R. Kwapisz, G. M. Weiss, and S. A. Moore, Activity recognition using cell phone accelerometers, *ACM SigKDD Explorations Newsletter*, vol. 12, no. 2, 74–82, 2011.
49. O. Banos, J.-M. Galvez, M. Damas, A. Guillen, L.-J. Herrera, H. Pomares, I. Rojas, C. Villalonga, C. S. Hong, and S. Lee, Multiwindow fusion for wearable activity recognition, in *International Work-Conference on Artificial Neural Networks*, 2015, pp. 290–297.
50. K. Maharatna, E. B. Mazomenos, J. Morgan, and S. Bonfiglio, Towards the development of next-generation remote healthcare system: Some practical considerations, in *2012 IEEE International Symposium on Circuits and Systems (ISCAS)*, 2012, pp. 1–4.
51. A. Mannini and A. M. Sabatini, Machine learning methods for classifying human physical activity from on-body accelerometers, *Sensors*, vol. 10, no. 2, 1154–1175, 2010.
52. D. Biswas, A. Cranny, A. Rahim, N. Gupta, N. Harris, K. Maharatna, and S. Ortmann, On the sensor choice and data analysis for classification of elementary upper limb movements, in *2014 IEEE-EMBS International Conference on Biomedical and Health Informatics (BHI)*, 2014, pp. 744–747.
53. S. Chernbumroong, S. Cang, A. Atkins, and H. Yu, Elderly activities recognition and classification for applications in assisted living, *Expert Systems with Applications*, vol. 40, no. 5, 1662–1674, 2013.
54. A. K. Jain, R. P. W. Duin, and J. Mao, Statistical pattern recognition: A review, *IEEE Transactions on Pattern Analysis and Machine Intelligence*, vol. 22, no. 1, 4–37, 2000.
55. S. Theodoridis, A. Pikrakis, K. Koutroumbas, and D. Cavouras, *Pattern Recognition*, Fourth edition. Academic Press, Cambridge, MA, 2009.
56. D. Biswas, A. Cranny, N. Gupta, K. Maharatna, and S. Ortmann, Recognition of elementary upper limb movements in an activity of daily living using data from wrist mounted accelerometers, in *2014 IEEE-Computer Society International Conference on Health Informatics (ICHI)*, 2014, pp. 232–237.
57. B. M. Abidine, L. Fergani, B. Fergani, and M. Oussalah, The joint use of sequence features combination and modified weighted SVM for improving daily activity recognition, *Pattern Analysis and Applications*, vol. 21, no. 1, 119–138, 2018.
58. Y. Kwon, K. Kang, and C. Bae, Unsupervised learning for human activity recognition using smartphone sensors, *Expert Systems with Applications*, vol. 41, no. 14, 6067–6074, 2014.
59. P. Natarajan and R. Nevatia, Coupled hidden semi markov models for activity recognition, in *2007 IEEE Workshop on Motion and Video Computing (WMVC'07)*, 2007, pp. 10–10.
60. C. Pham, T. Plötz, and P. Olivier, A dynamic time warping approach to real-time activity recognition for food preparation, in B. de Ruyter, R. Wichert, D.V. Keyson, P. Markopoulos, N. Streitz, M. Divitini, N. Georgantas and A. Mana Gomez (eds.) *Ambient Intelligence*, Berlin, Heidelberg: Springer, 2010, pp. 21–30.
61. Y. LeCun, Y. Bengio, and G. Hinton, Deep learning, *Nature*, vol. 521, no. 7553, 436, 2015.

62. A. Fischer and C. Igel, Training restricted Boltzmann machines: An introduction, *Pattern Recognition*, vol. 47, no. 1, 25–39, 2014.
63. S. S. Khan and B. Taati, Detecting unseen falls from wearable devices using channel-wise ensemble of autoencoders, *Expert Systems with Applications*, vol. 87, 280–290, 2017.
64. X. Li, Y. Zhang, J. Zhang, S. Chen, I. Marsic, R. A. Farneth, and R. S. Burd, Concurrent activity recognition with multimodal CNN-LSTM structure, arXiv preprint arXiv:1702.01638, 2017.
65. Y. Bengio, Learning deep architectures for AI, *Foundations and Trends\Textregistered in Machine Learning*, vol. 2, no. 1, 1–127, 2009.
66. Y. LeCun, L. Bottou, Y. Bengio, and P. Haffner, Gradient-based learning applied to document recognition, *Proceedings of the IEEE*, vol. 86, no. 11, 2278–2324, 1998.
67. S. Hochreiter and J. Schmidhuber, Long short-term memory, *Neural Computation*, vol. 9, no. 8, 1735–1780, 1997.
68. L. Buitinck, G. Louppe, M. Blondel, F. Pedregosa, A. Mueller, O. Grisel, V. Niculae, P. Prettenhofer, A. Gramfort, J. Grobler, and R. Layton, API design for machine learning software: experiences from the scikit-learn project, arXiv preprint arXiv:1309.0238, 2013.
69. F. Ordóñez and D. Roggen, Deep convolutional and lstm recurrent neural networks for multimodal wearable activity recognition, *Sensors*, vol. 16, no. 1, 115, 2016.
70. S. Yao, S. Hu, Y. Zhao, A. Zhang, and T. Abdelzaher, Deepsense: A unified deep learning framework for time-series mobile sensing data processing, in *Proceedings of the 26th International Conference on World Wide Web*, 2017, pp. 351–360.
71. C. A. Ronao and S.-B. Cho, Human activity recognition with smartphone sensors using deep learning neural networks, *Expert Systems with Applications*, vol. 59, 235–244, 2016.
72. S. Ma, L. Sigal, and S. Sclaroff, Learning activity progression in lstms for activity detection and early detection, in *Proceedings of the IEEE Conference on Computer Vision and Pattern Recognition*, 2016, pp. 1942–1950.
73. J. A. Ward, P. Lukowicz, and H. W. Gellersen, Performance metrics for activity recognition, *ACM Transactions on Intelligent Systems and Technology (TIST)*, vol. 2, no. 1, 6, 2011.
74. J. C. F. Caballero, F. J. Martínez, C. Hervás, and P. A. Gutiérrez, Sensitivity versus accuracy in multiclass problems using memetic pareto evolutionary neural networks, *IEEE Transactions on Neural Networks*, vol. 21, no. 5, 750–770, 2010.
75. S. Meairs, N. Wahlgren, U. Dirnagl, O. Lindvall, P. Rothwell, J.-C. Baron, K. Hossmann, B. Engelhardt, J. Ferro, J. McCulloch, and M. Kaste, Stroke research priorities for the next decade-a representative view of the European scientific community, *Cerebrovascular Diseases*, vol. 22, no. 2–3, 75–82, 2006.
76. C. J. Murray and A. D. Lopez, Alternative projections of mortality and disability by cause 1990–2020: Global Burden of Disease Study, *The Lancet*, vol. 349, no. 9064, 1498–1504, 1997.
77. S. Wilson, R. Davies, T. Stone, J. Hammerton, P. Ware, S. Mawson, N. Harris, C. Eccleston, H. Zheng, N. Black, and G. Mountain, Developing a telemonitoring system for stroke rehabilitation, *Contemporary Ergonomics*, vol. Year 2, 505–512, 2007.
78. S. R. Levine and M. Gorman, 'Telestroke' the application of telemedicine for stroke, *Stroke*, vol. 30, no. 2, 464–469, 1999.
79. S. V. Adamovich, G. G. Fluet, A. Mathai, Q. Qiu, J. Lewis, and A. S. Merians, Design of a complex virtual reality simulation to train finger motion for persons with hemiparesis: A proof of concept study, *Journal of Neuroengineering and Rehabilitation*, vol. 6, no. 1, 28, 2009.
80. S. L. Wolf, C. J. Winstein, J. P. Miller, P. A. Thompson, E. Taub, G. Uswatte, D. Morris, S. Blanton, D. Nichols-Larsen, and P. C. Clark, Retention of upper limb function in stroke survivors who have received constraint-induced movement therapy: The EXCITE randomised trial, *The Lancet Neurology*, vol. 7, no. 1, 33–40, 2008.
81. E. Taub and D. M. Morris, Constraint-induced movement therapy to enhance recovery after stroke, *Current Atherosclerosis Reports*, vol. 3, no. 4, 279–286, 2001.
82. D. M. Morris, G. Uswatte, J. E. Crago, E. W. Cook III, and E. Taub, The reliability of the wolf motor function test for assessing upper extremity function after stroke, *Archives of Physical Medicine and Rehabilitation*, vol. 82, no. 6, 750–755, 2001.

83. T. Platz, C. Pinkowski, F. van Wijck, I.-H. Kim, P. di Bella, and G. Johnson, Reliability and validity of arm function assessment with standardized guidelines for the Fugl-Meyer Test, Action Research Arm Test and Box and Block Test: A multicentre study, *Clinical Rehabilitation*, vol. 19, no. 4, 404–411, 2005.
84. A. Heller, D. Wade, V. A. Wood, A. Sunderland, R. L. Hewer, and E. Ward, Arm function after stroke: Measurement and recovery over the first three months, *Journal of Neurology, Neurosurgery & Psychiatry*, vol. 50, no. 6, 714–719, 1987.
85. J. Mao and A. K. Jain, A self-organizing network for hyperellipsoidal clustering (HEC), *IEEE Transactions on Neural Networks*, vol. 7, no. 1, 16–29, 1996.
86. M. Panwar, D. Biswas, H. Bajaj, M. Jobges, R. Turk, K. Maharatna, and A. Acharyya, Rehab-net: Deep learning framework for arm movement classification using wearable sensors for stroke rehabilitation, *IEEE Transactions on Biomedical Engineering*, 2019. doi: 10.1109/TBME.2019.2899927.
87. J. Salamon and J. P. Bello, Deep convolutional neural networks and data augmentation for environmental sound classification, *IEEE Signal Processing Letters*, vol. 24, no. 3, 279–283, 2017.
88. M. Atzori, M. Cognolato, and H. Müller, Deep learning with convolutional neural networks applied to electromyography data: A resource for the classification of movements for prosthetic hands, *Frontiers in Neurorobotics*, vol. 10, 9, 2016.
89. G. Klambauer, T. Unterthiner, A. Mayr, and S. Hochreiter, Self-normalizing neural networks, in I. Guyon, U.V. Luxburg, S. Bengio, H. Wallach, R. Fergus, S. Vishwanathan and R. Garnett (eds.) *Advances in Neural Information Processing Systems*, Long Beach, California, United States: Long Beach Convention & Entertainment Center, 2017, pp. 971–980.

11

Blood Pressure Monitoring

Rajarshi Gupta
University of Calcutta

CONTENTS

11.1 Introduction

Arterial pressure is defined as the hydrostatic pressure exerted by the circulating blood over the arteries as a result of the pumping action of the heart. Systolic pressure (SBP) is the highest pressure in a cardiac systole (ventricular contraction), while diastolic pressure (DBP) refers to the lowest (ventricular relaxation) one. Mean arterial pressure (MAP) is the algebraic difference between the SBP and DBP determined by dividing the area under the BP curve of one cardiac cycle by its period [1]. Pulse pressure (PP) is the algebraic difference between SBP and DBP. Periodic BP measurements are essential for cardiovascular patients, especially those with hypertension. Under normal activity of a patient, non-invasive and automatic electrical type BP monitors have become popular since many years due to their portability, quick response, and moderate accuracy level. The ambulatory blood pressure (BP) measurement devices are lightweight (often less than 300 g) can be worn for 24 h by a patient, and can be used to record SBP, DBP, and MAP at regular intervals, with each record estimated over several cardiac periods (normally less than 30 s). These can be operated by the patients themselves and do not require any 'clinical skill'.

Worldwide, there are few apex professional bodies that provide recommendations for standardization of automatic BP monitoring devices. These include the Association for Advancement of Medical Instrumentation (AAMI), British Hypertension Society (BHI), and European Society for Hypertension. The first protocol by the American National Standards Institute (ANSI)/AAMI SP10 was published in 1987 and subsequently updated in 1993 and 2002. BHS published a protocol, which was again updated in 1993. The latest ANS/AAMI standard was published in 2009 and revised in 2013 (ISO 81060-2:2013) together with the International Organization for Standardization (ISO) [2]. To validate an automated BP monitor with auscultatory reference sphygmomanometer, both the BHS and ANSI/AAMI/ISO protocols recommend a minimum of 255 measurements, with minimum 3 measurements collected from at least 85 individuals. The minimum time interval between two successive measurements should be 1 min. The parameters evaluated for the accuracy of measurements are mean error (ME), standard deviation error (SDE), and mean absolute error (MAE). According to the AAMI recommendations [3], the device must not differ from the mercury sphygmomanometers (considered as gold standard for non-invasive blood pressure (NIBP) measurements) by a mean difference greater than 5 mmHg, or a standard deviation not greater than 8 mmHg for SBP and DBP.

The five functional components of any automatic NIBP monitoring device are: (a) a pneumatic compression cuff; (b) a cuff inflation and control mechanism, normally an electrically operated small pump; (c) a cuff deflation mechanism; (d) a pressure transduction mechanism for the cuff; and, (e) sensor signal processing and indicator mechanism for direct readout. The three most popular NIBP monitors that are available in commercial standalone units are oscillometric, plethysmograph, and cuffless-type monitors. In these methods, a proportional electrical signal is generated either out of the inflated cuff pressure (CP) or volume change of blood in peripheral body parts. This electrical signal is suitably conditioned and digitally processed for direct computation of SBP, DBP, and MAP. A direct digital readout is provided for commercial table-top units after every measurement cycle. With the help of information and communication technology, the conditioned signal may be suitably telemetered to a remote end user for continuous monitoring.

11.2 Oscillometric NIBP Measurement Techniques

Oscillometric method is the most popular NI technique used in commercial ambulatory BP measurement instruments [4]. Oscillometric BP measurements use an air-inflated cuff, an automatic pressure control device, a pressure sensing transduction mechanism and associated signal conditioning circuits, a digital controller with an embedded algorithm for BP estimation, and a display/communication system for recording. The basic principle of oscillometry relies on measuring the small oscillations of intra-cuff pressure caused by heart-beat induced pulse volume changes by a pressure sensor. A proprietary algorithm in the embedded controller then calculates the SBP, DBP, and MAP [5].

The complete schematic diagram of an oscillometric BP monitor is shown in Figure 11.1. An air-inflatable cuff is wound around the subject's left upper arm at the same height of

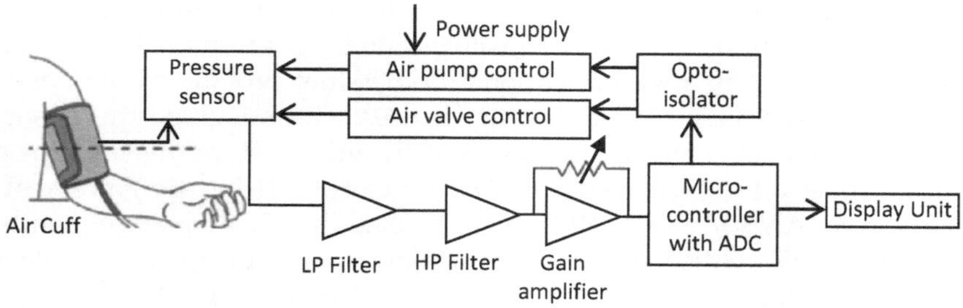

FIGURE 11.1
Schematic diagram of a complete oscillometric BP monitor.

the heart. Normally, the initial target pressure in oscillometry is 180 mmHg, which completely occludes the blood flow in the brachial artery. Figure 11.2a shows the arm cross section for oscillometry setup. During cuff inflation, the pressure (p_c) is applied through the soft tissue to compress the artery against the bone, gradually decreasing the lumen area to flat. From this pressure level, the cuff is slowly deflated either by a linear or step deflation valve to a value below the DBP. During this period, the cuff oscillations due to the gradual release of blood through the brachial artery are measured by a pressure sensor embedded within the cuff. Initially, the microcontroller provides the control for air cuff inflations, followed by the bleed valve control. With the gradual deflation of the cuff, the composite signal consisting of pulsations due to artery and cuff pressure is picked up by a peizoresistive sensor. The conditioned signal is filtered, amplified, and fed to the analog-to-digital converter of the microcontroller. The algorithms embedded in the microcontroller extract the oscillometric waveform envelope (OMWE) from the acquired data; determine the SBP, DBP, and MAP; and display these values through the liquid crystal display (LCD) unit [6].

A typical oscillometric cuff pressure waveform, also called the cuff deflation curve (CDC), shows pressure pulsations induced by the artery superimposed on the slowly decreasing deflated curve pressure, as shown in Figure 11.2b. From this, the oscillometric waveform (OMW), shown in Figure 11.2c, is extracted by a band-pass filter, with cutoff frequency between 0.3 and 20 Hz [7]. This eliminates the low-frequency component of the deflating cuff pressure and the high-frequency components that occasionally creep in due to hand movements, muscle contraction, etc. Another technique called detrending [8] involves subtracting a line of best fit (which represents the decreasing cuff pressure during deflation) from the CDC to generate the OMW. This requires locating the beginning of each pulse on the CDC and joining these points through a smooth curve.

The next job is to extract the OMWE. It can be extracted from the OMW by subtracting the amplitudes of the peaks from the corresponding trough and plotting them against time, as shown in Figure 11.2d. There are two ways to calculate the SBP, DBP, and MAP from the OMWE. The first algorithm is the maximum amplitude algorithm (MAA) [9], which determines MAP as the peak of the OMW. It also fixes the SBP and DBP as two fixed ratios of the MAP. As described in Ref. [10], the model predicted the SBP and DBP as 0.593 and 0.717, respectively, of the MAP. These are marked in Figure 11.2d as Ar_s and Ar_d, respectively. However, it has been found that although the MAP is accurately determined by MAA, the ratios of SBP and DBP are sensitive to the variation of BP waveform, PP, and age. Hence, the MAA is unable to provide accurate SBP and DBP. The second algorithm which estimates the SBP and DBP is maximum/minimum slope algorithm (MMSA) [10,11]. This technique estimates the SBP and DBP as the CP at which the first derivative of the OMWE, when plotted against the time, reaches its maximum and minimum, respectively. As shown in Figure 11.2e, when this derivative is plotted against the cuff pressure, it reaches a maximum positive value when cuff pressure equals DBP. The minimum negative value was found to occur at SBP, while at zero differential oscillations correspond to MAP. In Ref. [12], a trusted boundary around the SBP and DBP is proposed to estimate the trustworthiness of the estimated BPs, either through MAA or MMSA. The authors propose a parameter, *Ratio2*, which provides an expected range of SBP and DBP values.

To overcome the inaccuracies that arise due to the linear and fixed coefficient relation between the SBP, DBP, and OMWE, neural network (NN)-based approaches were proposed by some researchers [13,14]. Other advantages of NN is that it is free from noise sensitivity. A set of features using the height, derivative, and width of the cuff pressure oscillations were extracted to form envelopes and fed to a feed-forward NN [15]. The technique

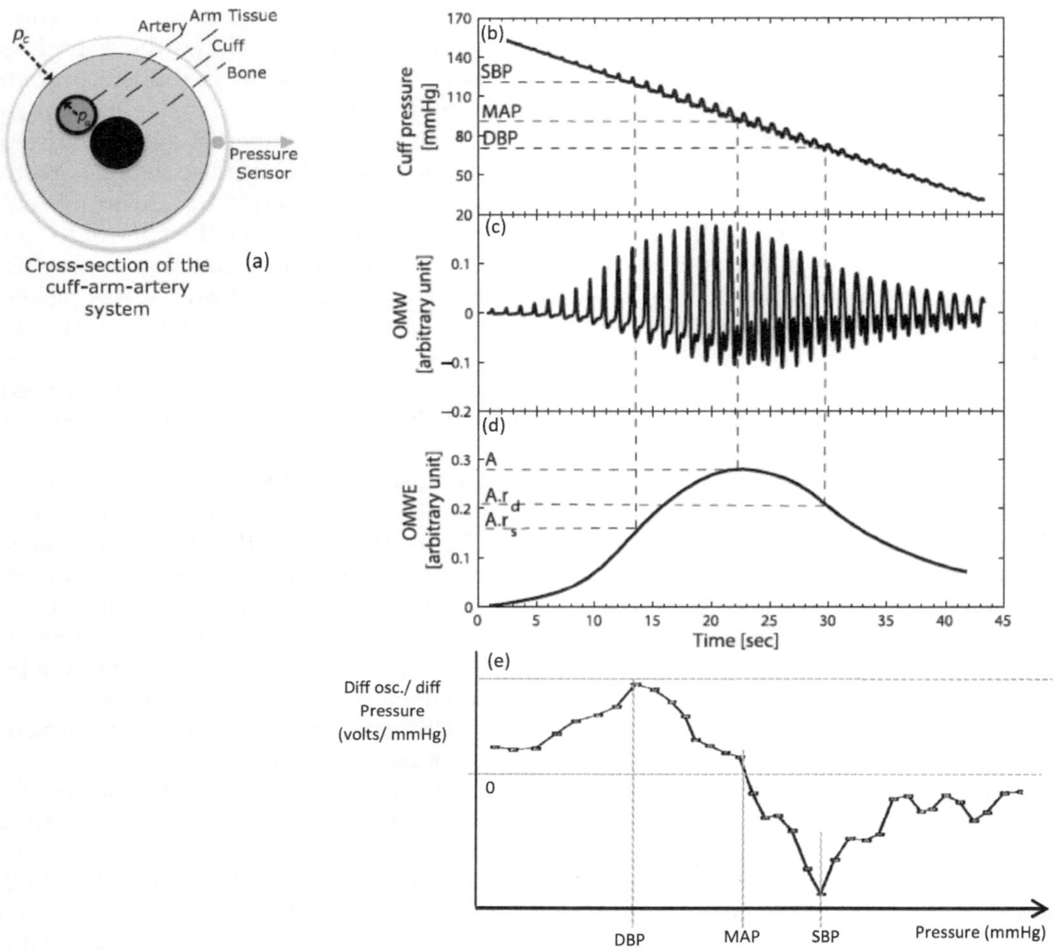

FIGURE 11.2
Oscillometric principle: (a) physical setup; (b) CDC; (c) OMW; (d and e) determination of SBP and DBP from OMWE using MAA and MMSA algorithms.

showed a very good agreement with actual BP, with a gain of 0.89 mmHg over MAA technique. A comparison of performance between different training algorithms of NN in terms of estimation error, training time, iterations, etc. were studied in Ref. [16] and compared with the MAA technique. The different training algorithms studied were steepest descent (with variable learning rate and momentum, resilient back propagation), quasi-Newton (Broyden–Fletcher–Goldfarb–Shanno, one step secant, Levenberg–Marquardt), and conjugate gradient (Fletcher–Reeves update, Polak–Ribiére update, Powell–Beale restart, scaled conjugate gradient). It was found that NN outperforms the MAA in terms of estimation error, with best results obtained using the steepest-descent resilient back propagation (SD-RBP) algorithm. The problems with these approaches are twofold. First, redundant OMWE data which degrades the NN performance in terms of generalization. Second, a larger network size that requires a large training data set. A natural extension of NN application for BP estimation is extracting features from the OMWE and feeding to NN for reduced computational load. In Ref. [17], the OMWE was modeled using a combination of two Gaussian functions and some features were extracted by minimizing the model error to feed a feed-forward NN. The method yielded a MAE and SDE of 23.33 and 28.68 mmHg respectively for SBP and 4.98 and 3.04 mmHg for DBP, respectively, over MAA technique.

There are some attempts to apply the oscillometric pulse morphology analysis for BP estimation [18,19]. Although these methods are not driven by definite theoretical or physiological background, some of them could achieve reasonable accuracy. In Ref. [19], the cardiovascular features like augmentation index, reflection index, and stiffness index, $\Delta T/T$ ratio (where, T is the pulse duration and the ΔT is the systolic peak to diastolic peak duration) were computed from the OMW during the cuff deflation process. The local maxima or minima of these features were estimated. The MAP was estimated as average of the selected pressure corresponding to maximum and minimum values. From the clusters of the features plot, the pulse corresponding to MAP and the regions of pulses corresponding to SBP and DBP were identified. The SBP and DBP were then calculated as the cuff pressure by averaging a group of pulses in SBP and DBP regions.

The last category of oscillometric NIBP measurement includes model-based techniques that utilize any of the following approaches: arterial BP pulse waveform model, transmural and cuff pressure model, cuff-pressure volume model, or arterial pressure-area model. However, a common limitation of many of these approaches is the dependence on the empirical coefficients, which may vary under different conditions like age, vascular compliance, and nervous system. In Ref. [20], the arterial BP and intracranial BP are modeled as harmonically related sinusoids modulated by respiration. The model parameters are determined using standard MATLAB® functions. In Ref. [21], the observation state-space model is considered as the function of respiratory rate, heart rate, and low frequency signal and use extended Kalman filter for state estimation. The arterial pressure-area model [22] considers the mechanics of stretching and collapsing the arterial wall with positive and negative transmural pressure, which is the differential pressure on the arterial wall due to cuff and arterial BP.

There have been many studies on the errors, repeatability, reproducibility, and calibration in oscillometric BP measurements [23]. The prime factors those limit the accuracy are: (a) the motion of the arm that introduces pressure oscillation noise; (b) cardiac diseases like arrhythmia; (c) variability due to the use of manufacturers' cuff. Some other sources of error are undesired external pressure applied on the cuff, tremor, or cardiovascular abnormalities during the measurement period. The commercial oscillometric BP devices with several implementations differ in their characteristics due to: (a) different cutoff frequencies for filtering pressure waveform, (b) different methods for averaging pulse size, (c) different methods and coefficients for extracting SBP and DBP from OMWE, and (d) different methods for controlling the cuff pressure during inflation and deflation. Due to many such factors, different manufacturers' devices achieved good accuracy over a certain group of patients. The research on repeatability addressed within-device and between-device repeatability [24]. Several simulators have been developed for the calibration of oscillometric BP devices; however, their accuracy in representing the OMW for different age groups and cardiovascular patients has been limited.

An extension of an oscillometric-type BP monitor implementing a smartphone is described in Ref. [25]. Here, the arm cuff is replaced by the index finger of the human subject, acting as the actuator. A photoplethysmogram (PPG) and force sensor are integrated on the backside of a smartphone. A visual guide is provided on the phone screen to alter the finger pressure on the transverse palmar arch artery. The oscillations due to pulsatile blood flow through the artery are sensed, and the oscillogram data is collected by the phone software. From this data, SBP, MAP, and DBP are computed similar to oscillometry.

11.3 Plethysmography-Based NIBP Measurement Technique

Plethysmography refers to volume changes in body tissue. The change in blood volume in the finger as a result of the pumping action of the heart has been utilized in plethysmography-based non-invasive arterial BP measurement. A popular and reliable method for continuous arterial BP measurement is vascular unloading, also known as volume clamping [26]. The commercial instrument that utilizes this technique is FINger Arterial PRESsure (FINAPRESS) NIBP [27], introduced in the early 1980s. This technique can be used for continuous 2-h non-invasive beat-by-beat BP measurement. This method can fairly estimate the MAP and systolic pressure (SP), while the diastolic pressure (DP) is calculated using a formula. The principle is based on measuring the local arterial volume pulsations using a PPG sensor placed downstream to the cuff (to occlude a finger artery) and detecting the systolic phase end point and maximum amplitude in the pulse waveform. The diameter of an artery under a cuff wrapped around the finger is kept constant (clamped) at a certain diameter, the 'set point', despite the changes in arterial pressure during each heartbeat.

The block schematic of the instrument is shown in Figure 11.3. An air inflatable cuff is wrapped around the finger phalange, embodying a PPG sensor. The PPG sensor output is read by a microcontroller unit to actuate a pump to inflate or deflate the cuff through a servo-nulling principle. During cardiac systole, the arterial pressure increases, detected by the PPG sensor. The servo mechanism rapidly increases the cuff pressure to keep the arterial diameter constant till the PPG oscillations attain maximum value. The servo system uses a position sensor on the actuator to keep the fluid-filled chamber at this pressure of maximum oscillations. At this point (zero transmural pressure), the arterial walls are maximally relieved (unloaded) of tension. At this point of 'unloaded' diameter of the artery, the cuff pressure equals the intra-arterial pressure. The unloaded diameter of the artery is the average diameter where the amplitude of the PPG oscillations are the largest. This diameter is not constant during a measurement cycle due to several factors, and, as a result, the calibration/reference value of the of the servo-control mechanism needs to be recalibrated periodically. Hence, an important aspect is the reference value of the servo-control mechanism, which is to be set based on the maximum amplitude of the PPG oscillations.

The method can be uncomfortable for prolonged recording since the cuff pressure is greater than the venous arm and often occludes the venous return path. To counteract this problem, some commercial units (Portapres and CNSystems) utilize alternate measurements on two fingers [28].

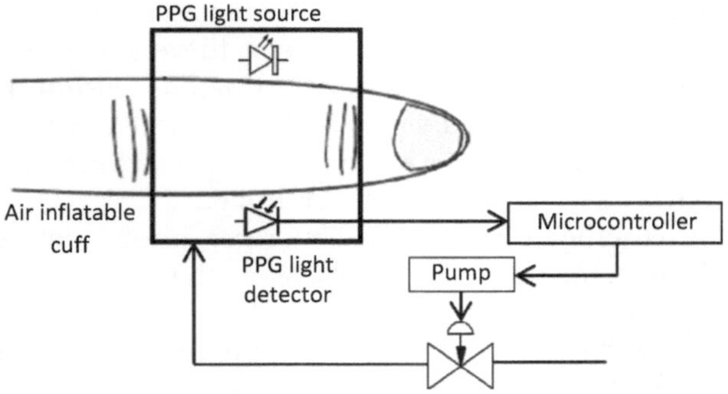

FIGURE 11.3
Schematic arrangement of vascular unloading technique.

An implementation of volume oscillometry for long-term NIBP measurement using a wearable instrument is illustrated [29,30]. The functional modules of this device are: (a) cuff pressure control and volume sensing unit from plethysmographic waveform, both interfaced to a microprocessor; (b) a microprocessor system with memory to estimate and store the cuff pressure (p_c); and (c) a personal computer for data reproduction and analysis. Among these, (a) and (b) are portable units and can be carried by the patient. The system configuration is similar to Figure 11.2 except the cuff is now applied to third/fourth left fingers. The pulse signal is measured by a PPG sensor, a combination of an infrared LED and a phototransistor. The sequence of operation is illustrated with reference to Figure 11.4.

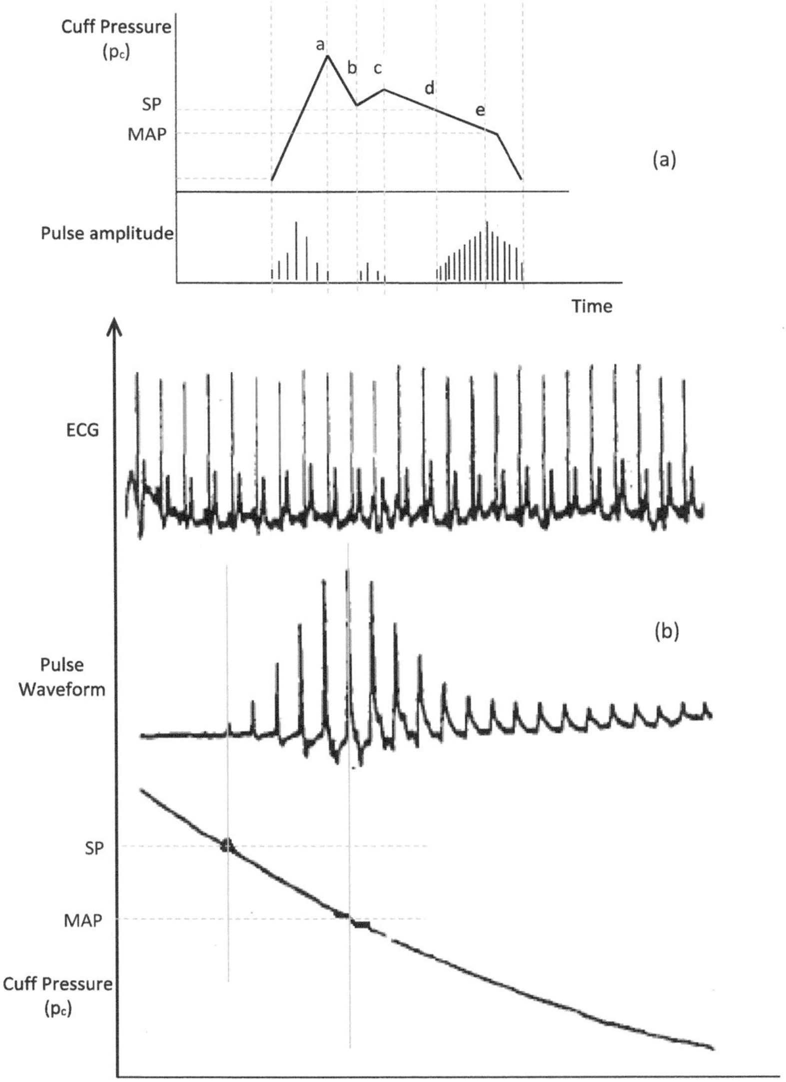

FIGURE 11.4
(a) Determination of SP and MAP from the pulse amplitude and cuff pressure applied; (b) Validation of the SP and MAP from the playback-recorded cuff-pressure-sensed pulse amplitude.

A start pulse is generated, following which the cuff pressure (p_c) is gradually increased from a preset value (20/30 mmHg) to an upper pressure limit (200/240 mmHg). On the detection of pulse peak at higher level, the p_c is further increased till pulse waveform completely disappears (point 'a' at Figure 11.4). From this point, a decrease and then an increase of p_c is repeated till the pulse waveform appears (point 'b') and then disappears (point 'c'). This is to ascertain that pulse waveforms are of sufficient shape and amplitude to estimate the SP. From point 'c', the p_c is gradually decreased at a rate of 2–4 mmHg per heartbeat till the pulsations reappear at point 'd'. This instant is considered the systolic end point, and the p_c at this is instant gives the SP value. With the further gradual decrease of p_c, the pulse amplitude gradually increases and, after point 'e', gradually decreases again. The p_c at the point 'e', maximum pulse amplitude, gives the MAP. If the discrimination between point 'd' and 'e' becomes difficult due to hand movement, the current measurement cycle is stopped, and p_c is set to the initial preset value. Normally, each measurement cycle is completed within 30 s. The DBP is calculated tentatively with the formula: DP w= (3 × (MAP-SP)/2.

The personal computer is used offline for data validation. For this, an additional electrocardiogram (ECG) measurement is performed to ascertain the pulse waveform reliability. The pulse, ECG, and p_c data stored in the RAM are transferred to the computer for a 'playback' mode analysis. The graphical representation in Figure 11.4b shows the SP, MAP in the p_c, and plethysmograph waveform. The SP, MAP, and p_c are marked in the pulse waveform.

11.4 Cuffless BP Monitoring Techniques Using Pulse Transit Time (PTT)

Despite good accuracy, oscillometric BP measurements are not the best choice for continuous, long-term, unconstrained BP measurements under ambulatory condition due to the uncomfortable cuff. Moreover, oscillometry is unreliable for subjects with atrial fibrillation, preeclampsia, and stiff arteries. Unconstrained BP monitoring has emerged as an alternative to facilitate user-friendly interfaces under normal activities of the user [31]. The basic theory of cuffless BP measurement is based on pulse wave velocity (PWV) [32]. The circulating blood exerts pressure on blood vessels at certain velocity, known as PWV. The PWV depends on the dimension of blood vessels and their elastic properties [33,34]. For measuring the SBP, the time interval is calculated from the opening of aortic valve, which initiates the pulse wave to travel through the artery, to a peripheral site. Technically this is sometimes referred to as the 'time interval that is taken by the pressure wave to travel between a proximal and a distal artery site'. There has been much research on the behavior and modeling of arterial walls [35,36]. The basic pulse wave propagation model relates vessel elasticity to the pulse arrival time (PAT). Since the arterial wall properties vary among different body sites, the reflections of the pulse wave takes place at bifurcations and branches and are most pronounced at the arterial terminations at body peripherals. The BP waveform at a given arterial site is the summation of the forward and backward traveling of the pulse wave. The model assumes that there is no time gap between the forward and backward traveling wave at body terminals; whereas, there is some time gap between those in the central/proximal arterial site. As shown in Figure 11.5, the pulse waveform is amplified as it travels along the artery towards the peripheral body site. Hence, wave reflection influences the magnitude of the shape and

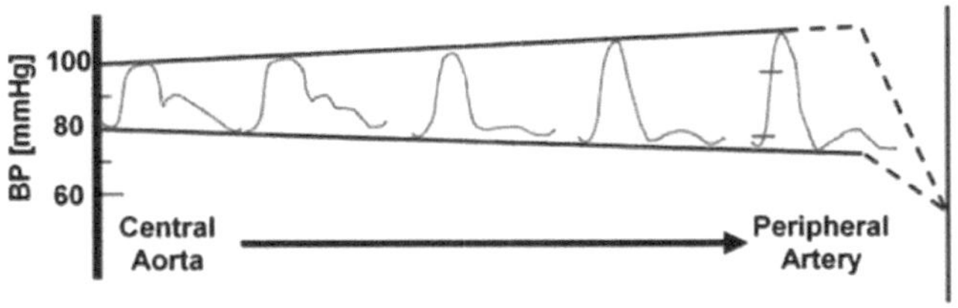

FIGURE 11.5
BP waves amplified with increasing distance from heart.

amplitude of BP. The well-known Moens–Korteweg equation [37] shows the PWV as a function of vessel and fluid parameters:

$$\mathrm{PWV} = \sqrt{\frac{E \cdot h}{2\rho \cdot r}} \tag{11.1}$$

where E: elastic modulus of the vessel, ρ: blood density in the vessel, h: wall thickness, and r: the inner radius of the vessel wall.

PAT is defined as the time interval between the peaks of R wave in an ECG signal and a fiducial point on a different peripheral pulse waveform indicating the pulse wave arrival within the same cardiac cycle. If changes in PWV are negligible in short-time interval, PAT (T) is inversely proportional to PWV per unit length of vessel and expressed as [38]:

$$T = \frac{L}{\mathrm{PWV}} \sqrt{\frac{\rho \cdot \Delta V}{V \cdot \Delta P}} \tag{11.2}$$

where L: the approximate length of the artery from heart to peripheral site, ΔV: change in blood volume corresponding to ΔP, the difference between SBP and DBP, and V: blood volume per unit length. PP is shown to be:

$$\mathrm{PP} = \mathrm{SBP} - \mathrm{DBP} = \sqrt{\frac{\rho \cdot \Delta V}{V \cdot T^2}} \tag{11.3}$$

The expressions for SBP and DBP are given by:

$$\begin{aligned} \mathrm{SBP} &= \mathrm{SBP}_0 - \frac{2}{\gamma T_0}(T - T_0) \\ \mathrm{DBP} &= \mathrm{SBP} - (\mathrm{PP}_0) \cdot \frac{T}{T_0} \end{aligned} \tag{11.4}$$

where suffix '0' indicates base values of the respective parameters [38], and γ is a coefficient representing vessel characteristics.

ECG, PPG, and impedance cardiography (ICG) are the three established techniques used for PPT/PAT estimation. The fiducial points detected on the ECG, PPG, and/or ICG waveforms for PAT and PTT measurements are indicated in Figure 11.6. The Q wave of ECG provides the most accurate measure of the start of ventricular systole. However, sometimes Q point is either difficult to detect, or not prominent, or buried under artifacts. Hence for

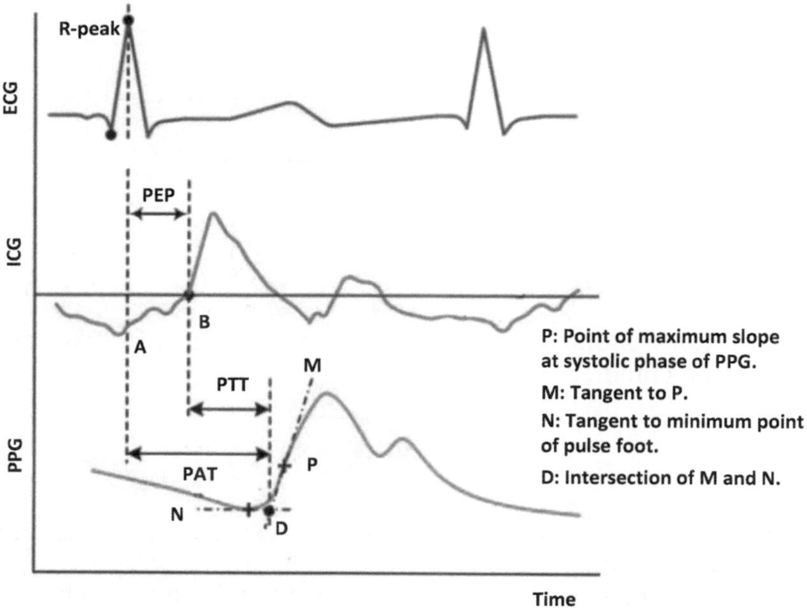

FIGURE 11.6
Definition of PTT and PAT.

all practical purposes, R peak, which is detected more easily can be taken as the start of a ventricular systole. The arrival of pulse wave at a peripheral site is marked by the PPG foot, which is also minimally disturbed by wave reflection. This point also indicates diastolic minimum time. The most accepted definition for PAT/PTT calculation is the point of intersection (D) between tangents at maximum slope (point P) and pulse minimum point, as shown in Figure 11.6. The ICG waveform represents the time derivative of the thoracic impedance (Z). The B point (Figure 11.6) in ICG waveform represents the onset of cardiac ejection and is detected by zero crossing before reaching to the maximum amplitude. The time duration between the R peak and B point represents the pre-ejection period (PEP). It is worth mentioning that the duration between the Q wave of ECG and B point in ICG could provide more precise PEP measurement. PTT is defined as the time between the aortic valve opening (B point on ICG) and a fiducial point on PPG. PAT is the sum of PEP and PTT.

There are a few measurement challenges for PAT/PTT estimation for NIBP measurement. The choice of PPG sensor modes among transmission and reflection type is important. Experimentally, the transmission-type PPG provides best estimation of PAT, operating in infrared range. The contact pressure for transmission-type PPG sensor should be near to mean BP to get optimal pulse amplitude. The problem with reflection-mode PPG is that, since the contact pressure is almost zero, the pulse amplitude is diminished. ICG is normally measured near the thorax, to measure the proximal waveform. However, it requires additional current (1–5 mA, 2–100 Hz) injection into the blood and, hence, is not very suitable for long-term BP measurements. Thus, many realizations for ambulatory BP measurement use ECG–PPG combination using PAT, instead of PTT [37–39]. Many other researchers considered PAT and PTT measures as same [40,41]. The second challenge is unreliable measurements due to artifacts; especially PPG since it is prone to corruption by motion artifacts. A plethora of research literature is available on the pre-processing and detection of fiducial points from ECG and PPG [42–45]. The third and biggest challenge is the calibration of PTT-based BP measurements. According to Equation (11.2), the BP is directly related to the personalized physiological features (like distance between measurement sites, blood volume per unit length, etc.), and hence the calibration is subject-dependent. The most standard calibration procedure adopts determining the unknown parameters in a linear model, the most common being the linear regression model [46]. A useful discussion on PAT/PTT to BP calibration is available in [47].

Although there are many studies on BP estimation from PTT, it was found unreliable for long-term monitoring under medication, especially during daytime due to variation in vascular tone [48].

11.5 NIBP Measurement Methods Based on PPG Only

An alternative way of measuring PTT is to engage two PPG sensors, one at proximal and the other at a distal body site, for example, ear and toe. The PTT so obtained is called PPG-PTT [49]. The wavelet-transform-based PPG-PTT estimation using dual PPG sensor is reported [50]. However, some distal sites like toes can be uncomfortable for continuous measurements. To alleviate this problem, a wearable eyeglass was developed for PPG-PTT measurement between either two of the three facial arteries, namely, angular, superficial, and occipital [51]. Some other successful implementation on NIBP measurement methods are available, which utilized multiple features from PPG and/or multiple physiological signals. In Ref. [52], total 14 time-domain features were extracted from simultaneously recorded ECG and PPG. The authors used a feature-selection process to use an optimal set of features using the genetic algorithm. Finally, the BP model was constructed using multivariate regression and support vector regression methods separately. Their results showed good short- and long-term performances with ME of −0.001 ± 3.1 mmHg for SBP and −0.004 ± 2.1 mmHg for DBP. There have been a few approaches for BP estimation from PPG signal only [53–56]. In Ref. [53], the authors used time-domain features like width at 1/3rd and 2/3rd of pulse amplitude and systolic and diastolic time to form a linear regression formula for SBP and DBP. To mitigate the effect of pulse propagation through atrial vessels during breathing cycle, which affects the PTT, a slope transit time (STT) was proposed in [54]. The authors showed that the slope of the aggregated pulse, resulting from initial and reflected pulses, varies during respiration and inverse of the slope in the systolic phase may be considered as an indication of the transit time of internal pulse wave, called the STT. In addition, this STT is independent of heart rate as well as ejection period. An application of mean STT, that is, STT calculated over a span of the PPG data, for reliable BP estimation is reported in Ref. [55]. An NN-based approach using multitaper method utilizing the PPG signal is described in Ref. [56]. After beat segmentation, two time-domain features, namely systolic upstroke time and diastolic time, along with the power spectral density at different frequency interval were extracted. They were fed to a multilayer perceptron neural net (MLP-NN), pre-trained with a back-propagation algorithm, to predict the SBP and DBP directly. The MD for SBP and DBP were −0.0217 ± 4.8950 mmHg and 0.0975 ± 2.9160 mmHg, respectively.

11.6 Conclusions

The number of devices and systems used for continuous BP measurement have been gradually increasing in the last decade. There has been much research on creating wearable and ambient assisted living environment around the patients to enable them to perform their normal activities. Low power sensing, intelligent computing, and short-range communication techniques have played a key role toward this approach. An important application has been the use of smartphones, most common personal gadgets. Even a moderately priced smartphone comes with useful in-built sensors, a moderate-level processor, and communication interfaces that can be exploited for NIBP measurements. A prime research using this has been the introduction of image PPG (*I*PPG) to use smartphone cameras, either for still images of facial segment or video streaming [57,58]. Another approach is the use of smart textiles and body sensor networks in conjunction for wearable sensing of physiological signals, particularly ECG and PPG in the context of BP measurements [59]. Nevertheless, most of the reported works have been validated on limited population groups with few postural activities. There are still ample research opportunities in the area of NIBP measurement in the coming years.

References

1. L. A. Geddes, *Handbook of Blood Pressure Measurement*. New York: Springer Sc.+Business Media, 1991.
2. ISO, ISO 16792:2015. Technical product documentation, in www.iso.org/standard/56865.html. Last Accessed: April, 2018.
3. E. O'Brien, B. Waeber, G. Parati, J. Staessen, and M. G. Myers, Blood pressure measuring devices: Recommendations of the European Society of Hypertension, *BMJ Br. Med. J.*, vol. 322, 531–536, 2001.
4. B. S. Alpert, D. Quinn, and D. Gallick, Oscillometric blood pressure: A review for clinicians, *J. Am. Soc. Hypertens.*, vol. 8, no. 12, 930–938, 2014.
5. M. Forouzanfar, H. R. Dajani, V. Z. Groza, M. Bolic, S. Rajan, and I. Batkin, Oscillometric blood pressure estimation: Past, present, and future, *IEEE Rev. Biomed. Eng.*, vol. 8, 44–63, 2015.
6. S. Lopez, Blood pressure monitor fundamentals and design, *Free Semicond.*, Application Note AN4328, 1–38, 2012.
7. S. Lee, J. H. Chang, S. W. Nam, C. Lim, S. Rajan, H. R. Dajani, and V. Z. Groza, Oscillometric blood pressure estimation based on maximum amplitude algorithm employing gaussian mixture regression, *IEEE Trans. Instrum. Meas.*, vol. 62, no. 12, 3387–3389, 2013.
8. J. N. Amoore, Extracting oscillometric pulses from the cuff pressure: Does it affect the pressures determined by oscillometric blood pressure monitors? *Blood Press. Monit.*, vol. 11, no. 5, 269–279, 2006.
9. P. D. Baker, D. R. Westenskow, and K. Kück, Theoretical analysis of non-invasive oscillometric maximum amplitude algorithm for estimating mean blood pressure, *Med. Biol. Eng. Comput.*, vol. 35, no. 3, 271–278, 1997.
10. G. Drzewiecki, R. Hood, and H. Apple, Theory of the oscillometric maximum and the systolic and diastolic detection ratios, *Ann. Biomed. Eng.*, vol. 22, no. 1, 88–96, 1994.
11. G. Drzewiecki, Noninvasive arterial blood pressure and mechanics, in *The Biomedical Engineering Handbook*, J. Bronzono, ed. Boca Raton, FL: CRC Press, 2000.

12. I. Koohi, S. Ahmad, I. Batkin, V. Z. Groza, S. Shirmohammadi, H. R. Dajani, and E. M. Petriu, Method for evaluation of trustworthiness of oscillometric blood pressure measurements, in *2015 IEEE International Symposium on Medical Measurements and Applications (MeMeA) Proceedings*, 2015, pp. 267–272.
13. S. Narus, T. Egbert, T. K. Lee, J. Lu, and D. Westenskow, Noninvasive blood pressure monitoring from the supraorbital artery using an artificial neural network oscillometric algorithm, *J. Clin. Monit.*, vol. 11, no. 5, 289–297, 1995.
14. M. Forouzanfar, H. R. Dajani, V. Z. Groza, M. Bolic, and S. Rajan, Adaptive neuro-fuzzy inference system for oscillometric blood pressure estimation, in *2010 IEEE International Workshop on Medical Measurements and Applications, MeMeA 2010-Proceedings*, 2010, pp. 125–129.
15. S. Colak and C. Isik, Blood pressure estimation using neural networks, in *2004 IEEE International Conference on Computational Intelligence for Measurement Systems and Applications*, July 2004, pp. 14–16.
16. M. Forouzanfar, H. R. Dajani, V. Z. Groza, M. Bolic, and S. Rajan, Comparison of Feed-Forward Neural Network training algorithms for oscillometric blood pressure estimation, in *4th International Workshop on Soft Computing Applications*, 2010, pp. 119–123.
17. M. Forouzanfar, H. R. Dajani, V. Z. Groza, M. Bolic, and S. Rajan, Feature-based neural network approach for oscillometric blood pressure estimation, *IEEE Trans. Instrum. Meas.*, vol. 60, no. 8, 2786–2796, 2011.
18. A. Sapinski, Standard algorithm for blood pressure measurement by sphygmo-oscillographic method, *Med. Biol. Eng. Comput.*, vol. 34, no. 1, 82–83, 1996.
19. M. Mafi, S. Rajan, M. Bolic, V. Z. Groza, and H. R. Dajani, Blood pressure estimation using oscillometric pulse morphology, in *Proceedings of the Annual International Conference of the IEEE Engineering in Medicine and Biology Society, EMBS*, 2011, pp. 2492–2496.
20. M. Aboy, J. McNames, R. Hornero, T. Thong, D. Cuesta, D. Novak, and B. Goldstein, A novel statistical model for simulation of arterial and intracranial pressure, in *Conference proceedings : Annual International Conference of the IEEE Engineering in Medicine and Biology Society*, 2004, pp. 129–132.
21. J. McNames and M. Aboy, Statistical modeling of cardiovascular signals and parameter estimation based on the extended kalman filter, *IEEE Trans. Biomed. Eng.*, vol. 55, no. 1, 119–129, 2008.
22. H. H. Hardy and R. E. Collins, On the pressure-volume relationship in circulatory elements, *Med. Biol. Eng. Comput.*, vol. 20, no. 5, 565–570, 1982.
23. E. Balestrieri and S. Rapuano, Instruments and methods for calibration of oscillometric blood pressure measurement devices, *IEEE Trans. Instrum. Meas.*, vol. 59, no. 9, 2391–2404, 2010.
24. A. J. Sims, C. A. Reay, D. R. Bousfield, J. A. Menes, and A. Murray, Low-cost oscillometric non-invasive blood pressure monitors: Device repeatability and device differences, *Physiol. Meas.*, vol. 26, no. 4, 441–445, 2005.
25. A. Chandrasekhar, C. S. Kim, M. Naji, K. Natarajan, J. O. Hahn, and R. Mukkamala, Smartphone-based blood pressure monitoring via the oscillometric finger-pressing method, *Sci. Transl. Med.*, vol. 10, no. 431, 1–11, 2018.
26. K. I. Yamakoshi, H. Shimazu, and T. Togawa, Indirect measurement of instantaneous arterial blood pressure in the human finger by the vascular unloading technique, *IEEE Trans. Biomed. Eng.*, vol. BME-27, no. 3, 150–155, 1980.
27. B. P. M. Imholz, W. Wieling, G. A. Van Montfrans, and K. H. Wesseling, Fifteen years experience with finger arterial pressure monitoring: Assessment of the technology, *Cardiovasc Res.*, vol. 38, no. 3. 605–616, 1998.
28. A. P. Avolio, M. Butlin, and A. Walsh, Arterial blood pressure measurement and pulse wave analysis-their role in enhancing cardiovascular assessment, *Physiol. Meas.*, vol. 31, no. 1, R1-R47, 2010.
29. H. I. K. Yamakoshi, A. Kawarada, A. Kamiya, H. Shimazu, Long-term ambulatory monitoring of indrect arterial blood pressure using a volume-oscillometric method, *Med. Biol. Eng. Comput.*, vol. 23, no. 5, 459–465, 1985.

30. Y. Imai, M. Nihei, K. Abe, S. Sasaki, N. Minami, M. Munakata, S. Yumita, Y. Onoda, H. Sekino, K. Yamakoshi, and K. Yoshinaga, A finger volume-oscillometric device for monitoring ambulatory blood pressure: Laboratory and clinical evaluations: Part A: Theory and practice, *Clin. Exp. Hypertens.*, vol. 9, no. 12, 2001–2025, 1987.
31. A. Stojanova, S. Koceski, and N. Koceska, Continuous blood pressure monitoring as a basis for Ambient Assisted Living (AAL) – Review of methodologies and devices, *J. Med. Syst.*, vol. 43, 1–12, 2019.
32. R. Mukkamala, J. O. Hahn, O. T. Inan, L. K. Mestha, C. S. Kim, H. Töreyin, and S. Kyal, Toward ubiquitous blood pressure monitoring via pulse transit time: Theory and practice, *IEEE Trans. Biomed. Eng.*, vol. 62, no. 8, 1879–1901, 2015.
33. A. C. Burton, Relation of structure to function of the tissues of the wall of blood vessels, *Physiol. Rev.*, vol. 34, no. 4, 619–642, 1954.
34. R. L. Armentano, J. G. Barra, F. M. Pessana, D. O. Craiem, S. Graf, D. B. Santana, and R. A. Sanchez, Smart smooth muscle spring-dampers, *IEEE Eng. Med. Biol. Mag.*, vol. 26, no. 1, 62–70, 2007.
35. M. F. O'Rourke and J. Hashimoto, Mechanical factors in arterial aging. A clinical perspective, *J. Am. Coll. Cardiol.*, vol. 50, no. 1, 1–13, 2007.
36. A. J. Bank, H. Wang, J. E. Holte, K. Mullen, R. Shammas, and S. H. Kubo, Contribution of collagen, elastin, and smooth muscle to in vivo human brachial artery wall stress and elastic modulus, *Circulation*, vol. 94, no. 12, 3263–3270, 1996.
37. W. Chen, T. Kobayashi, S. Ichikawa, Y. Takeuchi, and T. Togawa, Continuous estimation of systolic blood pressure using the pulse arrival time and intermittent calibration, *Med. Biol. Eng. Comput.*, vol. 38, no. 5, 569–574, 2000.
38. Z. Tang, T. Tamura, M. Sekine, M. Huang, W. Chen, M. Yoshida, K. Sakatani, H. Kobayashi, and S. Kanaya, A chair-based unobtrusive cuffless blood pressure monitoring system based on pulse arrival time, *IEEE J. Biomed. Heal. Inf.*, vol. 21, no. 5, 1194–1205, 2017.
39. F. S. Cattivelli and H. Garudadri, Noninvasive cuffless estimation of blood pressure from pulse arrival time and heart rate with adaptive calibration, in *2009 6th International Workshop on Wearable and Implantable Body Sensor Networks*, 2009, pp. 114–119.
40. J. Allen, Photoplethysmography and its application in clinical physiological measurement, *Physiol. Meas.*, vol. 28, no. 3, R1-R39, 2007.
41. J. E. Naschitz, S. Bezobchuk, R. Mussafia-Priselac, S. Sundick, D. Dreyfuss, I. Khorshidi, A. Karidis, H. Manor, M. Nagar, E. R. Peck, and S. Peck, Pulse transit time by r-wave-gated infrared photoplethysmography: Review of the literature and personal experience, *J. Clin. Monit.Comput.*, vol. 18, no. 5–6. 333–342, 2004.
42. J. Pan and W. J. Tompkins, A real-time QRS detection algorithm, *IEEE Trans. Biomed. Eng.*, vol. BME-32, no. 3, 230–236, 1985.
43. B. U. Kohler, C. Hennig, and R. Orglmeister, The principles of software QRS detection, *IEEE Eng. Med. Biol.*, vol. 6, no. 1, 42–57, 2002.
44. M. Elgendi, Detection of c, d, and e waves in the acceleration photoplethysmogram, *Comput. Methods Programs Biomed.*, vol. 117, no. 2, 125–136, 2014.
45. B. T. Ricardo Ferro, A. Ramírez Aguilera, and R. R. Fernández De La Vara Prieto, Automated detection of the onset and systolic peak in the pulse wave using Hilbert transform, *Biomed. Signal Process. Control*, vol. 20, 78–84, 2015.
46. Y. Choi, Q. Zhang, and S. Ko, Noninvasive cuffless blood pressure estimation using pulse transit time and Hilbert–Huang transform, *Comput. Electr. Eng.*, vol. 39, no. 1, 103–111, 2013.
47. D. Buxi, J. M. Redoutïe, and M. R. Yuce, A survey on signals and systems in ambulatory blood pressure monitoring using pulse transit time, *Physiol. Meas.*, vol. 36, no. 3, R1–R26, 2015.
48. R. A. Payne, Pulse transit time measured from the ECG: An unreliable marker of beat-to-beat blood pressure, *J. Appl. Physiol.*, vol. 100, no. 1, 136–141, 2006.
49. Y. Chen, C. Wen, G. Tao, and M. Bi, Continuous and noninvasive measurement of systolic and diastolic blood pressure by one mathematical model with the same model parameters and two separate pulse wave velocities, *Ann. Biomed. Eng.*, vol. 40, no. 4, 871–882, 2012.

50. A. Pielmuş, M. Pflugradt, T. Tigges, M. Klum, A. Feldheiser, O. Hunsicker, and R. Orglmeister, Novel computation of pulse transit time from multi-channel PPG signals by wavelet transform towards continuous, non-invasive blood pressure estimation, *Curr. Dir. Biomed. Eng.*, vol. 2, no. 1, 209–213, 2016.
51. C. Holz and E. J. Wang, Glabella: Continuously sensing blood pressure behavior using an unobtrusive wearable device, *Proc. ACM Interactive, Mobile, Wearable Ubiquitous Technol.*, vol. 1, no. 3, 1–23, 2017.
52. F. Miao, N. Fu, Y. T. Zhang, X. R. Ding, X. Hong, Q. He, and Y. Li, A novel continuous blood pressure estimation approach based on data mining techniques, *IEEE J. Biomed. Health Inf.*, vol. 21, no. 6, 1730–1740, 2017.
53. X. F. Teng and Y. T. Zhang, Continuous and noninvasive estimation of arterial blood pressure using a photoplethysmographic approach, in *Proceedings of the 25th Annual International Conference of the IEEE Engineering in Medicine and Biology Society*, vol. 4, 2003, pp. 3153–3156.
54. P. S. Addison, Slope Transit Time (STT): A pulse transit time proxy requiring only a single signal fiducial point, *IEEE Trans. Biomed. Eng.*, vol. 63, no. 11, 2441–2444, 2016.
55. Y. Chen, S. Member, S. Cheng, T. Wang, and T. Ma, Novel blood pressure estimation method using single photoplethysmography feature, in *39th Annual International Conference of the IEEE Engineering in Medicine and Biology Society* (EMBC), 2017, pp. 1712–1715.
56. L. Wang, W. Zhou, Y. Xing, and X. Zhou, A novel neural network model for blood pressure estimation using photoplethesmography without electrocardiogram, *J. Healthcare Eng.*, vol. 2018, 1–9, 2018.
57. H. Liu, K. Ivanov, Y. Wang, and L. Wang, Toward a smartphone application for estimation of pulse transit time, *Sensors*, vol. 15, no. 10, 27303–27321, 2015.
58. Y. C. Lin, Y. J. Wang, J. C. H. Cheng, and Y. H. Lin, Contactless monitoring of pulse rate and eye movement for uveal melanoma patients undergoing radiation therapy, *IEEE Trans. Instrum. Meas.*, vol. 68, no. 2, 474–482, 2018.
59. J. Espina, T. Falck, J. Muehlsteff, Y. Jin, M. A. Ad??n, and X. Aubert, Wearable body sensor network towards continuous cuff-less blood pressure monitoring, in *Proceedings of 5th International Workshop on Wearable and Implantable Body Sensor Networks, BSN2008, in conjunction with the 5th International Summer School and Symposium on Medical Devices and Biosensors, ISSS-MDBS 2008*, 2008, pp. 28–32.

12

Wireless Telecardiology

Luca Monzo, Michele Schiariti, and Paolo Emilio Puddu
Sapienza University

CONTENTS

12.1 Introduction

Telemedicine is the ability to provide interactive healthcare utilizing modern technology and telecommunications [1]. The terms "telemedicine" or "remote healthcare" and "e-Health" encompass both "telemonitoring" and telephone support. With telemonitoring, patients transmit data on their vital signs for real-time monitoring via a communication link or by store and forward systems. With telephone support, healthcare providers support patients or caregivers via the standard telephone system, which may involve monitoring of vital signs reported by patients. Methods involving standard telephone are relatively old-fashioned and present several disadvantages. It is important to imagine new methods, such as mobile phone communication [2], to enter a more nomadic era whereby monitoring is not just home confined.

On the other hand, telemedicine allows patients to visit with physicians live over video for immediate care or capture video/still images and patient data are stored and sent to physicians for diagnosis and follow-up treatment at a later time. Whether a patient lives in the center of London or Rome or deep in Sahara, telemedicine is an invaluable tool in healthcare. Instead of traveling to the nearest specialist, which depending where the patient lives could be anywhere between a 45-min drive and an 18-h car ride up sanded roads, the patient's service provider connects the patient directly to any medical specialist via telemedicine. The specialist then may hear the medical history and current condition directly from the patient instead of reading a written account dictated by the first health provider. There might be medical peripherals (such as electrocardiographic, echocardiographic signals, or nasopharyngoscope probes) that might be handled by the first-line health provider to allow the specialist receiving important direct elements for diagnosis. There might be direct questions from the specialist and immediate replies from the patient. At the end of the teleconsultation, the specialist can diagnose and recommend treatment immediately.

In general, the existing remote healthcare or telemedicine systems can be classified in three broad categories: real-time telemedicine, asynchronous telemedicine, and home care. The real-time telemedicine is the most common type of remote healthcare approach. Like the example above, live video allows the provider, patient, and specialist to communicate together to achieve the optimal outcome for the patient. It might be used in outpatient specialty consultation, for physician supervision of non-MD first-line health providers. It requires large bandwidths (>256 kB), although new technologies allow to overcome this technical constraint [3]. Asynchronous telemedicine is used when both the clinician and health service provider are not available or not required at the same time. The provider's voice or text dictation on the patient's history, current affliction including pictures and/or video, radiology images, electro-, and echocardiograms are attached for diagnosis. This record is either emailed or placed on a server for the clinician's access to follow up with his/her diagnosis and treatment plan. On the other hand, it is with home care technologies that the future approaches. When a patient is in the hospital and he/she is placed under general observation after a surgery or other medical procedure, the hospital is usually losing a valuable bed and the patient would rather not be there as well. Home health allows the remote observation and care of a patient. Home health equipment consists of vital signs capture and video conferencing capabilities; patient stats can be reviewed and alarms set from the hospital nurse's station, depending on the specific home health device. There are benefits in the telemedicine system for both the spoke sites and the patients. The formers receive education from the providers and the specialists, in particular. There is a better health outcome for the patients around the spoke site. The community around the spoke site enhances confidence in the local healthcare since they know that there are opportunities of continuing medical education for the local health providers via the telemedicine system.

There are important advantages for the patients as well. The loved ones remain in their community with close family support and cost savings from not having to travel extensively. When urgent care is needed, there is the possibility to look for immediate consultation. Confidentiality is increased since the patients receive consultation from the specialist without anyone else knowing apart from the general practitioner (GP) with whom the patient looks for tele-health. Finally, early diagnosis prior to escalated medical episodes is more frequently obtained, and in urgent situations the patient is more adequately stabilized prior to transport. All these have economic consequences: if patients remain in their communities, they can save money and no expenses will occur looking for medical advice in the absence of telemedicine. In general, the implementation of telemedicine systems was shown to substantially reduce health costs [4].

Among the vast range of medical disciplines in which telemedicine has been successfully applied (i.e., psychiatry, dermatology, radiology, neurology, ophthalmology, otolaryngology, rheumatology, pulmonary, urology, wound care, obstetrics, pediatrics and neonatology, pathology, emergency medicine, and trauma), telecardiology is certainly one of the most highly developed technology [5]. Within cardiology, a wide variety of both noninvasive and invasive medical signals are available for telehealthcare. Most are recorded manually by the patients via a device (e.g., systolic and diastolic blood pressure, pulse, 3-lead electrocardiogram (ECG), heart rate variability, body-weight, oxygen saturation, blood glucose, natriuretic peptides). Others obtained from invasive devices are recorded automatically (e.g., impedance, incidence of arrhythmias, pulmonary, and left atrial pressure) [6–8]. In addition to the provision of care to patients with heart disease, it has a vital role in educating these patients on the nature of their conditions, improving their compliance to medical therapy, and guiding them in practicing healthy life habits. The benefit of telecardiology in rural communities is especially important because of its capability of overcoming the obstacle of the large distances that would have to be covered in order to access medical assistance. As such, hazardous and even unnecessary transportation of critically ill patients for the purpose of diagnosis can be avoided by remote expert counseling. Finally, patients can receive second opinion and physicians can consult experts, capabilities that have proven to have a beneficial effect on both patient survival and recovery.

Telecardiology is used in a wide range of cardiovascular diseases (i.e., acute coronary syndromes (ACS), chronic heart failure, hypertension, etc.) and in various settings (i.e., pre-hospital, in-hospital, and after hospital discharge telecardiology) with beneficial results.

12.2 Telecardiology in the Management of Acute Coronary Syndromes

A particularly attractive application for telecardiology in the pre-hospitalization phase is in diagnosing and treating patients with ACS. The first data on the implementation of telemedicine for the diagnosis and treatment of acute myocardial infarction date from almost 20 years ago [9]. Studies on the transmission of ECGs from a moving ambulance were reported even before [10]. Patients with ST-elevation myocardial infarction (STEMI) should receive reperfusion therapy as soon as possible to reduce mortality and morbidity. Guidelines recommend primary percutaneous coronary intervention (PCI) as the preferable strategy, if this can be performed in a timely manner [11]. Thus, to treat STEMI patients within the recommended time limits, early diagnosis is essential to allow a timely transportation to a PCI center in due time. This has led to an increased interest in pre-hospital diagnosis. The diagnostic modality of choice has been the ECG. The ECG is recorded in the ambulance and transmitted to a PCI center. The ECG recording is interpreted by the on-call cardiologist and after consulting the patient and/or paramedic a decision is made to either direct the patient for primary PCI or carry him/her to the nearest local hospital for further diagnosis and initial pharmacological therapy (Figure 12.1).

Increasing the treatment efficacy in STEMI and reducing the door-to-balloon time, defined as the time between the arrival at the hospital and the first balloon inflation during PCI, are major targets of contemporary patient management [12]. Only a minority of hospitals treats patients with STEMI within 90 min after their arrival, and hospital performance has not improved substantially in recent years [13,14]. During the last two decades, many attempts have been made to implement telecardiology applications in order to reduce this crucial time period [9,15–18], and new devices capable to transmit 12-lead ECG tracings via regular and mobile telephones and even some that have the capability of interpreting the ECG findings were tested [19,20]. All studies prove that ECG-teleconsultation is technically

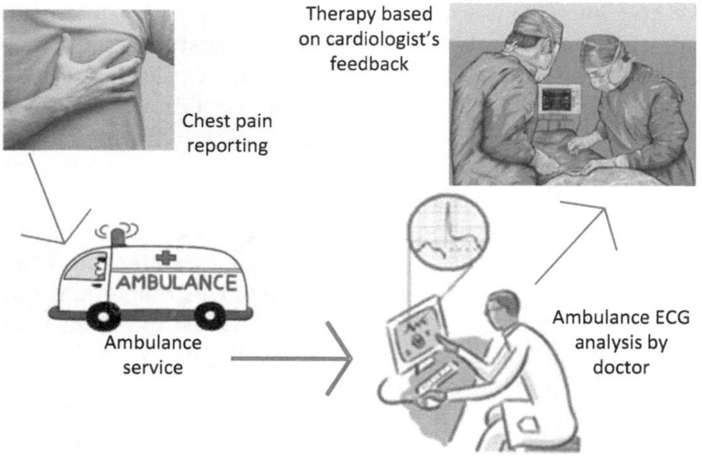

FIGURE 12.1
The set-up for prehospital ECG diagnosis in ACS. The ECG is recorded in the ambulance and sent wirelessly to a primary PCI center. The ECG is interpreted by the on-call cardiologist and, after talking to the patient and/or paramedic, a decision is made to either redirect the patient for primary PCI (in case of STEMI) or send the patient to the nearest local hospital for further diagnosis and therapy. ACS, acute coronary syndromes; ECG, electrocardiogram; PCI, percutaneous coronary intervention; STEMI, ST-elevation myocardial infarction.

feasible either from a moving ambulance or from remote places. The agreement between tele- and standard-ECG concerning alterations of the ST segment is usually very good [21]. ECG-teleconsultation allows faster triage, shortens door-to-needle time (just slightly earlier than time-to-balloon), and reduces in-hospital delays in STEMI patients compared with a prospective control group [9,10,15,16]. Transmission of a pre-hospital 12-lead ECG directly to the attending cardiologist's mobile telephone decreased door-to-PCI (coronary angioplasty) time by more than 1 h when patients were transported directly to PCI centers, bypassing local hospitals [15].

In a large cohort of patients, Brunetti et al. demonstrated that telecardiology evaluation and direct referral for primary PCI enable STEMI patients living far from a PCI center and in rural areas to achieve a system delay comparable with patients living in close vicinity of a PCI center. The routine use of prehospital ECG diagnosis and field triage was evaluated in a larger region comprising both urban and rural surroundings [22]. The authors found that pre-hospital diagnosis and triage reduces the importance of distance to primary PCI center, with a difference of only 9 minutes between patients belonging to urban versus rural hospital, but with a median distance from the PCI center of 10 km for the former versus 40 km for the latter. Moreover, the study showed a decreased mortality in patients with prehospital diagnosis and triage. A recent Dutch study further emphasized the importance and value of prehospital diagnosis for patients living far from a PCI center [23]. Telemetry may also enable the use of prehospital thrombolytic treatment in STEMI patients, thus reducing the call to treatment times in non-PCI capable settings. However, this benefit must be balanced against the very small proportion of eligible patients identified as suitable for prehospital thrombolysis [24].

In addition, ECG-teleconsultation screening may lower the rates of false negative diagnosis in the case of STEMI with atypical presentation [17,18]. Lastly, the pre-hospital diagnosis of STEMI has been associated with other favorable outcomes such as a reduction of the infarct size, a lesser impairment of the ejection fraction, a shorter length of stay, and a reduction in early and late mortality [25,26]. Current Acute Cardiac Care Association of the European Society of Cardiology guidelines on the pre-hospital management of chest pain and dyspnea, therefore, mandate pre-hospital ECG by telemedicine in the absence of a physician skilled in ECG interpretation [22].

In conclusion, telecardiology in ACS enables a more widespread access to rapid reperfusion therapy (by either pharmacological or mechanical intervention), thereby reducing treatment delay, morbidity, and mortality [27].

12.3 Telecardiology in the Management of Chronic Heart Failure

Approximately 6.5 million people in the USA have clinically manifest chronic heart failure (HF), and the projections show that the prevalence of HF will increase by 46% from 2012 to 2030, resulting in more than 8 million people above 18 years of age with HF [28]. Hospitalization for HF is a growing and major public health issue. Presently, HF is the leading cause of hospitalization among patients above 65 years of age and HF patients who have frequent hospitalizations are at risk for poor outcomes. Furthermore, the largest percentage of expenditures related to HF is directly attributable to hospital costs [29]. Guidelines of both American and European scientific societies recommended a multidisciplinary care approach that coordinates care along the continuum of HF and throughout the chain of care delivery by various services within healthcare systems [30,31].

Effective use of telemedicine in managing HF patients has been expanding considerably worldwide. A substantial reduction in hospital admission, in the length of hospitalization, and an increase in survival among HF patients who were managed by telemedicine were shown in randomized controlled trials [32–38] and in cohort studies [39–41]. Typically, a device is installed in the patient's home and another at the HF clinic. Preferably, the patient's device is portable and capable of both videoconferencing and data transmission, so that it can be managed by the healthcare professional using a remote control. According to individual arrangements, patients could be instructed to record blood pressure, heart rhythm, and weight and provide a subjective description of the symptoms every morning, then transmit data daily, just prior to or during a scheduled remote follow-up. The patients' vital signs and weight are transmitted daily to the system's monitor center; the monitor center's nurse telephones the patient to assess a number of parameters, including well-being, frequency of specific symptoms, adherence to treatment, and to remind him/her of the regular use of medications. The nurse instructs patients with "red alarms" (e.g., weight increase, modification of diastolic, and/or systolic blood pressures) and alerts the doctor (Figure 12.2). Data can then be reviewed before the videoconference with the patient, so treatment can be changed if needed. It is important that the healthcare professional be empowered to change medication regimens so that delay is diminished and the correct treatment is established in a timely manner. In addition to objective technological monitoring, most telemonitoring programs include subjective questionnaire regarding the patient's health and comfort. This questioning can take place automatically over the phone, or telemonitoring software can help to keep the patient in touch with the healthcare provider. The provider can then make decisions about the patient's treatment based on a combination of subjective and objective information similar to what would be shown during an on-site appointment.

Several advantages of the use of telemonitoring in HF have been reported [42]. Patients get a better understanding of their disease by being responsible for monitoring and being

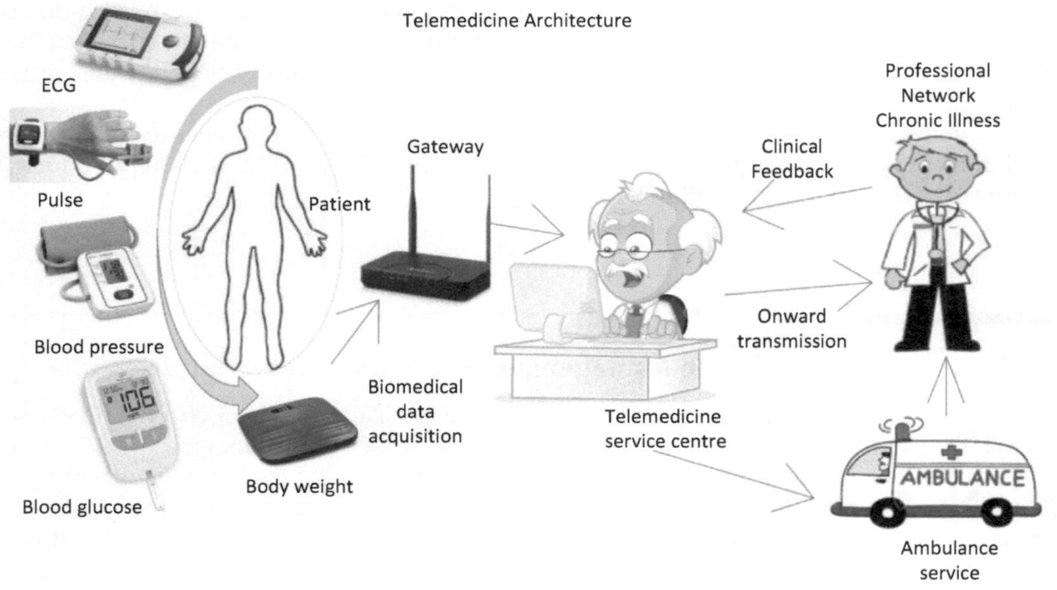

FIGURE 12.2
Telemedicine architecture in heart failure. Multiple non-invasive devices collect the patient's clinical data that are transmitted to a telemedicine service center, where trained healthcare professionals may refer red alerts to a doctor or, if needed, ask for emergency service.

involved in some self-managing, which enhances patients' education and empowerment [43]. In particular, psychological interventions such as cognitive behavioral therapy and mindfulness exercises have already been shown to be successful in changing lifestyle behavior [44,45] and to significantly reduce anxiety, depression, and clinical symptoms which are common in HF patients [46]. In the personal decision support system for Heart Failure Management (HeartMan) study, developed from the preliminary experience of the Chiron project [47], wireless monitoring of patients' physical condition and psychological state is integrated by a decision support system that suggests the most appropriate intervention (including exercise, nutrition, medication, and mental support) to modify and manage patient lifestyle to increase adherence to the medical advice [48]. This trial is still ongoing, and results are expected within the end of 2019. Telemonitoring also ensures preventive care with early detection of disease exacerbations and timely management [49]. Travel and waiting time are reduced or even abolished for the patient, so quality of life improves. Additionally, patients are at home in comfortable and safe surroundings and still in close contact with the hospital. An advantage for the hospital is that the technology leaves room for new and unstable patients. The results provide evidence that remote monitoring combined with discharge planning could reduce family caregiver burden, improve stress mastery, and family function during the first 30 days at home after HF patients are discharged from the hospital.

It is important to acknowledge that although previous studies have suggested that post-hospital telecardiology improves outcomes of HF patients, the effectiveness of tele-healthcare by reducing morbidity and mortality in HF patients is still controversial because of recent conflicting results. Three recent randomized controlled trials, (a) telemonitoring to improve heart failure outcomes (Tele-HF); (b) telemedical interventional monitoring in heart failure (TIM-HF); and (c) better effectiveness after transition-heart failure (BEAT-HF) do not support those findings [50–52] and conclude that, when compared with usual care, telecardiology is not associated with a significant reduction of cardiovascular death and hospitalization or an improvement in health-related quality of life.

However, in a recent meta-analysis and systematic review of 39 randomized clinical trials (11,758 patients), remote patient monitoring was associated with a reduction of 20% in all-cause mortality and of 37% in hospitalizations for HF, as well as to shorten the HF-related length of hospital stay [53]. In a budget impact analysis, the adoption of a telemedicine-based strategy entailed a progressive and linear increase in costs saved. In an other meta-analysis including 6,000 patients from 25 trials, case management type interventions led by a HF specialist nurse reduce HF-related readmissions after 12 months of follow-up, all-cause readmissions and all-cause mortality [54].

In the above-mentioned studies, telehealthcare was based on non-invasive medical data, but other studies have investigated the management of HF patients using data from implantable devices [55,56]. The efficacy of an implantable monitor lies in the fact that it continuously measures and stores hemodynamic information of HF patients that can be reviewed remotely. An increase in pulmonary vascular congestion is reflected by increased impedance [57], thus intra-thoracic impedance monitoring may be a valuable tool in the management of HF in patients, such as those with an implantable cardioverter defibrillator (ICD) or a cardiac resynchronization therapy device (CRT) [58]. This is supported by a study that found significantly greater sensitivity to predict worsening HF events for intra-thoracic impedance monitoring compared to acute weight increase [59]. Likewise, left ventricular filling pressures and pulmonary artery pressures are correlated with clinical congestion, functional limitation, and poor prognosis in HF patients [60]. Since the intracardiac and pulmonary artery pressures increase several days to weeks before the onset of clinical symptoms of congestion, close monitoring may provide early warning and assist appropriate management, including changes in medication. Abraham et al. [61] used a wireless implantable hemodynamic monitoring system to measure pulmonary artery pressure and reported that the use of the system significantly reduced the duration of hospital stay and the rate of hospital admissions related to HF, the latter by 30%. In a study that monitored left atrial pressure, this monitoring was associated with the reduced risk of acute decompensation and death [62].

In conclusion, the majority of evidences show that telemonitoring in HF not only reduces costs and increases the revenue but also saves lives and prevents complications by also improving patient safety and quality of care [54]. Nevertheless, additional randomized clinical trials using third- or fourth-generation telemedical systems that combine noninvasive and invasive variables with daily monitoring and management are needed to fully clarify the value of telehealthcare in cardiology.

12.4 Telecardiology in the Management of Cardiac Rehabilitation (CR) Program

Accessibility to the available traditional forms of cardiac rehabilitation programs in patients after a cardiac event (i.e., ACS, PCI, post-cardiac surgery, and HF hospitalization) is not adequate and adherence to the programs remains unsatisfactory [63]. Telecardiology support may, thus, be especially useful in rehabilitation programs. The literature reporting on telerehabilitation is rather scarce and mainly focuses on low-risk patients [64,65]. These studies have shown favorable effects resulting from telemonitored CR.

Telemedicine-based rehabilitation after cardiac surgery has been demonstrated to be feasible and safe. Scalvini et al. [66] demonstrated that patients who underwent cardiac surgery (EuroSCORE 0–10) followed a 1-month home rehabilitation program supervised by a nurse-tutor and a physiotherapist, and they had a significant increase in the 6-min walking test distance at the end of the program compared to the baseline (404 vs 307 m, $p<0.001$). When two models of assistance (telecardiology versus usual care) for patients discharged after ACS were compared in the assessment of angina, telecardiology slightly reduced hospital readmissions (44% vs 56%) in a short-term follow-up [67]. Giallauria et al. assessed the efficacy of telecardiology in improving the effects of CR during three sessions weekly for 8 weeks in post-MI patients with reduced ejection fraction. Physical capacity and exercise duration improvement were comparable in patients trained in an out-patient center and at home with ECG monitoring, but patients who trained at home without telemonitoring failed to obtain favorable effects [68]. In a large population who survived hospitalization after sustaining an acute myocardial infarction, subjects followed by telemedicine support had significantly higher survival rates at 1 year compared to usual care (4.4% vs 9.7%; $p<0.0001$) [69]. The direct 12-week comparison of a conventional and a telemedicine approach showed the suitability of telemedicine for delivering cardiac rehabilitation for risk factor modification and exercise monitoring to patients who otherwise would not have access to it [70].

Studies wholly dedicated to home-based telerehabilitation of HF patients showed the high impact of this intervention on health status. One of the first studies to assess telerehabilitation in HF patients specifically was that of Smart et al. who described a 35-week home-based CR model initiated 16 weeks after a hospital-based exercise training program. Patients were provided with heart rate monitors and exercise diaries. They were also offered scheduled telephone and e-mail consultations. This study showed that home rehabilitation based on heart rate monitoring maintained the oxygen consumption improvement achieved during previous rehabilitation in hospital settings, only in patients with HF adherent to the program [71]. Two randomized studies evaluated telerehabilitation in HF patients. The first one showed that an 8-week home-based telerehabilitation program provided improvements in physical capacity and quality of life similar to that of a standard outpatient rehabilitation program [72]. The second one demonstrated that a home-based telemonitored Nordic walking training program was well accepted, safe, and effective among HF patients, including those with implanted devices [73].

Telerehabilitation may also serve as a helpful strategy for reducing depression that are common in HF patients and that work as a barrier for exercise training. Allowing a continuous monitoring and management of patients in a familiar environment and as part of the everyday way of life, telerehabilitation may enhance exercise adherence and ultimately health-related quality of life [74]. Moreover, telerehabilitated patients might receive a persistent psychological telesupport and teleassistance from the telemonitoring team (nurse, physician, physiotherapist) [71,73]. In the Personal Decision Support System for Heart Failure Management (HeartMan) study, wireless monitoring of patients' physical condition and psychological state is integrated by a decision support system to allow personalized lifestyle advice [48]. This system might increase adherence and efficacy of physical exercise both supporting physical activity by a psychological support and adapting the strength of exercise to the current physical condition of the patient (based on health wireless telemonitoring). The trial is still ongoing and results are expected within the end of 2019.

A crucial issue regarding home-based telerehabilitation is the safety of HF patients who participate, mainly because of the lack of a close and direct contact with medical staff that in-hospital or outpatient programs can offer [75]. Overall, home-based telerehabilitation was generally reported as safe, with only minor events, such as atrial arrhythmias or skin reactions caused by the electrodes [72,73]. Overall, the data available so far suggest that the benefits of regular physical training outweigh the risks it may entail.

In conclusion, current data show that telemedicine-based rehabilitation is an effective and safe way of cardiac rehabilitation, with a strong beneficial impact on health status and quality of life.

12.5 Telecardiology in the Diagnosis and Management of Hypertension

Hypertension affects almost one in three adults in the United States [29] and is a serious concern in the healthcare service worldwide as it is one of the major risk factors for developing cardiovascular diseases including coronary heart disease, stroke, and HF or renal failure. In addition, it is one of the most frequent reasons for accessing to medical care [76]. Inadequate blood pressure (BP) control and poor adherence to treatment remain among the major limitations in the management of hypertensive patients, particularly of those at high risk of cardiovascular events. In fact, no more than 30% of treated hypertensive patients maintain satisfactory BP control [77]. Compliance to therapy seems to be related to many factors including the high number of pills to be taken daily, the lack of motivation and the patient's poor involvement in the management of this clinical condition, the absence of symptoms, the inability of the patient to understand the real long-term consequences of high BP, and finally the psychological and personality traits of the patient [78].

The use of telecardiology may help empowering hypertensive patients, influencing their attitudes and behaviors, and improving their medical condition. Telecardiology can reinforce and empower the physician–patient relationship, may even try to individualize it, and thus improve BP and cardiovascular risk control. Moreover, it allows physicians and health facilities to expand their reach, beyond their own offices, and to easily provide services to an increased number of patients, with consistent savings in time demand, and with the same quality content of traditional in-person consultations [79]. Telecardiology services also allow hypertensive patients to easily and rapidly communicate to their doctors the occurrence of acute symptoms or sudden BP raises, thus enhancing the monitoring, tracking, and communication of various biometric information, which enable greater engagement and partnership of patients in their care. Finally, telecardiology may offer hypertensive patients the access to diagnostic procedures that might not be available otherwise, without the need to travel long distances [80].

A specific application of telecardiology in hypertension management is represented by BP telemonitoring (BPT), which allows remote data transmission of BP and additional non-vital parameters, from the patient's living site or from a community setting (e.g., primary care clinic or pharmacy store) to the doctor's office or to a hospital [80,81]. BP readings are collected in a professional healthcare environment, or at home over several days or over the 24h, in ambulatory conditions, through electronic automated BP monitors. Measurements are stored in the device memory and then forwarded to a remote computer host (usually wireless system mostly based on Bluetooth, Wi-Fi, ZigBee, or near field communication technologies) where they are reviewed by the referring physician for treatment adjustments.

BPT is a well-established practice that has been shown to improve patient adherence to treatment regimens and to achieve target BP levels. It has the potential for the improvement of hypertension control and associated healthcare outcomes, as documented in numerous meta-analyses [79,82–84]. In 7,037 hypertensive patients enrolled in 23 selected high-quality randomized controlled studies, a regular BPT at home was associated with a significantly larger reduction in both office and ambulatory BP as compared with usual care [85], with a significantly larger proportion of BPT patients achieving office BP normalization. Another interesting feature about telecardiology is its ability to provide measurements of BP without provoking the "white coat" effect [86]. Finally, BPT has been demonstrated to reduce the total cost of hypertension care compared with usual care [87].

Web-based telemedicine platforms may also be used to provide home or 24-h ambulatory BP monitoring through community pharmacies, extending the screening for hypertension and providing a quick, accurate, and professional feedback and adjustment of care plans in treated hypertensive patients, with the support and supervision of the general practitioner or the specialist [88]. The synergy between BPT and pharmacist case management of hypertensive patients may facilitate high BP screening and detection. Furthermore, adding web-based pharmacist care to BPT and web-provided education on lifestyle may be particularly effective for improving BP control in treated hypertensives. The effectiveness of such a telemedicine-based approach for hypertension management has been shown in a number of randomized or observational studies [89–91].

In conclusion, current evidence suggests that telecardiology, and in particular BPT, may be useful in hypertensive patients needing a tighter BP control (such as those at high-risk). Moreover, BPT may support doctors for a closer and continuous follow-up of hypertensive patients as well as in situations that require the monitoring of multiple vital signs. The diffusion of telecardiology solutions among the general public will help patients' caregivers to improve the quality and the effectiveness of the delivered care.

12.6 Telecardiology for Arrhythmias and Cardiovascular Implantable Electronic Devices (CIED)

The number of patients with CIEDs, such as pace-makers, ICDs, and CRT, has increased tremendously in the modern era, with more than three million patients with existing CIEDs and approximately 400,000 new devices estimated to be implanted worldwide each year [92]. According to current guidelines, patients with CIEDs need a follow-up in a range from 3 to 12 months [93]. Today, CIEDs are managed through various mechanisms. Conventional follow-up for these patients has been in-person during office- or hospital-based visits, which requires patients to travel to offices for assessments and optimization of device function. In-person device interrogations are performed by clinically employed nurses or allied professionals, CIED-trained physicians, CIED-trained advanced practice nurses or physician assistants, and/or industry representatives. Usually the follow-up visits include both an evaluation of the device function (including battery status and capture thresholds) and a clinical examination, which may result in reprogramming of the device or changes in medical treatment. These time-consuming visits are a burden to the patient and his relatives, as they often require long transportation, waiting time, and

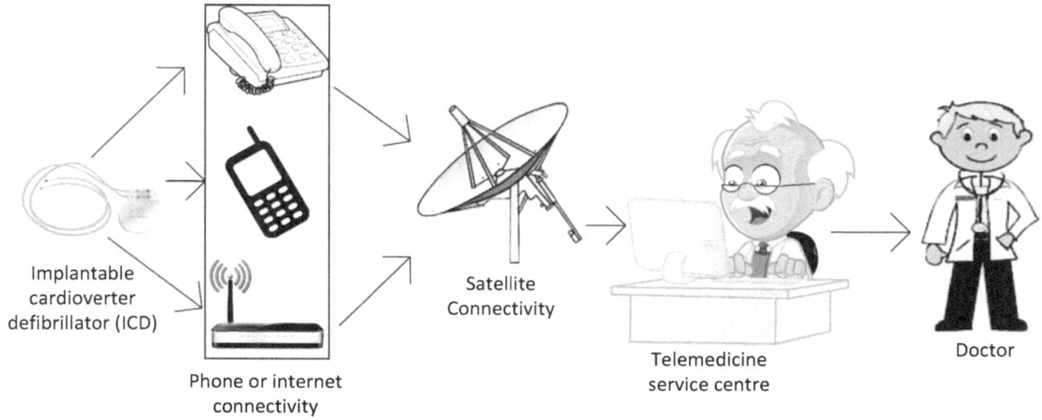

FIGURE 12.3
Remote monitoring system applied to CIEDs. A device equipped with a micro-antenna automatically communicates with a transmitter located close to the patient. Data are then received and analyzed in a telemedicine service center by trained staff and then, if necessary, referred to a dedicated cardiologist.

time off work. Moreover, they require space and dedicated staff in the outpatient clinic because of the workload associated with every visit.

Remote monitoring and remote follow-up by the use of telecardiology offers an alternative management strategy [94]. Preferably, the system should be able to automatically transmit data stored in the device to the outpatient clinic using the wireless global system for mobile communication (GSM) network or a landline. Automatic wireless data transmission requires a pace-maker or ICD/CRT equipped with a micro-antenna for communication with a transmitter located close to the patient as shown in Figure 12.3. This setup diminishes the need for patient compliance and increases the frequency of data transmissions. Several companies provide systems for automatic wireless data transmission. Almost all data stored in the device's memory (e.g., battery voltage, lead characteristics, arrhythmias, alerts) can be transmitted. The alerts may be based on a change in device performance (battery status, lead impedance), programming (disabling of ventricular fibrillation therapy, insufficient safety margins for sensing or capture), or medical data (arrhythmias, indication of lung fluid accumulation). Data are usually transmitted to a central database, where they are processed and made available to the physician on a secure webpage. Additional notifications by e-mail, SMS, fax, or phone messages may be valuable when critical data are available for consultation. Daily monitoring allows transmission of data with any predefined alerts to a physician. Telemonitoring also reduces the amount of follow-up visits for checking the device's parameters (e.g., battery status, leads impedance, etc.). Several reports showed home-monitoring as a safe alternative to conventional care and significantly lowered the number of ambulatory visits [95,96], increasing efficiency for healthcare providers, and improving quality of care for patients [97].

The safety and clinical benefits conferred by telehealthcare in patients with implantable devices have been validated in several studies [95,98,99]. In the COMPAS trial [95], long-term remote monitoring of pace-maker users was a safe substitute for conventional follow-up, decreased the number of ambulatory visits, and enabled early detection of important clinical and device-related adverse events. Another study showed that half of the regularly scheduled visits could be avoided by the use of remote monitoring, without impairing patient safety [98]. Additionally, only 6% of patients undergo device reprogramming or admittance to hospital during clinic visits, and thus 94% of these visits might as well be executed by remote monitoring [99]. Remote monitoring could diagnose 99% of arrhythmia- or device-related problems in patients with an ICD, when combined with a clinical follow-up. In addition to fewer scheduled clinic visits, unscheduled visits following an ICD shock might be prevented by remote follow-up; after such an event, the patient may perform manual upload of data, transmitted to the healthcare provider for immediate determination of whether the shock was appropriate or not, and whether device reprogramming or medical modifications are needed [100]. Finally, previous observations showed that remote monitoring reduces the time to event recognition and diagnosis and the time from the event to a clinical decision for the individual patient [55,101].

Remote monitoring has also been showed to have a beneficial impact on survival. As identified in the long-term outcome after ICD and CRT implantation and influence of remote device follow-up (ALTITUDE) survival study [102], a 50% higher survival rate was found in patients with ICD and CRT-D devices followed with remote monitoring than in those followed only with in-person visits. The survival benefit did not appear to be related to patient age, gender, device type, or how long the device had been implanted. Also, no differences were found in survival rate related to a patient's economic or educational status. Although the ALTITUDE Survival Study included only patients with ICDs and CRT-Ds, similar results were replicated in a study designed to include pace-maker and CRT pace-makers in addition to ICD and CRT-D. Interestingly, Varma et al. [103] discovered that survival rate improvement correlated with adherence to remote monitoring prescription. The reported 50% higher survival rate was seen in patients who demonstrated 75% or more adherence to the remote monitoring protocol. However, the group that participated in remote monitoring with a protocol adherence rate less than 75% also gained a survival benefit when compared with those study participants who did not participate in remote monitoring. Similarly, a reduction in all-cause mortality rate greater than 50% was demonstrated by a group of remotely monitored ICD or CRT-D patients with chronic HF on optimal medical therapy when compared with the control group that received standard in-office follow-up [104]. Anyway, to date it is not known with certainty whether remote monitoring independently increases survival rate or whether it is associated with other factors that increase survival rate, such as more compliant patients and health care providers who practice the most up-to-date optimal medical care.

Remote monitoring is associated with a reduction in inappropriate ICD shocks. In the Effectiveness and Cost of ICD follow-up Schedule with Telecardiology (ECOST) trial, the number of inappropriate shocks in the remotely followed group was 52% lower than that in the control group. Remote monitoring expedites provider notification of ICD shocks, which in turn enables a CIED follow-up clinic to assess the rhythm together with the appropriateness and effectiveness of the shock. As needed, a clinic visit could then take place with the goal of preventing further inappropriate shocks through device reprogramming or prescription of other therapies aimed at management of arrhythmias and other comorbidities [105]. This is of particular interest because the reduction in ICD shocks plays an important role in the physical and psychological well-being of patients with CIEDs and it is well documented that ICD shocks are linked to incidence of anxiety, depression, and sometimes traumatic syndromes in patients having an implanted defibrillator [106], as well as to an increase in mortality [107].

Remote monitoring may finally be a valuable tool to assist in the diagnosis of arrhythmias or ECG abnormalities [108,109]. Symptoms secondary to arrhythmias, such as palpitations and syncope, can be documented on ECG tracings, but many ECG changes are transient or paroxysmal, and the search for corroboratory evidence of these arrhythmias can be lengthy and problematic and missed even by long-term Holter ECG recordings [110]. The detection of these arrhythmias has crucial therapeutic implications, such as the provision of anti-arrhythmic and anticoagulation treatment for high-risk atrial fibrillation (AF) patients, permanent pace-makers for patients suffering from high-degree atrioventricular nodal block, ablation for patients suffering from recurrent supraventricular tachycardia, and others. The diagnostic yield increases substantially with the use of patient-activated short-term ECG recordings [110]. The use of a remote device that enables real-time cardiac arrhythmic monitoring and able to provide a specific diagnosis and recommendations of action has been demonstrated to raise the rate of diagnosis and the efficacy of treatment [111]. In addition, it enabled patients to undergo dose titration or change of their medication in the outpatient setting, thus reducing the rate of hospitalization [112].

AF remote detection and management deserve a particular mention in this setting. Remote monitoring is advantageous in patients with paroxysmal AF because alerts could identify the start-point of the arrhythmia, resulting in an unscheduled follow-up either in office or by phone, and appropriate interventions (e.g., antiarrhythmic medication, anticoagulation or antiplatelet medication, external cardioversion, device reprogramming) [113]. AF may trigger inappropriate therapy or promote loss of effective cardiac resynchronization and may places patients at risk for stroke, especially if they are not receiving anticoagulation therapy, or for HF decompensation. Almost all modern CIEDs have AF detection criteria that will trigger a remote alert when the CIED and/or the alert notification criteria in the remote monitoring web portal are programmed to do so. Timely access to arrhythmia data, including electrogram, rate, onset, and duration of episodes, enables clinicians to assess the appropriateness of anticoagulation therapy and rate control in accordance with evidence-based AF management guidelines.

12.7 Impact of Telecardiology on Health Costs

The cost-effectiveness of telecardiology is still a matter of debate. Besides several studies where telecardiology reduced costs and hospitalizations, other studies failed to do so.

In the TARIFF study, remote monitoring of patients with CIEDs was cost saving from the perspectives of the healthcare system, patients, and caregivers [114], thus reducing routine device checks in hospital. Similarly, in the HomeGuide registry home monitoring was highly effective in detecting and managing clinical events in CIED patients with remarkably low manpower and resource consumption [115]. In a randomized comparison, the costs and benefits of technology-enabled, home-based cardiac rehabilitation were slightly lower than for patients attending a rehabilitation service in person [116]. In a regional registry for pre-hospital ECG diagnosis of STEMI, a cost per quality-adjusted life year gain of €1,927 was calculated [117]. Roth et al. found a reduction of emergency department visits in 1,600 subscribers to a telemedicine system and found the approximate savings to the national economy were $830,000 per 10,000 members per year [118]. Home-based telecardiology management of chronic HF lowered costs by 24% [119]. In a decision analysis modeling to examine the cost-effectiveness of different remote monitoring technologies, a structured telephone support via human to machine was the most cost-effective strategy and yielded an estimated incremental cost-effectiveness ratio compared to usual care [120]. In a UK study in the primary care setting, telecardiology prevented secondary care referrals in 65.8% of cases, with an estimated savings for the NHS of about £300,000 [121]. Finally, self-monitoring with self-titration of anti-hypertensives and telemonitoring of BP measurements not only was shown to reduce blood pressure, compared with usual care, but also to represent a cost-effective use of healthcare resources [122].

Opposite, a comprehensive review of the literature suggests that there is a lack of concrete evidence to assess the economic impact of telecardiology [123]. Comparing cost-effectiveness of telemonitoring versus usual care in patients with HF in the TEHAF study, there were no significant differences in annual costs per patient between groups [124]. In another study, telecardiology was not cost effective after recent discharge in patients with HF, with a similar gain in terms of quality adjusted life years by patients using telehealth in addition to usual care or usual care only [125].

In conclusion, although there is a lack of concrete evidence to assess the economic impact of telecardiology, mainly due to different technologies used (first vs second generation telecardiology tools, namely telephone follow-up versus modern communication technologies, Web 2.0 tools, or even artificial intelligence) and settings (rural versus urban, HF versus rehabilitation, etc.), the majority of the recent evidence shows some potential cost reduction associated with the implementation of telecardiology supporting good agreement with some clinical hopes and/or expectations expressed earlier [126].

12.8 Conclusions

By adapting to the population scale the potential of technology and intervening at the level of primary and secondary prevention, wireless telecardiology could become an essential tool for delivery of effective health care in patients with a broad range of cardiovascular diseases, in turn reducing the cardiovascular disease burden in a society-wide manner and alleviating the tremendous pressure (and costs) the national health services are under.

The next-generation of remote telecardiology systems should aim to identify those individuals (in primary prevention) or patients (in secondary or tertiary care) who may need clinical attention in short-term even before the symptoms are manifested and thereby allowing the clinicians to move towards a proactive care pathway. Finally, telecardiology data will help clinicians to study the characteristic trends of physiological parameters in a person-centric way, targeting diagnosis and treatment on the patient and no more on the disease.

Funding

The research described in this paper partially was carried out in the Chiron project (https://artemis-ia.eu/project/17-chiron.html), which was co-funded by the ARTEMIS Joint Undertaking (grant agreement # 2009-1-100228) and by national authorities, and partially in the HeartMan project (http://cordis.europa.eu/project/rcn/199014_en.html), which received funding from the European Union's Horizon 2020 research and innovation program under grant agreement No 689660. The funders had no role in study design, data collection and analysis, decision to publish, or preparation of the manuscript.

Competing Interests

The authors have declared that no competing interests exist.

Acknowledgments

The consortium of the HeartMan project, which partially funded this research, consists of Jožef Stefan Institute, Sapienza University, Ghent University, Italian National Research Council, ATOS Spain SA, SenLab, KU Leuven, MEGA Electronics Ltd., and European Heart Network.

References

1. WHO, A Report of the WHO group consultation on health telematics. Health telematics policy in support of WHO's Health-For-All strategy for global health development, pp. 11–16, 1997.
2. S. Bonfiglio, Fostering a continuum of care. In: Donnelly M., Paggetti C., Nugent C., Mokhtari M. (eds.) *Impact Analysis of Solutions for Chronic Disease Prevention and Management: 10th International Conference on Smart Homes and Health Telematics, ICOST 2012, Artimino, Tuscany, Italy, June 12-15, Proceedings*, Lecture Notes in Computer Science, vol. 7251, Springer, Berlin, Heidelberg, 2012, pp. 35–41.
3. S. Maheshwari, A. Acharyya, M. Schiariti, and P. E. Puddu, Personalized reduced 3-lead system formation methodology for remote health monitoring applications and reconstruction of standard 12-lead system, *International Archives of Medicine*, vol. 8, 2015, doi: 10.3823/1661.
4. C. Henderson, et al., Cost effectiveness of telehealth for patients with long term conditions (Whole Systems Demonstrator telehealth questionnaire study): Nested economic evaluation in a pragmatic, cluster randomised controlled trial, *BMJ*, vol. 346, f1035, 2013, doi: 10.1136/bmj.f1035.
5. D. Hailey, A. Ohinmaa, and R. Roine, Published evidence on the success of telecardiology: A mixed record, *J Telemed Telecare*, vol. 10, Suppl 1, 36–8, 2004, doi: 10.1258/1357633042614195.
6. A. Acharyya, P. E. Puddu, and M. Schiariti, Reduced lead system selection methodology for reliable standard 12-lead reconstruction targeting personalised remote health monitoring applications AU - Maheshwari, Sidharth, *Comput Methods Biomech Biomed Eng: Imaging Visualization*, vol. 2, no. 2, 107–20, 2014, doi: 10.1080/21681163.2013.859097.
7. A. Acharyya, P. E. Puddu, E. B. Mazomenos, M. Schiariti, and K. Maharatna, Robust and accurate personalised reconstruction of standard 12-lead system from Frank vectorcardiographic system AU - Maheshwari, Sidharth, *Comput Methods Biomech Biomed Eng: Imaging Visualization*, vol. 4, no. 3–4, 183–92, 2016, doi: 10.1080/21681163.2014.931029.
8. S. D. Anker, F. Koehler, and W. T. Abraham, Telemedicine and remote management of patients with heart failure, *Lancet*, vol. 378, no. 9792, 731–9, 2011, doi: 10.1016/S0140-6736(11)61229-4.
9. C. J. Terkelsen, et al., Telemedicine used for remote prehospital diagnosing in patients suspected of acute myocardial infarction, *J Intern Med*, vol. 252, no. 5, 412–20, 2002. [Online]. Available: www.ncbi.nlm.nih.gov/pubmed/12528759.
10. P. Giovas, et al., Transmission of electrocardiograms from a moving ambulance, *J Telemed Telecare*, vol. 4, Suppl 1, 5–7, 1998, doi: 10.1258/1357633981931533.

11. B. Ibanez, et al., 2017 ESC Guidelines for the management of acute myocardial infarction in patients presenting with ST-segment elevation: The Task Force for the management of acute myocardial infarction in patients presenting with ST-segment elevation of the European Society of Cardiology (ESC), *Eur Heart J*, vol. 39, no. 2, 119–77, 2018, doi: 10.1093/eurheartj/ehx393.
12. E. H. Bradley, et al., Strategies for reducing the door-to-balloon time in acute myocardial infarction, *N Engl J Med*, vol. 355, no. 22, 2308–20, 2006, doi: 10.1056/NEJMsa063117.
13. S. F. Jencks, E. D. Huff, and T. Cuerdon, Change in the quality of care delivered to Medicare beneficiaries, 1998–1999 to 2000–2001, *JAMA*, vol. 289, no. 3, 305–12, 2003. [Online]. Available: www.ncbi.nlm.nih.gov/pubmed/12525231.
14. R. L. McNamara, et al., Hospital improvement in time to reperfusion in patients with acute myocardial infarction, 1999 to 2002, *J Am Coll Cardiol*, vol. 47, no. 1, 45–51, 2006, doi: 10.1016/j.jacc.2005.04.071.
15. M. Sejersten, et al., Effect on treatment delay of prehospital teletransmission of 12-lead electrocardiogram to a cardiologist for immediate triage and direct referral of patients with ST-segment elevation acute myocardial infarction to primary percutaneous coronary intervention, *Am J Cardiol*, vol. 101, no. 7, 941–6, 2008, doi: 10.1016/j.amjcard.2007.11.038.
16. B. J. Carmody, A novel approach to transmission of the out-of-hospital EKG in patients with ST segment elevation myocardial infarction, *Ann Emerg Med*, vol. 52, no. 2, 183–4, 2008, doi: 10.1016/j.annemergmed.2008.01.339.
17. N. D. Brunetti, et al., Telecardiology improves quality of diagnosis and reduces delay to treatment in elderly patients with acute myocardial infarction and atypical presentation, *Eur J Cardiovasc Prev Rehabil*, vol. 17, no. 6, 615–20, 2010, doi: 10.1097/HJR.0b013e328331f9e5.
18. N. D. Brunetti, et al., Pre-hospital electrocardiogram triage with tele-cardiology support is associated with shorter time-to-balloon and higher rates of timely reperfusion even in rural areas: Data from the Bari-Barletta/Andria/Trani public emergency medical service 118 registry on primary angioplasty in ST-elevation myocardial infarction, *Eur Heart J Acute Cardiovasc Care*, vol. 3, no. 3, 204–13, 2014, doi: 10.1177/2048872614527009.
19. J. C. Hsieh, K. C. Yu, and C. C. Yang, The realization of ubiquitous 12-lead ECG diagnosis in emergency telemedicine, *Telemed J E Health*, vol. 15, no. 9, 898–906, 2009, doi: 10.1089/tmj.2009.0006.
20. A. Roth, Y. Bloch, Y. Villa, Z. Schlesinger, S. Laniado, and E. Kaplinsky, The CB-12L: A new device for transtelephonic transmission of a 12-lead electrocardiogram, *Pacing Clin Electrophysiol*, vol. 20, no. 9 Pt 1, 2243–7, 1997. [Online]. Available: www.ncbi.nlm.nih.gov/pubmed/9309750.
21. B. Schwaab, A. Katalinic, J. Riedel, and A. Sheikhzadeh, Pre-hospital diagnosis of myocardial ischaemia by telecardiology: Safety and efficacy of a 12-lead electrocardiogram, recorded and transmitted by the patient, *J Telemed Telecare*, vol. 11, no. 1, 41–4, 2005, doi: 10.1177/1357633X0501100109.
22. F. Beygui, et al., Pre-hospital management of patients with chest pain and/or dyspnoea of cardiac origin. A position paper of the Acute Cardiovascular Care Association (ACCA) of the ESC, *Eur Heart J Acute Cardiovasc Care*, 2015, doi: 10.1177/2048872615604119.
23. J. Z. Atary, et al., Standardised pre-hospital care of acute myocardial infarction patients: MISSION! guidelines applied in practice, *Neth Heart J*, vol. 18, no. 9, 408–15, 2010. [Online]. Available: www.ncbi.nlm.nih.gov/pubmed/20862235.
24. M. Woollard, K. Pitt, A. J. Hayward, and N. C. Taylor, Limited benefits of ambulance telemetry in delivering early thrombolysis: A randomised controlled trial, *Emerg Med J*, vol. 22, no. 3, 209–15, 2005, doi: 10.1136/emj.2003.013482.
25. M. Sanchez-Ross, et al., The STAT-MI (ST-Segment Analysis Using Wireless Technology in Acute Myocardial Infarction) trial improves outcomes, *JACC Cardiovasc Interventions*, vol. 4, no. 2, 222–7, 2011, doi: 10.1016/j.jcin.2010.11.007.
26. A. W. Chan, et al., Improved survival associated with pre-hospital triage strategy in a large regional ST-segment elevation myocardial infarction program, *JACC Cardiovasc Interv*, vol. 5, no. 12, 1239–46, 2012, doi: 10.1016/j.jcin.2012.07.013.
27. C. Sable, Telecardiology: Potential impact on acute care, *Crit Care Med*, vol. 29, no. 8 Suppl, N159–65, 2001. [Online]. Available: www.ncbi.nlm.nih.gov/pubmed/11496038.

28. P. A. Heidenreich, et al., Forecasting the impact of heart failure in the United States: A policy statement from the American Heart Association, *Circ Heart Fail*, vol. 6, no. 3, 606–19, 2013, doi: 10.1161/HHF.0b013e318291329a.
29. E. J. Benjamin, et al., Heart disease and stroke statistics-2017 update: A report from the American Heart Association, *Circulation*, vol. 135, no. 10, e146–e603, 2017, doi: 10.1161/CIR.0000000000000485.
30. P. Ponikowski, et al., 2016 ESC Guidelines for the diagnosis and treatment of acute and chronic heart failure: The Task Force for the diagnosis and treatment of acute and chronic heart failure of the European Society of Cardiology (ESC). Developed with the special contribution of the Heart Failure Association (HFA) of the ESC, *Eur J Heart Fail*, vol. 18, no. 8, 891–975, 2016, doi: 10.1002/ejhf.592.
31. C. W. Yancy, et al., 2017 ACC/AHA/HFSA focused update of the 2013 ACCF/AHA guideline for the management of heart failure: A report of the American College of Cardiology/American Heart Association Task Force on Clinical Practice Guidelines and the Heart Failure Society of America, *Circulation*, vol. 136, no. 6, e137–e61, 2017, doi: 10.1161/CIR.0000000000000509.
32. J. G. Cleland, A. A. Louis, A. S. Rigby, U. Janssens, A. H. Balk, and T.-H. Investigators, Noninvasive home telemonitoring for patients with heart failure at high risk of recurrent admission and death: The Trans-European Network-Home-Care Management System (TEN-HMS) study, *J Am Coll Cardiol*, vol. 45, no. 10, 1654–64, 2005, doi: 10.1016/j.jacc.2005.01.050.
33. W. C. Dunagan, et al., Randomized trial of a nurse-administered, telephone-based disease management program for patients with heart failure, *J Card Fail*, vol. 11, no. 5, 358–65, 2005. [Online]. Available: www.ncbi.nlm.nih.gov/pubmed/15948086.
34. A. Kashem, M. T. Droogan, W. P. Santamore, J. W. Wald, and A. A. Bove, Managing heart failure care using an internet-based telemedicine system, *J Card Fail*, vol. 14, no. 2, 121–6, 2008, doi: 10.1016/j.cardfail.2007.10.014.
35. B. Riegel, B. Carlson, D. Glaser, and T. Romero, Randomized controlled trial of telephone case management in Hispanics of Mexican origin with heart failure, *J Card Fail*, vol. 12, no. 3, 211–9, 2006, doi: 10.1016/j.cardfail.2006.01.005.
36. A. D. Galbreath, et al., Long-term healthcare and cost outcomes of disease management in a large, randomized, community-based population with heart failure, *Circulation*, vol. 110, no. 23, 3518–26, 2004, doi: 10.1161/01.CIR.0000148957.62328.89.
37. L. R. Goldberg, et al., Randomized trial of a daily electronic home monitoring system in patients with advanced heart failure: The Weight Monitoring in Heart Failure (WHARF) trial, *Am Heart J*, vol. 146, no. 4, 705–12, 2003, doi: 10.1016/S0002-8703(03)00393-4.
38. E. K. Kasper, et al., A randomized trial of the efficacy of multidisciplinary care in heart failure outpatients at high risk of hospital readmission, *J Am Coll Cardiol*, vol. 39, no. 3, 471–80, 2002. [Online]. Available: www.ncbi.nlm.nih.gov/pubmed/11823086.
39. A. Roth, I. Kajiloti, I. Elkayam, J. Sander, M. Kehati, and M. Golovner, Telecardiology for patients with chronic heart failure: The 'SHL' experience in Israel, *Int J Cardiol*, vol. 97, no. 1, 49–55, 2004, doi: 10.1016/j.ijcard.2003.07.030.
40. A. J. Morguet, P. Kuhnelt, A. Kallel, M. Jaster, and H. P. Schultheiss, Impact of telemedical care and monitoring on morbidity in mild to moderate chronic heart failure, *Cardiology*, vol. 111, no. 2, 134–9, 2008, doi: 10.1159/000119701.
41. M. Gambetta, P. Dunn, D. Nelson, B. Herron, and R. Arena, Impact of the implementation of telemanagement on a disease management program in an elderly heart failure cohort, *Prog Cardiovasc Nurs*, vol. 22, no. 4, 196–200, 2007. [Online]. Available: www.ncbi.nlm.nih.gov/pubmed/18059196.
42. M. Marschollek, Decision support at home (DS@HOME)–System architectures and requirements, *BMC Med Inform Decis Mak*, vol. 12, 43, 2012, doi: 10.1186/1472-6947-12-43.
43. S. Spinsante, R. Antonicelli, I. Mazzanti, and E. Gambi, Technological approaches to remote monitoring of elderly people in cardiology: A usability perspective, *Int J Telemed Appl*, vol. 2012, 104561, 2012, doi: 10.1155/2012/104561.

44. C. B. Becker, C. Wilson, A. Williams, M. Kelly, L. McDaniel, and J. Elmquist, Peer-facilitated cognitive dissonance versus healthy weight eating disorders prevention: A randomized comparison, *Body Image*, vol. 7, no. 4, 280–8, 2010, doi: 10.1016/j.bodyim.2010.06.004.
45. L. Carim-Todd, S. H. Mitchell, and B. S. Oken, Mind-body practices: An alternative, drug-free treatment for smoking cessation? A systematic review of the literature, *Drug Alcohol Depend*, vol. 132, no. 3, 399–410, 2013, doi: 10.1016/j.drugalcdep.2013.04.014.
46. M. J. Sullivan, et al., The Support, Education, and Research in Chronic Heart Failure Study (SEARCH): A mindfulness-based psychoeducational intervention improves depression and clinical symptoms in patients with chronic heart failure, *Am Heart J*, vol. 157, no. 1, 84–90, 2009, doi: 10.1016/j.ahj.2008.08.033.
47. M. Mlakar, P. E. Puddu, M. Somrak, S. Bonfiglio, and M. Lustrek, Mining telemonitored physiological data and patient-reported outcomes of congestive heart failure patients, *PLoS One*, vol. 13, no. 3, e0190323, 2018, doi: 10.1371/journal.pone.0190323.
48. A. Baert, et al., A personal decision support system for Heart Failure Management (HeartMan): Study protocol of the HeartMan randomized controlled trial, *BMC Cardiovasc Disord*, vol. 18, no. 1, 186, 2018, doi: 10.1186/s12872-018-0921-2.
49. N. Varma, B. Stambler, and S. Chun, Detection of atrial fibrillation by implanted devices with wireless data transmission capability, *Pacing Clin Electrophysiol*, vol. 28, Suppl 1, S133–6, 2005, doi: 10.1111/j.1540-8159.2005.00083.x.
50. S. I. Chaudhry, et al., Telemonitoring in patients with heart failure, *N Engl J Med*, vol. 363, no. 24, 2301–9, 2010, doi: 10.1056/NEJMoa1010029.
51. F. Koehler, et al., Impact of remote telemedical management on mortality and hospitalizations in ambulatory patients with chronic heart failure: The telemedical interventional monitoring in heart failure study, *Circulation*, vol. 123, no. 17, 1873–80, 2011, doi: 10.1161/CIRCULATIONAHA.111.018473.
52. M. K. Ong, et al., Effectiveness of remote patient monitoring after discharge of hospitalized patients with heart failure: The better effectiveness after transition – Heart Failure (BEAT-HF) randomized clinical trial, *JAMA Intern Med*, vol. 176, no. 3, 310–8, 2016, doi: 10.1001/jamainternmed.2015.7712.
53. M. H. Lin, W. L. Yuan, T. C. Huang, H. F. Zhang, J. T. Mai, and J. F. Wang, Clinical effectiveness of telemedicine for chronic heart failure: A systematic review and meta-analysis, *J Investig Med*, vol. 65, no. 5, 899–911, 2017, doi: 10.1136/jim-2016-000199.
54. A. Takeda, S. J. Taylor, R. S. Taylor, F. Khan, H. Krum, and M. Underwood, Clinical service organisation for heart failure, *Cochrane Database Syst Rev*, no. 9, CD002752, 2012, doi: 10.1002/14651858.CD002752.pub3.
55. N. Varma, A. E. Epstein, A. Irimpen, R. Schweikert, C. Love, and T. Investigators, Efficacy and safety of automatic remote monitoring for implantable cardioverter-defibrillator follow-up: The Lumos-T Safely Reduces Routine Office Device Follow-up (TRUST) trial, *Circulation*, vol. 122, no. 4, 325–32, 2010, doi: 10.1161/CIRCULATIONAHA.110.937409.
56. C. M. Yu, et al., Intrathoracic impedance monitoring in patients with heart failure: Correlation with fluid status and feasibility of early warning preceding hospitalization, *Circulation*, vol. 112, no. 6, 841–8, 2005, doi: 10.1161/CIRCULATIONAHA.104.492207.
57. R. S. Small, et al., Changes in intrathoracic impedance are associated with subsequent risk of hospitalizations for acute decompensated heart failure: Clinical utility of implanted device monitoring without a patient alert, *J Card Fail*, vol. 15, no. 6, 475–81, 2009, doi: 10.1016/j.cardfail.2009.01.012.
58. D. Catanzariti, et al., Monitoring intrathoracic impedance with an implantable defibrillator reduces hospitalizations in patients with heart failure, *Pacing Clin Electrophysiol*, vol. 32, no. 3, 363–70, 2009, doi: 10.1111/j.1540-8159.2008.02245.x.
59. W. T. Abraham, et al., Intrathoracic impedance vs daily weight monitoring for predicting worsening heart failure events: Results of the Fluid Accumulation Status Trial (FAST), *Congest Heart Fail*, vol. 17, no. 2, 51–5, 2011, doi: 10.1111/j.1751-7133.2011.00220.x.

60. M. R. Zile, et al., Transition from chronic compensated to acute decompensated heart failure: Pathophysiological insights obtained from continuous monitoring of intracardiac pressures, *Circulation*, vol. 118, no. 14, 1433–41, 2008, doi: 10.1161/CIRCULATIONAHA.108.783910.
61. W. T. Abraham, et al., Wireless pulmonary artery haemodynamic monitoring in chronic heart failure: A randomised controlled trial, *Lancet*, vol. 377, no. 9766, 658–66, 2011, doi: 10.1016/S0140-6736(11)60101-3.
62. P. B. Adamson, et al., Ongoing right ventricular hemodynamics in heart failure: Clinical value of measurements derived from an implantable monitoring system, *J Am Coll Cardiol*, vol. 41, no. 4, 565–71, 2003. [Online]. Available: www.ncbi.nlm.nih.gov/pubmed/12598066.
63. B. Bjarnason-Wehrens, et al., Cardiac rehabilitation in Europe: Results from the European Cardiac Rehabilitation Inventory Survey, *Eur J Cardiovasc Prev Rehabil*, vol. 17, no. 4, 410–8, 2010, doi: 10.1097/HJR.0b013e328334f42d.
64. G. F. Fletcher, A. J. Chiaramida, M. R. LeMay, B. L. Johnston, J. E. Thiel, and M. C. Spratlin, Telephonically-monitored home exercise early after coronary artery bypass surgery, *Chest*, vol. 86, no. 2, 198–202, 1984. [Online]. Available: www.ncbi.nlm.nih.gov/pubmed/6611245.
65. E. Piotrowicz, et al., Feasibility of home-based cardiac telerehabilitation: Results of TeleInterMed study, *Cardiol J*, vol. 21, no. 5, 539–46, 2014, doi: 10.5603/CJ.a2014.0005.
66. S. Scalvini, E. Zanelli, L. Comini, M. D. Tomba, G. Troise, and A. Giordano, Home-based exercise rehabilitation with telemedicine following cardiac surgery, *J Telemed Telecare*, vol. 15, no. 6, 297–301, 2009, doi: 10.1258/jtt.2009.090208.
67. A. Chiantera, et al., Role of telecardiology in the assessment of angina in patients with recent acute coronary syndrome, *J Telemed Telecare*, vol. 11, Suppl 1, 93–4, 2005, doi: 10.1258/1357633054461967.
68. F. Giallauria, et al., Efficacy of telecardiology in improving the results of cardiac rehabilitation after acute myocardial infarction, *Monaldi Arch Chest Dis*, vol. 66, no. 1, 8–12, 2006, doi: 10.4081/monaldi.2006.536.
69. A. Roth, et al., Telemedicine for post-myocardial infarction patients: An observational study, *Telemed J E Health*, vol. 15, no. 1, 24–30, 2009, doi: 10.1089/tmj.2008.0068.
70. L. C. Dalleck, L. K. Schmidt, and R. Lueker, Cardiac rehabilitation outcomes in a conventional versus telemedicine-based programme, *J Telemed Telecare*, vol. 17, no. 5, 217–21, 2011, doi: 10.1258/jtt.2010.100407.
71. N. Smart, B. Haluska, L. Jeffriess, and T. H. Marwick, Predictors of a sustained response to exercise training in patients with chronic heart failure: A telemonitoring study, *Am Heart J*, vol. 150, no. 6, 1240–7, 2005, doi: 10.1016/j.ahj.2005.01.035.
72. E. Piotrowicz, et al., A new model of home-based telemonitored cardiac rehabilitation in patients with heart failure: Effectiveness, quality of life, and adherence, *Eur J Heart Fail*, vol. 12, no. 2, 164–71, 2010, doi: 10.1093/eurjhf/hfp181.
73. E. Piotrowicz, et al., Home-based telemonitored Nordic walking training is well accepted, safe, effective and has high adherence among heart failure patients, including those with cardiovascular implantable electronic devices: A randomised controlled study, *Eur J Prev Cardiol*, vol. 22, no. 11, 1368–77, 2015, doi: 10.1177/2047487314551537.
74. K. A. Barbour and N. H. Miller, Adherence to exercise training in heart failure: A review, *Heart Fail Rev*, vol. 13, no. 1, 81–9, 2008, doi: 10.1007/s10741-007-9054-x.
75. C. M. O'Connor, et al., Efficacy and safety of exercise training in patients with chronic heart failure: HF-ACTION randomized controlled trial, *JAMA*, vol. 301, no. 14, 1439–50, 2009, doi: 10.1001/jama.2009.454.
76. D. R. Berlowitz, et al., Inadequate management of blood pressure in a hypertensive population, *N Engl J Med*, vol. 339, no. 27, 1957–63, 1998, doi: 10.1056/NEJM199812313392701.
77. K. Wolf-Maier, et al., Hypertension treatment and control in five European countries, Canada, and the United States, *Hypertension*, vol. 43, no. 1, 10–7, 2004, doi: 10.1161/01.HYP.0000103630.72812.10.
78. G. Bobrie, N. Postel-Vinay, J. Delonca, P. Corvol, and S. Investigators, Self-measurement and self-titration in hypertension: A pilot telemedicine study, *Am J Hypertens*, vol. 20, no. 12, 1314–20, 2007, doi: 10.1016/j.amjhyper.2007.08.011.

79. C. Weston, A. Gilkes, S. Durbaba, P. Schofield, P. White, and M. Ashworth, Long term condition morbidity in English general practice: A cross-sectional study using three composite morbidity measures, *BMC Fam Pract*, vol. 17, no. 1, 166, 2016, doi: 10.1186/s12875-016-0563-3.
80. S. Omboni and R. Ferrari, The role of telemedicine in hypertension management: Focus on blood pressure telemonitoring, *Curr Hypertens Rep*, vol. 17, no. 4, 535, 2015, doi: 10.1007/s11906-015-0535-3.
81. G. Parati and S. Omboni, Role of home blood pressure telemonitoring in hypertension management: An update, *Blood Press Monit*, vol. 15, no. 6, 285–95, 2010, doi: 10.1097/MBP.0b013e328340c5e4.
82. S. Liu, et al., Reducing blood pressure with internet-based interventions: A meta-analysis, *Can J Cardiol*, vol. 29, no. 5, 613–21, 2013, doi: 10.1016/j.cjca.2013.02.007.
83. L. L. Zullig, S. D. Melnyk, K. Goldstein, R. J. Shaw, and H. B. Bosworth, The role of home blood pressure telemonitoring in managing hypertensive populations, *Curr Hypertens Rep*, vol. 15, no. 4, 346–55, 2013, doi: 10.1007/s11906-013-0351-6.
84. A. AbuDagga, H. E. Resnick, and M. Alwan, Impact of blood pressure telemonitoring on hypertension outcomes: A literature review, *Telemed J E Health*, vol. 16, no. 7, 830–8, 2010, doi: 10.1089/tmj.2010.0015.
85. S. Omboni, T. Gazzola, G. Carabelli, and G. Parati, Clinical usefulness and cost effectiveness of home blood pressure telemonitoring: Meta-analysis of randomized controlled studies, *J Hypertens*, vol. 31, no. 3, 455–67; discussion 467–8, 2013, doi: 10.1097/HJH.0b013e32835ca8dd.
86. S. Wagner, Blood pressure self-measurement, *Adv Exp Med Biol*, vol. 956, 97–107, 2017, doi: 10.1007/5584_2016_151.
87. G. Parati, et al., Blood pressure control and treatment adherence in hypertensive patients with metabolic syndrome: Protocol of a randomized controlled study based on home blood pressure telemonitoring vs. conventional management and assessment of psychological determinants of adherence (TELEBPMET Study), *Trials*, vol. 14, 22, 2013, doi: 10.1186/1745-6215-14-22.
88. S. Omboni and E. Sala, The pharmacist and the management of arterial hypertension: The role of blood pressure monitoring and telemonitoring, *Expert Rev Cardiovasc Ther*, vol. 13, no. 2, 209–21, 2015, doi: 10.1586/14779072.2015.1001368.
89. B. B. Green, et al., Effectiveness of home blood pressure monitoring, Web communication, and pharmacist care on hypertension control: A randomized controlled trial, *JAMA*, vol. 299, no. 24, 2857–67, 2008, doi: 10.1001/jama.299.24.2857.
90. B. B. Green, M. L. Anderson, J. D. Ralston, S. L. Catz, and A. J. Cook, Blood pressure 1 year after completion of web-based pharmacist care, *JAMA Intern Med*, vol. 173, no. 13, 1250–2, 2013, doi: 10.1001/jamainternmed.2013.1037.
91. L. S. Robins, J. E. Jackson, B. B. Green, D. Korngiebel, R. W. Force, and L. M. Baldwin, Barriers and facilitators to evidence-based blood pressure control in community practice, *J Am Board Fam Med*, vol. 26, no. 5, 539–57, 2013, doi: 10.3122/jabfm.2013.05.130060.
92. E. Buch, N. G. Boyle, and P. H. Belott, Pacemaker and defibrillator lead extraction, *Circulation*, vol. 123, no. 11, e378–80, 2011, doi: 10.1161/CIRCULATIONAHA.110.987354.
93. D. Slotwiner, et al., HRS Expert Consensus Statement on remote interrogation and monitoring for cardiovascular implantable electronic devices, *Heart Rhythm*, vol. 12, no. 7, e69–100, 2015, doi: 10.1016/j.hrthm.2015.05.008.
94. J. E. Poole, et al., Prognostic importance of defibrillator shocks in patients with heart failure, *N Engl J Med*, vol. 359, no. 10, 1009–17, 2008, doi: 10.1056/NEJMoa071098.
95. P. Mabo, et al., A randomized trial of long-term remote monitoring of pacemaker recipients (the COMPAS trial), *Eur Heart J*, vol. 33, no. 9, 1105–11, 2012, doi: 10.1093/eurheartj/ehr419.
96. L. Guedon-Moreau, et al., A randomized study of remote follow-up of implantable cardioverter defibrillators: Safety and efficacy report of the ECOST trial, *Eur Heart J*, vol. 34, no. 8, 605–14, 2013, doi: 10.1093/eurheartj/ehs425.
97. M. Landolina, et al., Remote monitoring reduces healthcare use and improves quality of care in heart failure patients with implantable defibrillators: The evolution of management strategies of heart failure patients with implantable defibrillators (EVOLVO) study, *Circulation*, vol. 125, no. 24, 2985–92, 2012, doi: 10.1161/CIRCULATIONAHA.111.088971.

98. P. Brugada, What evidence do we have to replace in-hospital implantable cardioverter defibrillator follow-up? *Clin Res Cardiol*, vol. 95, Suppl 3, III3–9, 2006, doi: 10.1007/s00392-006-1302-x.
99. H. Heidbuchel, et al., Potential role of remote monitoring for scheduled and unscheduled evaluations of patients with an implantable defibrillator, *Europace*, vol. 10, no. 3, 351–7, 2008, doi: 10.1093/europace/eun010.
100. M. Santini, et al., Remote monitoring of patients with biventricular defibrillators through the CareLink system improves clinical management of arrhythmias and heart failure episodes, *J Interv Card Electrophysiol*, vol. 24, no. 1, 53–61, 2009, doi: 10.1007/s10840-008-9321-3.
101. G. H. Crossley, A. Boyle, H. Vitense, Y. Chang, R. H. Mead, and Connect Investigators, The CONNECT (Clinical Evaluation of Remote Notification to Reduce Time to Clinical Decision) trial: The value of wireless remote monitoring with automatic clinician alerts, *J Am Coll Cardiol*, vol. 57, no. 10, 1181–9, 2011, doi: 10.1016/j.jacc.2010.12.012.
102. L. A. Saxon, et al., Long-term outcome after ICD and CRT implantation and influence of remote device follow-up: The ALTITUDE survival study, *Circulation*, vol. 122, no. 23, 2359–67, 2010, doi: 10.1161/CIRCULATIONAHA.110.960633.
103. N. Varma, J. P. Piccini, J. Snell, A. Fischer, N. Dalal, and S. Mittal, The relationship between level of adherence to automatic wireless remote monitoring and survival in pacemaker and defibrillator patients, *J Am Coll Cardiol*, vol. 65, no. 24, 2601–10, 2015, doi: 10.1016/j.jacc.2015.04.033.
104. G. Hindricks, et al., Implant-based multiparameter telemonitoring of patients with heart failure (IN-TIME): A randomised controlled trial, *Lancet*, vol. 384, no. 9943, 583–90, 2014, doi: 10.1016/S0140-6736(14)61176-4.
105. L. Guedon-Moreau, et al., Decreased delivery of inappropriate shocks achieved by remote monitoring of ICD: A substudy of the ECOST trial, *J Cardiovasc Electrophysiol*, vol. 25, no. 7, 763–70, 2014, doi: 10.1111/jce.12405.
106. S. F. Sears, J. D. Hauf, K. Kirian, G. Hazelton, and J. B. Conti, Posttraumatic stress and the implantable cardioverter-defibrillator patient: What the electrophysiologist needs to know, *Circ Arrhythm Electrophysiol*, vol. 4, no. 2, 242–50, 2011, doi: 10.1161/CIRCEP.110.957670.
107. J. P. Daubert, et al., Inappropriate implantable cardioverter-defibrillator shocks in MADIT II: Frequency, mechanisms, predictors, and survival impact, *J Am Coll Cardiol*, vol. 51, no. 14, 1357–65, 8 2008, doi: 10.1016/j.jacc.2007.09.073.
108. S. Maheshwari, A. Acharyya, P. E. Puddu, and M. Schiariti, Methodology for automated detection of fragmentation in QRS complex of Standard 12-lead ECG, *Conf Proc IEEE Eng Med Biol Soc*, vol. 2013, 3789–92, 2013, doi: 10.1109/EMBC.2013.6610369.
109. S. Maheshwari, et al., An automated algorithm for online detection of fragmented QRS and identification of its various morphologies, *J R Soc Interface*, vol. 10, no. 89, 20130761, 2013, doi: 10.1098/rsif.2013.0761.
110. G. Kaleschke, et al., Prospective, multicentre validation of a simple, patient-operated electrocardiographic system for the detection of arrhythmias and electrocardiographic changes, *Europace*, vol. 11, no. 10, 1362–8, 2009, doi: 10.1093/europace/eup262.
111. S. S. Singh and H. S. Hsiao, Development of a remote handheld cardiac arrhythmia monitor, *Conf Proc IEEE Eng Med Biol Soc*, vol. 1, 3608–11, 2006, doi: 10.1109/IEMBS.2006.260096.
112. J. A. Olson, A. M. Fouts, B. J. Padanilam, and E. N. Prystowsky, Utility of mobile cardiac outpatient telemetry for the diagnosis of palpitations, presyncope, syncope, and the assessment of therapy efficacy, *J Cardiovasc Electrophysiol*, vol. 18, no. 5, 473–7, 2007, doi: 10.1111/j.1540-8167.2007.00779.x.
113. R. P. Ricci, L. Morichelli, and M. Santini, Remote control of implanted devices through Home Monitoring™ technology improves detection and clinical management of atrial fibrillation, *Europace*, vol. 11, no. 1, 54–61, 2009, doi: 10.1093/europace/eun303.
114. R. P. Ricci, et al., Economic analysis of remote monitoring of cardiac implantable electronic devices: Results of the Health Economics Evaluation Registry for Remote Follow-up (TARIFF) study, *Heart Rhythm*, vol. 14, no. 1, 50–7, 2017, doi: 10.1016/j.hrthm.2016.09.008.

115. R. P. Ricci, et al., Effectiveness of remote monitoring of CIEDs in detection and treatment of clinical and device-related cardiovascular events in daily practice: The HomeGuide Registry, *Europace*, vol. 15, no. 7, 970–7, 2013, doi: 10.1093/europace/eus440.
116. F. Whittaker and V. Wade, The costs and benefits of technology-enabled, home-based cardiac rehabilitation measured in a randomised controlled trial, *J Telemed Telecare*, vol. 20, no. 7, 419–22, 2014, doi: 10.1177/1357633X14552376.
117. N. D. Brunetti, et al., Prehospital telemedicine electrocardiogram triage for a regional public emergency medical service: Is it worth it? A preliminary cost analysis, *Clin Cardiol*, vol. 37, no. 3, 140–5, 2014, doi: 10.1002/clc.22234.
118. A. Roth, et al., Potential reduction of costs and hospital emergency department visits resulting from prehospital transtelephonic triage–The Shahal experience in Israel, *Clin Cardiol*, vol. 23, no. 4, 271–6, 2000. [Online]. Available: www.ncbi.nlm.nih.gov/pubmed/10763075.
119. S. Scalvini, et al., Effect of home-based telecardiology on chronic heart failure: Costs and outcomes, *J Telemed Telecare*, vol. 11, Suppl 1, 16–8, 2005, doi: 10.1258/1357633054461688.
120. P. Thokala, et al., Telemonitoring after discharge from hospital with heart failure: Cost-effectiveness modelling of alternative service designs, *BMJ Open*, vol. 3, no. 9, e003250, 2013, doi: 10.1136/bmjopen-2013-003250.
121. Lancashire and South Cumbria Cardiac Network. NHS North West. Delivering Benefits for Patients and the NHS in Lancashire & Cumbria. A report for commissioners. Manchester (UK) 2009. https://joinup.ec.europa.eu/collection/ehealth/document/uk-cardiac-telemedicine-primary-care-aeu-delivering-benefits-patients-and-nhs-lancashire-and-cumbria.
122. B. Kaambwa, et al., Telemonitoring and self-management in the control of hypertension (TASMINH2): A cost-effectiveness analysis, *Eur J Prev Cardiol*, vol. 21, no. 12, 1517–30, 2014, doi: 10.1177/2047487313501886.
123. M. E. Davalos, M. T. French, A. E. Burdick, and S. C. Simmons, Economic evaluation of telemedicine: Review of the literature and research guidelines for benefit-cost analysis, *Telemed J E Health*, vol. 15, no. 10, 933–48, 2009, doi: 10.1089/tmj.2009.0067.
124. J. J. Boyne, et al., Cost-effectiveness analysis of telemonitoring versus usual care in patients with heart failure: The TEHAF-study, *J Telemed Telecare*, vol. 19, no. 5, 242–8, 2013, doi: 10.1177/1357633X13495478.
125. A. Pandor, et al., Home telemonitoring or structured telephone support programmes after recent discharge in patients with heart failure: Systematic review and economic evaluation, *Health Technol Assess*, vol. 17, no. 32, 1–207, v-vi, 2013, doi: 10.3310/hta17320.
126. P. E. Puddu, et al., A Clinician's view of next-generation remote healthcare system, In: K. Maharatna, S. Bonfiglio (eds.) *Systems Design for Remote Healthcare*, Springer, New York, NY, USA, 2013, pp. 1–30.

13

Diabetes Monitoring System

Saptarshi Das
University of Exeter

CONTENTS

13.1 Measurements, Model Estimation, and Control of Diabetes

Diabetes is a metabolic disorder characterized by chronic hyperglycemia leading to microvascular and macrovascular complications. Diabetes can be divided into two categories, namely type 1 and type 2, with different pathogenesis. Type 1 diabetes is caused due to immune-mediated destruction of the beta cells at the site of insulin secretion and production. Although not related to obesity, this can cause other cardiac complications. Type 1 diabetes is usually treated by insulin therapy to control hyperglycemia and sustain life; whereas in type 2 diabetes, inadequate insulin secretion results in hyperglycemia. Type 2 diabetes is generally associated with increasing age (usually after 40), excess caloric intake, less physical activity, and obesity which plays a big role in its development. It leads to cardiovascular risks like dyslipidemia and hypertension. It has been estimated that almost 90% of the world's diabetes population is of type 2 and 5% to 10% is of type 1. Over the years, diabetes has led to many other complications like diabetic retinopathy (leading to blindness), diabetic neuropathy (leading to limb loss), kidney failure, heart disease, and stroke which doubles the risk of dying. Diabetes control is inherently an interdisciplinary field of research. Over the last 50 years, many models and closed-control strategies have been proposed to develop an artificial pancreas.

The World Health Organization has estimated that more than 180 million people in the world have diabetes which is likely to double by 2020 [1]. Diabetes is the fifth-highest cause of death after communicable diseases, cardiovascular diseases, cancer, and injury and the primary cause of death in low and middle-income countries. Research and study on diabetes encompass many interdisciplinary fields like physiology and pathophysiology, drugs, artificial pancreas, transplantation, patient management, healthcare, and systems biology.

The glucose-insulin control system in the human body is regulated by a complex neuro-hormonal control system. Insulin is the primary regulator of glucose homeostasis that promotes glucose utilization and inhibits glucose production from stored resources in the body. Working at different scales, counter-regulatory hormones like glucagon, epinephrine, cortisol, and growth hormone defend the body from life-threatening hypoglycemia. Insulin control and hypoglycemia counter-regulation are balanced by neuro-modulation. Glucose in the body is produced by the liver and utilized in both insulin-dependent (e.g., central nervous system and red blood cells) and insulin-independent tissues (e.g., muscle and adipose tissues). Insulin is secreted by the beta cells in the pancreas, then reaches the system circulation after liver degradation, and is finally cleared by the kidney. The glucose and insulin systems interact by feedback control signals; for example, during glucose perturbation after a meal, the beta cells secrete more insulin after sensing high plasma glucose concentration, thereby promoting glucose utilization and inhibiting glucose production to balance the plasma glucose. Insulin sensitivity and beta-cell responsitivity progressively deteriorate in type 2 diabetes. In type 1 diabetes, the beta cells don't respond to glucose perturbation, and therefore, insulin must be provided exogenously into the patient's body to compensate for hyperglycemia. However, insulin treatment is often risky and may lead to severe hypoglycemia. Therefore, for type 1 diabetic patients it is a challenge to maintain reduced hyperglycemia without risking hypoglycemia; blood glucose level is the measured quantity in such optimization. Several control models have been developed to assist diabetes control, which are reviewed in the following section.

13.2 Minimal Models

Minimal models describe the key functionality rather than the detailed substrate/hormone interactions. For large detailed models, it is usually difficult to estimate all model parameters from *in-vivo* dynamic data; whereas minimal models are much simpler. The desirable features of a minimal model include:

- Physiologically motivated
- Parameter estimation with good precision is possible through a single dynamic response of the system.
- The model parameters should vary within physiologically plausible ranges.
- The whole system dynamics can be described with a minimum number of parameters.

The glucose system is divided into many parts of a compartmental model [1]:

- Steady-state insulin action.
- Non-steady-state insulin action.
- Dynamic perturbation.
- Dynamic perturbation with tracer.

Plasma glucose and insulin's response to an oral or intravenous administration of glucose is given by the following system of ordinary differential equations (ODEs):

$$\begin{aligned}\frac{dG(t)}{dt} &= -a_1G(t) - a_2I(t) + J(t)\\ \frac{dI(t)}{dt} &= a_3G(t) - a_4I(t)\end{aligned} \tag{13.1}$$

where the system parameters are given as:

G: plasma glucose,

I: plasma insulin,

J: glucose input either as an intravenous injection, or the absorption rate of glucose during a meal, or an oral glucose tolerance test.

The minimal model assumes that glucose kinetics can be described by one compartment and both the remote insulin controls, the net hepatic glucose balance (NHGB) and peripheral glucose disposal:

$$\begin{aligned}&\frac{dQ_1(t)}{dt} = \mathrm{NHGB}\big(Q(t), I'(t)\big) - R_\mathrm{d}\big(Q(t), I'(t)\big) + D\cdot\delta(t), \quad Q(0) = Q_\mathrm{b}\\ &\frac{dI'(t)}{dt} = -k_3\cdot I'(t) + k_2\cdot\big[I(t) - I_\mathrm{b}\big], \quad I'(0) = 0\\ &G(t) = \frac{Q(t)}{V}\end{aligned} \tag{13.2}$$

where the parameters are given as:

Q: plasma glucose mass,

Q_b: the basal value of the plasma glucose mass,

I: plasma insulin concentration,

I_b: the basal value of the plasma insulin concentration,

D: glucose dose,

V: glucose distribution volume,

k_2, k_3: the rate parameters.

The NHGB depends on plasma glucose and remote insulin (I'):

$$\mathrm{NHGB}\big(Q(t), I'(t)\big) = \mathrm{NHGB}_0 - \big[k_5 + k_6\cdot I'(t)\big]\cdot Q(t) \tag{13.3}$$

where R_d is the rate of glucose disappearance from the peripheral tissues, which is given as:

$$R_\mathrm{d}\big(Q(t), I'(t)\big) = R_{\mathrm{d}0} + \big[k_1 + k_4\cdot I'(t)\big]\cdot Q(t) \tag{13.4}$$

This non-linear model is usually reparametrized to be uniquely identifiable as:

$$\frac{dQ(t)}{dt} = -\left[p_1 + X(t)\right]\cdot Q(t) + p_1\cdot Q_b + D\cdot\delta(t), \quad Q(0) = Q_b$$
$$\frac{dX(t)}{dt} = -p_2\cdot X(t) + p_3\cdot\left[I(t) - I_b\right], \quad X(0) = 0 \tag{13.5}$$
$$G(t) = \frac{Q(t)}{V}$$

with the following relations:

$$X(t) = (k_4 + k_6)\cdot I'(t)$$
$$p_1 = k_1 + k_5$$
$$p_2 = k_3 \tag{13.6}$$
$$p_3 = k_2\cdot(k_4 + k_6)$$
$$p_4 = \text{NHGB}_0 - R_{d0} = p_1\cdot Q_b.$$

The parameters are:

X: insulin action,

p_1: fractional (i.e., per unit distribution volume) glucose effectiveness,

p_2: the rate constant of the remote insulin compartment the insulin action emanates from,

p_3: scale factor which governs the amplitude of insulin action.

This model allows the estimation of insulin sensitivity as:

$$S_I^{\text{IVGTT}} = \frac{p_3}{p_2}\cdot V \quad (\text{dL/kg/min per } \mu\text{U/mL}), \tag{13.7}$$

where S_I^{IVGTT} is a steady-state measure which means it does not account for how fast or slow the insulin action takes place.

Another mass balance equation is used by describing the rate of appearance of glucose into plasma (Ra) as:

$$\frac{dQ(t)}{dt} = -\left[p_1 + X(t)\right]\cdot Q(t) + p_1\cdot Q_b + \text{Ra}(t, \boldsymbol{\alpha}) \quad Q(0) = Q_b \tag{13.8}$$

with $\alpha = [\alpha_1, \alpha_2, \ldots, \alpha_N]$ being the parameter vector describing Ra.

In the dynamic perturbation with tracer category, another important quantity is the endogenous glucose production (EGP) responsible for negligible or negative action of insulin on the liver and is given as:

$$\text{EGP}(t) = \text{EGP}_b - \text{GE}^L\cdot\left[G(t) - G_b\right] - X^L(t)\cdot G(t), \quad \text{EGP}(0) = \text{EGP}_b \tag{13.9}$$

where the parameters are given as:

$\mathrm{EGP_b}$: basal EGP,

$\mathrm{GE^L}$: liver glucose effectiveness,

G: glucose concentration and

G_b: the basal value of G,

X^L: the liver insulin action or the deviation from insulin.

The dynamical system this follows is:

$$\frac{dX^L(t)}{dt} = -p_2 \cdot X^L(t) + p_3^L\left[I(t) - I_b\right], \quad X^L(0) = 0, \tag{13.10}$$

where

p_2: rate constant describing the dynamics of insulin action on glucose production,

p_3^L: scale factor governing the amplitude of hepatic insulin action.

An improved model has been suggested in Ref. [1] as:

$$\mathrm{EGP}(t) = \mathrm{EGP_b} - k_G \cdot \left[G(t) - G_b\right] - X^L(t) - X^{\mathrm{Der}}(t), \quad \mathrm{EGP}(0) = \mathrm{EGP_b} \tag{13.11}$$

where X^L can be described by the system of ODEs:

$$\begin{aligned} \frac{dX_1(t)}{dt} &= -p_2^L \cdot X_1(t) + p_3^L \cdot \left[I(t) - I_b\right], \quad X_1(0) = 0 \\ \frac{dX^L(t)}{dt} &= -p_2^L \cdot X^L(t) + p_3^L \cdot X_1(t), \quad X^L(0) = 0 \end{aligned} \tag{13.12}$$

and

$$X^{\mathrm{Der}}(t) = \begin{cases} k_{\mathrm{GR}} \cdot \dfrac{dG(t)}{dt}, & \text{if } \dfrac{dG(t)}{dt} \geq 0 \\ 0, & \text{if } \dfrac{dG(t)}{dt} < 0 \end{cases} \tag{13.13}$$

where the two parameters are:

p_2^L: rate constant describing the dynamics of insulin action on glucose production,

p_3^L: scale factor governing the amplitude of hepatic insulin action,

k_{GR}: parameter governing the magnitude of glucose derivative control.

EGP can be calculated using the endogenous glucose concentration (G_{end}) which indicates the compartment of total glucose concentration measured in the plasma due to glucose production. The quantity G_{end} is related to EGP by the integral equation:

$$G_{\mathrm{end}}(t) = \int_0^t h(t,\tau) \cdot \mathrm{EGP}(\tau) \cdot d\tau + G_b \cdot h(t,0) \tag{13.14}$$

where $h(t,\tau)$ is the time-varying impulse response of glucose system given by tracer minimal model, and G_b is the basal glucose.

The whole-body-to-tissue model usually employs tracer elements and is described as the following ODEs:

$$\begin{aligned}
\frac{dC_c(t)}{dt} &= K_1 C_p(t) - (k_2 + k_3)C_c(t) + k_4 C_e(t), \quad C_c(0) = 0 \\
\frac{dC_e(t)}{dt} &= k_3 C_c(t) - (k_4 + k_5)C_e(t), \quad C_e(0) = 0 \\
\frac{dC_m(t)}{dt} &= k_5 C_e(t), \quad C_m(0) = 0
\end{aligned} \tag{13.15}$$

and

$$C(t) = (1 - V_b)(C_c(t) + C_e(t) + C_m(t)) + V_b C_b(t) \tag{13.16}$$

where the parameters are given as:

C_p: fluorodeoxyglucose (FDG) plasma arterial concentration,

C_c: extra-cellular concentration of FDG normalized to tissue volume,

C_e: FDG tissue concentration,

C: Total ^{18}F activity concentration in the region of interest,

K_1 [mL/mL/min] and k_2 [min^{-1}]: the exchange between plasma and extracellular space,

k_3 [min^{-1}] and k_4 [min^{-1}]: the rates of transport in and out of the cell,

k_5 [min^{-1}]: the rate of phosphorylation,

V_b: fractional blood volume in the region of interest,

C_b: whole blood tracer concentration.

One of these models can calculate the fractional uptake of the FDG and K [mL/mL/min] as:

$$K = \frac{K_1 k_3 k_5}{k_2 k_4 + k_2 k_5 + k_3 k_5} \tag{13.17}$$

More details on this model can be found in Ref. [1].

For an insulin system, firstly, an insulin kinetics model in steady state and insulin secretion can be investigated. The C-peptide concentration measurements (C) are linear in a wide range of concentration and related to the basal pancreatic secretion by the following convolution integral:

$$C(t) = \int_0^t h(t-\tau) \cdot \mathrm{SR}(\tau) \cdot d\tau \tag{13.18}$$

where h is the impulse response function of this system.

As described before, during the intravenous glucose tolerance test, the basal insulin secretion model is given by the pancreatic secretion rate (SR) as:

$$\text{SR}(t) = m \cdot F(t) \tag{13.19}$$

with F as the ready releasable insulin given by the ODE:

$$\frac{dF}{dt} = -m \cdot F(t) + Y(G,t), \quad F(0) = F_0 \tag{13.20}$$

with F_0 as the amount of insulin released immediately after the glucose stimulus and $Y(G,t)$ as the provision of new insulin which depends on the glucose level:

$$\frac{dY(G,t)}{dt} = -\frac{1}{T} \cdot [Y(G,t) - Y(G,\infty)], \quad Y(0) = 0. \tag{13.21}$$

13.3 Maximal Models

Compared to the minimal models, the maximal models are more detailed or fine-grained, non-linear, and higher-order, with a large number of parameters to estimate. Usually such models are not estimable without running large experimental investigations. However, they have been widely used for simulation to check model validity. A healthy state simulator, described in Ref. [1], has been usedon 204 non-diabetic subjects with simulation models of plasma glucose, plasma insulin, EGP, glucose rate of appearance, glucose utilization, insulin secretion, etc. Maximal models have also been used for pre-diabetes simulator, type 2 diabetes simulator, and type 1 diabetes simulator with unit process models and forcing functions for the liver, gastrointestinal tract, muscle and adipose tissue, beta cell, etc. *In-silico* subject simulation has been reported in Ref. [1] with a feedback controller and simulated insulin pump for type 1 diabetes. For insulin secretion, the following mass balance equation is used for the intermediate pool (I):

$$\frac{dI(t)}{dt} = M(g,t) - r \cdot I(t) - p^+ \cdot I(t) + p^- \cdot \int_0^\infty h(g,t) \cdot dg \tag{13.22}$$

where M is the mobilization flux and r is the rate of re-internalization.

The readily releasable pool (RRP) can be described by the time-varying density function $h(g,t)$ that indicates the amount of insulin in the beta cells within the pool with a threshold between g and $(g+dg)$. The granules are primed with rate p^+ and assumed to lose the capacity with rate p^-. In addition, the granule is considered to fuse with rate f^+ if it is in a triggered beta cell, which leads to the following equation:

$$\frac{dh(g,t)}{dt} = p^+ \cdot I(t) \cdot j(g) - p^- \cdot h(g,t) - f^+ \cdot h(g,t) \cdot \theta(G-g) \tag{13.23}$$

Here, $\theta(G-g)$ is the heaviside step function which takes the value of unity for $G > g$ or zero and I is the total intermediate pool. The primary flux p^+I distributes among cells with threshold g, described the time constant function $j(g)$.

13.4 Diabetes Monitoring Signals and Controls

Since the 1970s, various signal processing techniques have been proposed during in-hospital monitoring of blood glucose concentration and other substances like insulin, C-peptide, glucagon, etc. In recent times, new continuous glucose monitoring (CGM) systems have emerged that can monitor glucose concentration as frequently as every 5 min. These CGM monitoring devices are minimally invasive and portable, can measure glucose subcutaneously, and assess blood glucose concentration indirectly through interstitial fluid sampling. Although the measurement accuracy is not fully solved yet, this technology shows high potential for real-time prevention and treatment of hypoglycemia and hyperglycemia. Among various signal processing methods such as glucose-insulin oscillations, peak detection, and spectral/correlation analysis, hormone pulsatility and the use of approximate entropy are notable. Other challenges of continuous glucose monitoring with time-series data include CGM sensor calibration, CGM versus blood glucose measurement, filtering, prediction, hypoglycemia versus hyperglycemia alert, etc.

Several architectures have been proposed for glucose control which include safety algorithms and real-time control. Among various real-time control schemes, the feedback and feedforward control, autoregressive moving average exogenous (ARMAX) and non-linear ARMAX system identification, proportional integral derivative (PID) control, model predictive control (MPC) with quadratic cost and constraints on the manipulated variables (insulin pump), and real-time detection and estimation using Kalman filter are notable [1]. Various strategies for tuning the control loops have been adopted, for example, control variability grid analysis, robustness versus personalization of the controller parameters, run-to-run control, behavioral analysis, etc. More in-depth models, signals, and control strategies for diabetes are discussed in Ref. [1] and the large number of references therein.

Recently, Turksoy et al. [2] introduced a multi-module multivariable adaptive control strategy for artificial pancreas for type 1 diabetes. The artificial pancreas collects information from many sensors, computes the optimal insulin amount to be infused, and then manipulates the infusion rate of the pump. A recursive model of glucose concentration dynamics is first estimated using the ARMAX method of system identification. The controller takes various inputs like:

- Glucose and activity feedback
- Hypoglycemia detection and carbohydrates suggestion
- Meal detection and hyperglycemia prevention
- Exercise classification
- Fault detection and diagnosis

Using these inputs, the multivariable adaptive controller drives the insulin infusion pump for the patient. The measurements used in this scheme are continuous glucose monitoring sensor and wearable biometric sensors. The study in Ref. [2] with clinical experiments also concluded that the multivariable approach provides better results than the single variable version using only CGM measurement.

Meal detection has been further researched in Ref. [3] for 9 patients with type 1 diabetes over 27 different main meals. The multivariable adaptive artificial pancreas system includes a minimal model similar to Equation (13.1), followed by the unscented Kalman filter for state estimation of the non-linear system. The CGM measurements are used for nine subjects during breakfast, lunch, and dinner to validate the algorithm. A similar analog PID-based glucose-control algorithm known as the Hill equation was implemented in Ref. [4] using a beta-cell model comprising an ODE and the sigmoid function. For type 1 diabetes control using artificial pancreas, PID and sliding-mode-reference-conditioning-based safety auxiliary feedback control has been implemented in Ref. [5] with enhanced robustness and fault-tolerance properties.

A model predictive iterative learning control scheme has been proposed in Wang et al. [6] for artificial pancreatic beta cells in type 1 diabetes. This involves a virtual patient which uses an autoregressive exogenous model. The robustness of the MPC and iterative learning control on repetitive and non-repetitive diets, robustness to subject variations, and set-point updating have also been investigated.

For type 1 diabetic patients, an improved overnight safety scheme has been proposed in Facchinetti et al. [7] for online failure detection of the glucose sensor and insulin pump system. The two cases considered are CGM sensor failure and continuous subcutaneous insulin infusion pump failure. The failure detection employs a Kalman predictor and online prediction and alert module. The method was validated on *in-silico* data of 100 virtual subjects and real type 1 diabetes mellitus data. The robustness of the failure detection monitoring system against noise and the domain of validity has also been investigated on these two databanks.

In the field of closed-loop control of diabetes, MPC has wide applicability over conventional therapy in the development of artificial pancreas with various modules, for example, data handling/filtering, state estimation/update, closed-loop control algorithm, safety supervision algorithm, actuation, data-logging and outcome measures, etc. [8]. The important issues for modular control are design flexibility, incremental testing, regulatory approval, and deployment. The safety supervision algorithm has various elements like insulin request classifier, correction filter, hypoglycemia indicators, etc. This system is based on a linear MPC with a linearized model of the non-linear insulin-glucose dynamics. Moreover, the MPC works on the difference between the CGM signal and the patient's nominal blood glucose profile with improved individualization capability and has been validated on *in-silico* experiments.

13.5 Non-Invasive Diabetes Monitoring

Recently, more works have been devoted to non-invasive monitoring of diabetes. For example, Pai et al. [9] used a cloud computing platform to implement a photoacoustic spectroscopy system with detailed calculation of the signal to noise ratio. In the calibration and testing phase, several well-known signal processing operations were performed, such as pre-processing, feature extraction (e.g., positive/negative peaks, max/min peak, peak to peak amplitudes) followed by *in-vitro* and *in-vivo* testing, glucose estimation using polynomial kernels. The estimation performances were compared with respect to three accuracy measures – root mean square error, mean absolute difference, and mean absolute relative difference which results in a kernel-based calibration model for continuous non-invasive glucose monitoring using photoacoustic measurements. The algorithm has been deployed in mobile cloud computing platform with Internet of Things devices, internet gateway, cloud services with dataflow pipelines like data processing, data analytics, data storage, application control, etc. The embedded back-end implementation was tested against power consumption, security, *in-vitro* testing, system stability, safety, *in-vivo* testing, etc.

Much of the future research in this domain is expected to have a few key functionalities – unobtrusive sensing, modeling the onset and progress of diabetes mellitus, and user-centered approach [10]. Zarkogianni et al. [10] reviewed various emerging technologies for the management of diabetes mellitus. Among the commercially available devices, only two use non-invasive methods like Raman spectroscopy (HG1-c by C8 Medisensors) and thermal ultrasound and electromagnetic (GlucoTrack by Integrity Applications Ltd.). These devices have been benchmarked against other invasive devices by Dexcom, Medtronic, and Abbott with detailed comparison of the sensor lifetime, sensor warm-up time, frequency of calibration, frequency of recording, accuracy, etc. In Ref. [10], several artificial intelligence models for risk prediction and early diagnosis of type 2 diabetes have been reviewed in, for example, fuzzy neural networks, adaptive neuro-fuzzy inference system, support vector machine, linear discriminant analysis, adaptive network-based fuzzy inference system. In addition, a number of experts and their performances are compared using classification accuracy, sensitivity, specificity, etc. For long-term risk modeling, several statistical models – Cox regression, Tobit survival regression, Cox proportional hazard model and fractional polynomials, univariate and multivariate logistic regression, survival analysis, Weibull proportional hazard regression, Markov model, etc. – have been used for different cohort sizes (55, >1,000, >5,000, >1.2 million, etc.). Moreover, several glucose prediction models are compared in Ref. [10] based on artificial intelligence and autoregressive models. Long-term implanted sensor/telemetry systems are studied in Ref. [11] for glucose monitoring in diabetic patients. Raman spectra of blood has also been used for glucose monitoring devices in Ref. [12].

13.6 Recent Research Trends in Diabetes Monitoring and Control

Diabetes monitoring and control is emerging as prominent research in the last decade. In Scopus, some patterns were retrieved from the first 2,000 highest cited papers on this topic. The word clouds in Figure 13.1 show that a few frequently used words in the papers' titles are diabetes, glucose, type 1 and 2, insulin, monitoring, along with some relatively less used words like cardiovascular, glycemic, close-loop, management, self-monitoring, pump, etc. These word frequencies show the relative research efforts and different sub-areas in this field which are actively growing. Figure 13.2 bar chart shows the type of recently published works, many of which were journal research articles than review articles, followed by conference papers, short surveys, and letters.

Similarly, Figure 13.3 shows the indexed and authors' keywords in the top 2,000 papers in this active research area. As opposed to the title-based text analysis, the keyword-based analysis is often capable of capturing finer technical details or emphasis on fewer most popular technologies. In both the word clouds, apart from diabetes, the most frequent words were glucose, insulin, blood, monitoring, continuous, control, etc. which represent the finer details of the technologies gradually increasing over the years. Figure 13.4, exploring the citation patterns in this research field, shows that the citations of the papers on diabetes monitoring have rather drooping characteristics, at least for the upper quartile. The gradual decrease in the median citation from the year 2000 almost follows an exponential trend because the citation dynamics are usually a cumulative phenomenon

FIGURE 13.1
Word-clouds of the titles of recent diabetes monitoring research papers. (Datasource: Scopus.)

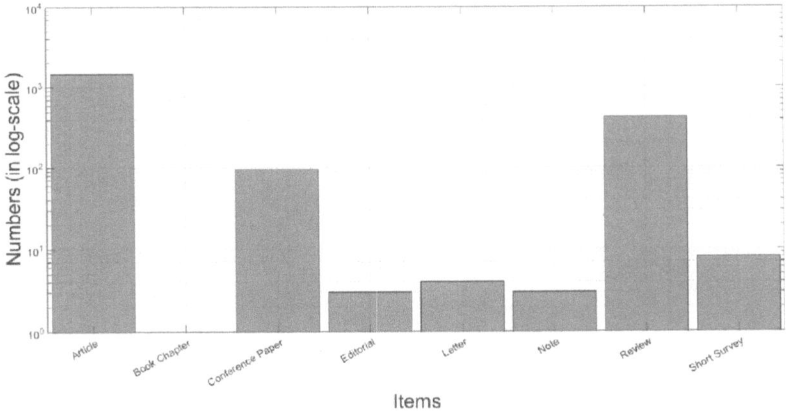

FIGURE 13.2
Type of recent papers on diabetes monitoring system. (Datasource: Scopus.)

and such recent technological advancements take years to get assimilated into clinical practice. Furthermore, there are more outliers in the papers between the years 2001 and 2013 as shown in Figure 13.4 which indicates that there are a few highly impactful papers compared to the citation of average papers published in recent years. Next, the source titles of the top 2,000 papers on this topic are explored in Figure 13.5. Most of the impactful papers were published in journals like Diabetes Care (American Diabetes Association), Diabetes Technology and Therapeutics (Liberto Pub.), Journal of Diabetes Science and Technology (Sage), Diabetic Medicines (Wiley), Diabetologia, Diabetes (American Diabetes Association), The Lancet, IEEE Transactions on Biomedical Engineering, etc. A more focused study on mobile health technologies for diabetes mellitus ranging from 2011 to 2017 and using 3 databases like ScienceDirect, SpringerLink, and IEEEXplore have been reported in Ref. [13].

With the advent of 5G and smart technologies, several personalized options have opened to analyze big healthcare data on clouds [14]. The five goals of technology using 5G

FIGURE 13.3
Indexed and author's keywords for recent papers on diabetes monitoring system. (Datasource: Scopus.)

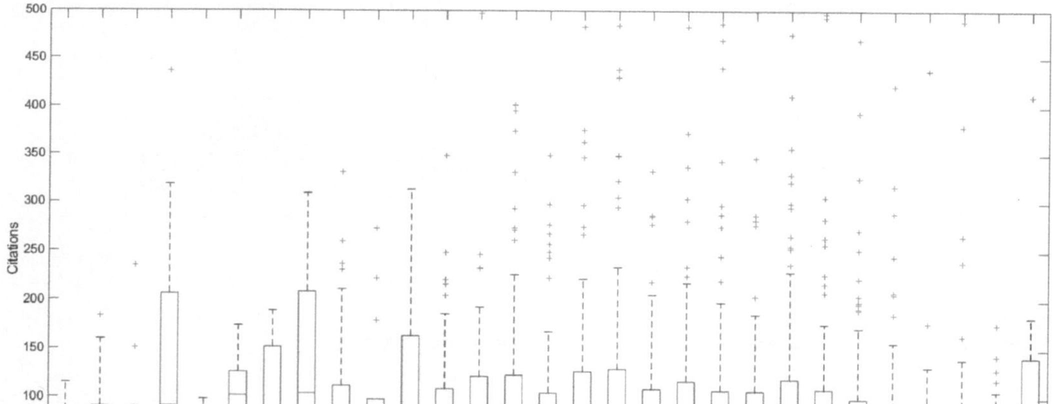

FIGURE 13.4
Year-wise citations of recent papers on diabetes monitoring system. (Datasource: Scopus.)

are – cost effectiveness, comfortability, personalization, sustainability, and smartness. The 5G-smart diabetes architecture in Ref. [14] has three layers – sensing layer, personalized diagnosis layer, and data sharing layer. The performance was validated using machine learning algorithms like decision tree, artificial neural network, support vector machine, ensemble, etc., which suggests the prevention and treatment of diabetes using recommendations like diet, sport, data sharing in a social network, etc.

FIGURE 13.5
Source titles for recent works on diabetes monitoring system. (Datasource: Scopus.)

13.7 Conclusion

This chapter reviews the recent trends in diabetes monitoring using the top 2,000 cited articles and reports text analytics results on past research activities and emerging trends in this domain. It also reviews various diabetes monitoring and control models available in the literature which are primarily divided into two categories – minimal and maximal models. Although the citation trends suggest that fundamental research in this domain is nearly saturated, there are many commercialization activities and patents issued recently on this topic [15]. This provides an opportunity to grow new industries in this domain which requires a good and affordable business model particularly for low and middle-income countries of the developing world where diabetes is emerging almost at the scale of a large epidemic.

References

1. Cobelli, C., Dalla Man, C., Sparacino, G., Magni, L., De Nicolao, G., and Kovatchev, B.P., 2009. Diabetes: Models, signals, and control. *IEEE Reviews in Biomedical Engineering*, 2, 54–96.
2. Turksoy, K., Littlejohn, E., and Cinar, A., 2018. Multimodule, multivariable artificial pancreas for patients with type 1 diabetes: Regulating glucose concentration under challenging conditions. *IEEE Control Systems Magazine*, 38(1), 105–124.
3. Turksoy, K., Samadi, S., Feng, J., Littlejohn, E., Quinn, L., and Cinar, A., 2016. Meal detection in patients with type 1 diabetes: A new module for the multivariable adaptive artificial pancreas control system. *IEEE Journal of Biomedical and Health Informatics*, 20(1), 47–54.
4. Pagkalos, I., Herrero, P., Toumazou, C., and Georgiou, P., 2014. Bio-inspired glucose control in diabetes based on an analogue implementation of a β-cell model. *IEEE Transactions on Biomedical Circuits and Systems*, 8(2), 186–195.

5. Revert, A., Garelli, F., Picó, J., De Battista, H., Rossetti, P., Vehí, J., and Bondia, J., 2013. Safety auxiliary feedback element for the artificial pancreas in type 1 diabetes. *IEEE Transactions on Biomedical Engineering*, 60(8), 2113–2122.
6. Wang, Y., Dassau, E., and Doyle III, F.J., 2010. Closed-loop control of artificial pancreatic β-cell in type 1 diabetes mellitus using model predictive iterative learning control. *IEEE Transactions on Biomedical Engineering*, 57(2), 211–219.
7. Facchinetti, A., Favero, S., Sparacino, G., and Cobelli, C., 2013. An online failure detection method of the glucose sensor-insulin pump system: Improved overnight safety of type-1 diabetic subjects. *IEEE Transactions on Biomedical Engineering*, 60(2), 406–416.
8. Patek, S.D., Magni, L., Dassau, E., Karvetski, C., Toffanin, C., De Nicolao, G., Del Favero, S., Breton, M., Dalla Man, C., Renard, E., and Zisser, H., 2012. Modular closed-loop control of diabetes. *IEEE Transactions on Biomedical Engineering*, 59(11), 2986–2999.
9. Pai, P.P., Sanki, P.K., Sahoo, S.K., De, A., Bhattacharya, S., and Banerjee, S., 2018. Cloud computing-based non-invasive glucose monitoring for diabetic care. *IEEE Transactions on Circuits and Systems I: Regular Papers*, 65(2), 663–676.
10. Zarkogianni, K., Litsa, E., Mitsis, K., Wu, P.Y., Kaddi, C.D., Cheng, C.W., Wang, M.D., and Nikita, K.S., 2015. A review of emerging technologies for the management of diabetes mellitus. *IEEE Transactions on Biomedical Engineering*, 62(12), 2735–2749.
11. Lucisano, J.Y., Routh, T.L., Lin, J.T., and Gough, D.A., 2017. Glucose monitoring in individuals with diabetes using a long-term implanted sensor/telemetry system and model. *IEEE Transactions on Biomedical Engineering*, 64(9), 1982–1993.
12. Mougiakakou, S.G., Bartsocas, C.S., Bozas, E., Chaniotakis, N., Iliopoulou, D., Kouris, I., Pavlopoulos, S., Prountzou, A., Skevofilakas, M., Tsoukalis, A., and Varotsis, K., 2010. SMARTDIAB: A communication and information technology approach for the intelligent monitoring, management and follow-up of type 1 diabetes patients. *IEEE Transactions on Information Technology in Biomedicine*, 14(3), 622–633.
13. El-Sappagh, S., Ali, F., El-Masri, S., Kim, K., Ali, A., and Kwak, K.S., 2019. Mobile health technologies for diabetes mellitus: Current state and future challenges. *IEEE Access*, 7, 21917–21947.
14. Chen, M., Yang, J., Zhou, J., Hao, Y., Zhang, J., and Youn, C.H., 2018. 5G-smart diabetes: Toward personalized diabetes diagnosis with healthcare big data clouds. *IEEE Communications Magazine*, 56(4), 16–23.
15. Mertz, L., 2018. Automated insulin delivery: Taking the guesswork out of diabetes management. *IEEE Pulse*, 9(1), 8–9.

Index

N

O

P

Q

R

S

T

V